AF885979

Studies in Computational Intelligence

Volume 948

Series Editor

Janusz Kacprzyk, Polish Academy of Sciences, Warsaw, Poland

The series "Studies in Computational Intelligence" (SCI) publishes new developments and advances in the various areas of computational intelligence—quickly and with a high quality. The intent is to cover the theory, applications, and design methods of computational intelligence, as embedded in the fields of engineering, computer science, physics and life sciences, as well as the methodologies behind them. The series contains monographs, lecture notes and edited volumes in computational intelligence spanning the areas of neural networks, connectionist systems, genetic algorithms, evolutionary computation, artificial intelligence, cellular automata, self-organizing systems, soft computing, fuzzy systems, and hybrid intelligent systems. Of particular value to both the contributors and the readership are the short publication timeframe and the world-wide distribution, which enable both wide and rapid dissemination of research output.

Indexed by SCOPUS, DBLP, WTI Frankfurt eG, zbMATH, SCImago.

All books published in the series are submitted for consideration in Web of Science.

More information about this series at http://www.springer.com/series/7092

Erik Cuevas · Alma Rodríguez ·
Avelina Alejo-Reyes · Carolina Del-Valle-Soto

Recent Metaheuristic Computation Schemes in Engineering

Springer

Erik Cuevas
CUCEI
Universidad de Guadalajara
Guadalajara, Mexico

Avelina Alejo-Reyes
Centro de Eneñanza Técnica Industrial
Universidad de Guadalajara
Guadalajara, Mexico

Alma Rodríguez
CUCEI
Universidad de Guadalajara
Guadalajara, Mexico

Facultad de Ingeniería
Universidad Panamericana
Zapopan, Jalisco, Mexico

Centro de Eneñanza Técnica Industrial
Universidad de Guadalajara
Guadalajara, Mexico

Carolina Del-Valle-Soto
Centro de Eneñanza Técnica Industrial
Universidad de Guadalajara
Guadalajara, Mexico

ISSN 1860-949X ISSN 1860-9503 (electronic)
Studies in Computational Intelligence
ISBN 978-3-030-66006-2 ISBN 978-3-030-66007-9 (eBook)
https://doi.org/10.1007/978-3-030-66007-9

© The Editor(s) (if applicable) and The Author(s), under exclusive license to Springer Nature Switzerland AG 2021
This work is subject to copyright. All rights are solely and exclusively licensed by the Publisher, whether the whole or part of the material is concerned, specifically the rights of translation, reprinting, reuse of illustrations, recitation, broadcasting, reproduction on microfilms or in any other physical way, and transmission or information storage and retrieval, electronic adaptation, computer software, or by similar or dissimilar methodology now known or hereafter developed.
The use of general descriptive names, registered names, trademarks, service marks, etc. in this publication does not imply, even in the absence of a specific statement, that such names are exempt from the relevant protective laws and regulations and therefore free for general use.
The publisher, the authors and the editors are safe to assume that the advice and information in this book are believed to be true and accurate at the date of publication. Neither the publisher nor the authors or the editors give a warranty, expressed or implied, with respect to the material contained herein or for any errors or omissions that may have been made. The publisher remains neutral with regard to jurisdictional claims in published maps and institutional affiliations.

This Springer imprint is published by the registered company Springer Nature Switzerland AG
The registered company address is: Gewerbestrasse 11, 6330 Cham, Switzerland

Preface

Many problems in engineering nowadays concern with the goal of an “optimal” solution. Several optimization methods have therefore emerged, being researched and applied extensively to different optimization problems.

Typically, optimization methods arising in engineering are computationally complex because they require evaluation of a quite complicated objective function which is often multimodal, non-smooth, or even discontinuous. The difficulties associated with using mathematical optimization on complex engineering problems have contributed to the development of alternative solutions. Metaheuristic computation techniques are stochastic optimization methods that have been developed to obtain near-optimum solutions in complex optimization problems, for which traditional mathematical techniques normally fail.

Metaheuristic methods consider as an inspiration our scientific understanding of biological, natural, or social systems, which at some level of abstraction can be represented as optimization processes. In their operation, searcher agents emulate a group of biological or social entities which interact with each other based on specialized operators that model a determined biological or social behavior. These operators are applied to a population (or several sub-populations) of candidate solutions (individuals) that are evaluated with respect to their fitness. Thus, in the evolutionary process, individual positions are successively approximated to the optimal solution of the system to be solved.

Due to their robustness, metaheuristic techniques are well-suited options for industrial and real-world tasks. They do not need gradient information and they can operate on each kind of parameter space (continuous, discrete, combinatorial, or even mixed variants). Essentially, the credibility of evolutionary algorithms relies on their ability to solve difficult real-world problems with the minimal amount of human effort.

There exist some common features clearly appear in most of the metaheuristic approaches, such as the use of diversification to force the exploration of regions of the search space, rarely visited until now, and the use of intensification or exploitation, to investigate thoroughly some promising regions. Another common feature is the use of memory to archive the best solutions encountered.

Metaheuristic schemes are used to estimate the solutions to complex optimization problems. They are often designed to meet the requirements of particular problems because no single optimization algorithm can solve all problems competitively. Therefore, in order to select an appropriate metaheuristic technique, its relative efficacy must be appropriately evaluated.

Metaheuristic search methods are so numerous and varied in terms of design and potential applications; however, for such an abundant family of optimization techniques, there seems to be a question which needs to be answered: Which part of the design in a metaheuristic algorithm contributes more to its better performance? One widely accepted principle among researchers considers that metaheuristic search methods can reach a better performance when an appropriate balance between exploration and exploitation of solutions is achieved. While there seems to exist a general agreement on this concept, in fact, there is barely a vague conception of what the balance of exploration and exploitation really represent. Indeed, the classification of search operators and strategies present in a metaheuristic method is often ambiguous, since they can contribute in some way to explore or exploit the search space.

Most of the problems in science, engineering, economics, and life can be translated as an optimization or a search problem. According to their characteristics, some problems can be simple that can be solved by traditional optimization methods based on mathematical analysis. However, most of the problems of practical importance in engineering represents complex scenarios so that they are very hard to be solved by using traditional approaches. Under such circumstances, metaheuristic has emerged as the best alternative to solve this kind of complex formulations. Therefore, metaheuristic techniques have consolidated as a very active research subject in the last ten years. During this time, various new metaheuristic approaches have been introduced. They have been experimentally examined on a set of artificial benchmark problems and in a large number of practical applications. Although metaheuristic methods represent one of the most exploited research paradigms in computational intelligence, there are a large number of open challenges in the area of metaheuristics. They range from premature convergence, inability to maintain population diversity, and the combination of metaheuristic paradigms with other algorithmic schemes, toward extending the available techniques to tackle ever more difficult problems.

Several works that compare the performance among metaheuristic approaches have been reported in the literature. Nevertheless, they suffer from one the following limitations:

(A) their conclusions are based on the performance of popular evolutionary approaches over a set of synthetic functions with exact solutions and well-known behaviors, without considering the application context or including recent developments.

(B) their conclusions consider only the comparison of their final results which cannot evaluate the nature of a good or bad balance between exploration and exploitation.

Numerous books have been published taking in account many of the most widely known methods, namely simulated annealing, tabu search, evolutionary algorithms, ant colony algorithms, particle swarm optimization, or genetic algorithms, but attempts to consider the discussion of alternative approaches are scarce. The excessive publication of developments based on the simple modification of popular metaheuristic methods presents an important disadvantage, in that it distracts attention away from other innovative ideas in the field of metaheuristics. There exist several alternative metaheuristic methods which consider very interesting concepts; however, they seem to have been completely overlooked in favor of the idea of modifying, hybridizing, or restructuring traditional metaheuristic approaches.

This book has two objectives. The first goal is to present advances that discuss new alternative metaheuristic developments which have proved to be effective in their application to several complex problems. The book considers different new metaheuristic methods and their practical applications. This structure is important to us, because we recognize this methodology as the best way to assist researchers, lecturers, engineers, and practitioners in the solution of their own optimization problems. The second objective of this book is to present the performance comparison of various metaheuristic techniques when they face complex optimization problems. In the comparisons, the following criteria have been adopted:

(A) special attention is paid to recently developed metaheuristic algorithms,
(B) the balance between exploration and exploitation has been considered to evaluate the search performance and,
(C) the use of demanding applications such as image processing, complex scheduling problems, supplier selection, and Internet of Things (IoT).

This book includes ten chapters. The book has been structured so that each chapter can be read independently from the others. Chapter 1 describes the main characteristics and properties of metaheuristic methods. This chapter analyzes the most important concepts of metaheuristic schemes.

The book is divided into three parts. The first part of the book that involves Chaps. 2 and 3 presents recent metaheuristic algorithms, their operators and characteristics.

In Chap. 2, a metaheuristic scheme is based on the hunting model of Yellow Saddle Goatfish. In this strategy, the complete group of fish is distributed in subpopulations to cover the whole hunting region. In each sub-population, all fish participate collectively in the hunt considering two different roles: chaser and blocker. In the hunt, a chaser fish actively tries to find the prey in a certain area, whereas a blocker fish moves spatially to avoid the escape of the prey. In the approach, different computational operators are designed in order to emulate this peculiar hunting behavior. With the use of this biological model, the new search strategy improves the optimization results in terms of accuracy and convergence in comparison to other popular optimization techniques. The performance of this method is tested by analyzing its results with other related evolutionary computation techniques. Several standard benchmark functions commonly used in the literature were considered to obtain optimization results. Furthermore, the proposed model is applied to solve certain engineering optimization problems. Analysis of the experimental results exhibits the efficiency, accuracy, and robustness of the proposed algorithm.

Chapter 3 presents a hybrid metaheuristic method for solving optimization problems. The presented approach combines (A) the explorative characteristics of the invasive weed optimization method, (B) the probabilistic models of the estimation distribution algorithms, and (C) the dispersion capacities of a mixed Gaussian–Cauchy distribution to produce its own search strategy. With these mechanisms, the method conducts an optimization strategy over search areas that deserve a special interest according to a probabilistic model and the fitness value of the existent solutions. In this method, each individual of the population generates new elements around its own location, dispersed according to the mixed distribution. The number of new elements depends on the relative fitness value of the individual regarding the complete population. After this process, a group of promising solutions are selected from the set compound by the (a) new elements and the (b) original individuals. Based on the selected solutions, a probabilistic model is built from which a certain number of members (c) are sampled. Then, all the individuals of the sets (a), (b), and (c) are joined in a single group and ranked in terms of their fitness values. Finally, the best elements of the group are selected to replace the original population. This process is repeated until a termination criterion has been reached. To test the performance of this method, several comparisons to other well-known metaheuristic methods have been made. The comparison consists of analyzing the optimization results over different standard benchmark functions within a statistical framework. Conclusions based on the comparisons exhibit the accuracy, efficiency, and robustness of the proposed approach.

The second part of the book which involves Chaps. 4–6 presents the use of recent metaheuristic algorithms in different domains. The idea is to compare the potential of new metaheuristic alternatives algorithms within a practical perspective.

In Chap. 4, a corner detector algorithm based on cellular nonlinear/neural networks (CNN) for grayscale images is presented. In the approach, the original processing scheme of the CNN is modified to include a nonlinear operation for increasing the contrast of the local information in the image. With this adaptation, the final CNN parameters that allow the appropriate detection of corner points are estimated through the differential evolution algorithm (DE) by using standard training images. Different test images have been used to evaluate the performance of the proposed corner detector. Its results are also compared with popular corner methods from the literature. Computational simulations demonstrate that the proposed CNN approach presents competitive results in comparison with other algorithms in terms of accuracy and robustness.

Chapter 5 discusses an accurate methodology for retinal vessel and optic disk segmentation. The scheme combines two different techniques: The lateral inhibition (LI) and the differential evolution (DE). The LI scheme produces a new image with enhanced contrast between the background and retinal vessels. Then, the DE algorithm is used to obtain the appropriate threshold values through the minimization of the cross-entropy function from the enhanced image. To evaluate the performance of the proposed approach, several experiments over images extracted from

STARE, DRIVE, and DRISHTI-GS databases have been conducted. Simulation results demonstrate a high performance of the proposed scheme in comparison with similar methods reported in the literature.

Chapter 6 presents a simple clustering model inspired by the way in which the human visual system associates patterns spatially. The model, at some abstraction level, can be characterized as a density grouping strategy. The approach is based on cellular neural networks (CNNs) which have demonstrated to be the best models for emulating the human visual system. In the proposed method, similar to the biological model, the CNN is used to build especially groups while an automatic mechanism tries different resolution scales to find the best possible data categorization. Different datasets have been adopted to evaluate the performance of the proposed algorithm. Their results are also compared with popular density clustering techniques from the literature. Computational results demonstrate that the proposed CNN approach presents competitive results in comparison with other algorithms regarding accuracy and robustness.

The third part of the book which involves Chaps. 7 and 8 presents two overviews. They consider the use of recent metaheuristic algorithms in the area of supplier selection and Internet of Things (IoT).

Chapter 7 presents the main concepts of metaheuristic schemes for wireless sensor networks (WSNs). WSNs are multi-functional, low-cost, and low-power networks and rely on communications among devices, from sensor nodes to one or more sink nodes. Sink nodes, sometimes called coordinator nodes or root nodes, may be more robust and have larger processing capacity than the other nodes. Sensor networks can be widely used in various environments, sometimes hostile. Some of the many applications of WSNs are in the medical field, agriculture, monitoring and detection, automation, and data mining.

Finally, in Chap. 8, the problem of inventory management through metaheuristic principles is considered. Mathematical models of the inventory management problem may be complex and NP-hard, and as a result, evaluating all possible solutions to find the cheapest one may be unfeasible, even with a computer. When that happens, metaheuristic algorithms may be used to find a reasonable solution in a reasonable amount of time.

As authors, we wish to thank many people who were somehow involved in the writing process of this book. We express our gratitude to Prof. Janusz Kacprzyk, who so warmly sustained this project. Acknowledgments also go to Dr. Thomas Ditzinger and Jennifer Sweety Johnson, who so kindly also support this book project.

Guadalajara, Mexico

Erik Cuevas
Alma Rodríguez
Avelina Alejo-Reyes
Carolina Del-Valle-Soto

Contents

Chapter 1
Introductory Concepts of Metaheuristic Computation

This chapter presents the main concepts of metaheuristic schemes. The objective of this chapter is to introduce the characteristics and properties of these approaches. An important propose of this chapter is also to recognize the importance of metaheuristic methods to solve optimization problems in the cases in which traditional techniques are not suitable.

1.1 Formulation of an Optimization Problem

Most of the industrial and engineering systems require the use of an optimization process for their operation. In such systems, it is necessary to find a specific solution that is considered the best in terms of a cost function. In general terms, an optimization scheme corresponds to a search strategy that has as an objective to obtain the best solution considering a set of potential alternatives. This bets solution represents the best possible solution that, according to the cost function, solves the optimization formulation appropriately [1].

Consider a public transportation system of a specific town, for illustration proposes. In this example, it is necessary to find the "best" path to a particular target destination. To assess each possible alternative and then get the best possible solution, an adequate criterion should be taken into account. A practical criterion could be the relative distances among all possible routes. Therefore, a hypothetical optimization scheme selects the option with the smallest distance as a final output. It is important to recognize that several evaluation elements are also possible, which could consider other important criteria such as the number of transfers, the time required to travel from a location to another or ticket price.

Optimization can be formulated as follows: Consider a function $f : S \rightarrow \Re$ which is called the cost function, find the argument that minimizes f:

$$x^* = \arg \min_{x \in S} f(\mathbf{x}) \tag{1.1}$$

© The Author(s), under exclusive license to Springer Nature Switzerland AG 2021

E. Cuevas et al., *Recent Metaheuristic Computation Schemes in Engineering*,
Studies in Computational Intelligence 948,
https://doi.org/10.1007/978-3-030-66007-9_1

S corresponds to the search space that refers to all possible solutions. In general terms, each possible solution solves in a different quality the optimization problem. Commonly, the unknown elements of x represent the decision variables of the optimization formulation. The cost function f determines the quality of each candidate solution. It evaluates the way in which a candidate element x solves the optimization formulation.

In the example of public transportation, S represents all subway stations, bus lines, etc., available in the database of the transportation system. $\mathbf{x}$ represents a possible path that links the start location with the final destination. $f(\mathbf{x})$ is the cost function that assesses the quality of each possible route. Some other constraints can be incorporated as a part of the problem definition, such as the ticket price or the distance to the destination (in different situations, it is taken into account the combination of both indexes, depending on our preferences).

When additional constraints exist, the optimization problem is called constrained optimization (different from unconstrained optimization where such restrictions are not considered). Under such conditions, an optimization formulation involves the next elements:

- One or several decision variables from $\mathbf{x}$, which integrate a candidate solution
- A cost function $f(\mathbf{x})$ that evaluates the quality of each solution $\mathbf{x}$
- A search space S that defines the set of all possible solutions to $f(\mathbf{x})$
- Constraints that represent several feasible regions of the search space S.

In practical terms, an optimization approach seeks within a search space S a solution for $f(\mathbf{x})$ in a reasonable period of time with enough accuracy. The performance of the optimization method also depends on the type of formulation. Therefore, an optimization problem is well-defined if the following conditions are established:

1. There is a solution or set of solutions that satisfy the optimal values.
2. There is a specific relationship between a solution and its position so that small displacements of the original values generate light deviations in the objective function $f(\mathbf{x})$.

1.2 Classical Optimization Methods

Once an engineering problem has been translated into a cost function, the next operation is to choose an adequate optimization method. Optimization schemes can be divided into two sets: classical approaches and metaheuristic methods [2].

Commonly, $f(\mathbf{x})$ presents a nonlinear association in terms of its modifiable decision variables $\mathbf{x}$. In classical optimization methods, an iterative algorithm is employed to analyze the search space efficiently. Among all approaches introduced in the literature, the methods that use derivative-descent principles are the most popular. Under such techniques, the new position x_{k+1} is determined from the current location x_k in a direction toward $\mathbf{d}$:

$$x_{k+1} = x_k + \alpha \mathbf{d}, \tag{1.2}$$

where α symbolizes the learning rate that determines the extent of the search step in the direction to **d**. The direction **d** in Eq. 1.2 is computed, assuming the use of the gradient (**g**) of the objective function $f(\cdot)$.

One of the most representative methods of the classical approaches is the steepest descent scheme. Due to simplicity, this technique allows us to solve efficiently objective functions. Several other derivative-based approaches consider this scheme as the basis for the construction of more sophisticated methods. The steepest descent scheme is defined under the following formulation:

$$x_{k+1} = x_k - \alpha \mathbf{g}(f(x)), \tag{1.3}$$

In spite of its simplicity, classical derivative-based optimization schemes can be employed as long as the cost function presents two important constraints:

(I) The cost function can be two-timed derivable.
(II) The cost function is unimodal; i.e., it presents only one optimal position.

The optimization problem defines a simple case of a derivable and unimodal objective function. This function fulfills the conditions (I) and (II):

$$f(x_1, x_2) = 10 - e^{-(x_1^2 + 3 \cdot x_2^2)} \tag{1.4}$$

Figure 1.1 shows the function defined by formulation 1.4.

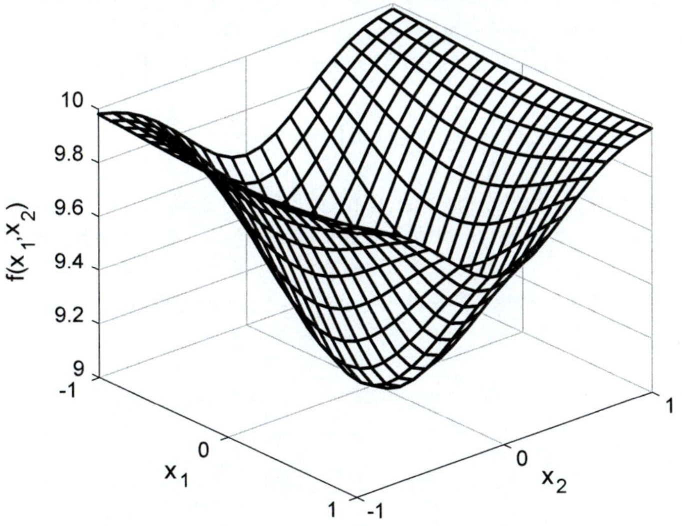

Fig. 1.1 Cost function with unimodal characteristics

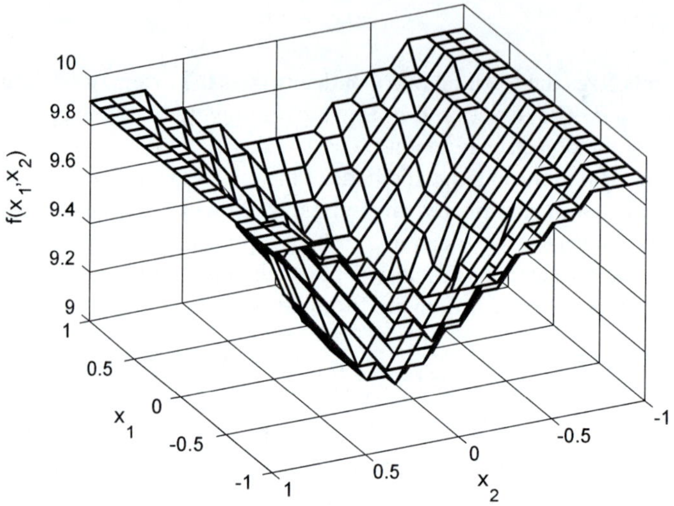

Fig. 1.2 A non-differentiable produced through the use of the floor function

Considering the current complexity in the design of systems, there are too few cases in which traditional methods can be applied. Most of the optimization problems imply situations that do not fulfill the constraints defined for the application of gradient-based methods. One example involves combinatorial problems where there is no definition of differentiation. There exist also many situations why an optimization problem could not be differentiable. One example is the "floor" function, which delivers the minimal integer number of its argument. This operation applied in Eq. 1.4 transforms the optimization problem from (Eq. 1.5) one that is differentiable to others not differentiable. This problem can be defined as follows (Fig. 1.2):

$$f(x_1, x_2) = \text{floor}\left(10 - e^{-(x_1^2 + 3 \cdot x_2^2)}\right) \quad (1.5)$$

Although an optimization problem can be differentiable, there exist other restrictions that can limit the use of classical optimization techniques. Such a restriction corresponds to the existence of only an optimal solution. This fact means that the cost function cannot present any other prominent local optima. Let us consider the minimization of the Griewank function as an example.

$$\begin{array}{ll} \text{minimize} & f(x_1, x_2) = \frac{x_1^2 + x_2^2}{4000} - \cos(x_1)\cos\left(\frac{x_2}{\sqrt{2}}\right) + 1 \\ \text{subject to} & -5 \le x_1 \le 5 \\ & -5 \le x_2 \le 5 \end{array} \quad (1.6)$$

A close analysis of the formulation presented in Eq. 1.6; it is clear that the optimal global solution is located in $x_1 = x_2 = 0$. Figure 1.3 shows the cost function established in Eq. 1.6. As can be seen from Fig. 1.3, the cost function presents many

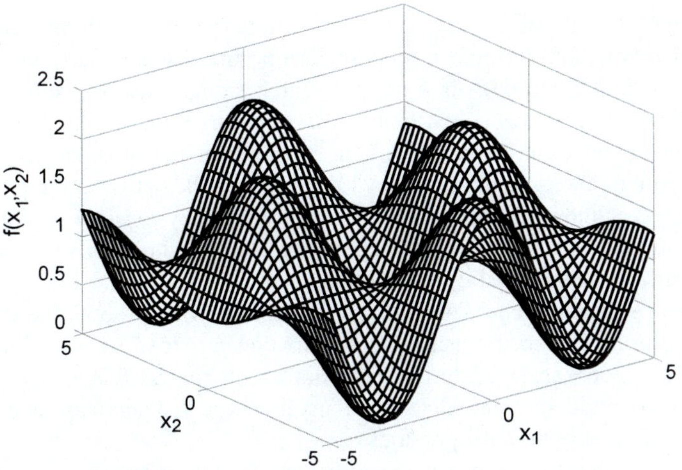

Fig. 1.3 The Griewank function with multimodal characteristics

local optimal solutions (multimodal) so that the gradient-based techniques with a randomly generated initial solution will prematurely converge to one of them with a high probability.

Considering the constraints of gradient-based approaches, it makes difficult their use to solve a great variety of optimization problems in engineering. Instead, some other alternatives which do not present restrictions are needed. Such techniques can be employed in a wide range of problems [3].

1.3 Metaheuristic Computation Schemes

Metaheuristic [4] schemes are derivative-free methods, which do not need that the cost function maintains the restrictions of being two-timing differentiable or unimodal. Under such conditions, metaheuristic methods represent global optimization methods which can deal with several types of optimization problems such as nonconvex, nonlinear, and multimodal problems subject to linear or nonlinear constraints with continuous or discrete decision variables.

The area of metaheuristic computation maintains a rich history. With the demands of more complex industrial processes, it is necessary for the development of new optimization techniques that do not require prior knowledge (hypotheses) on the optimization problem. This lack of assumptions is the main difference between classical gradient-based methods. In fact, the majority of engineering system applications are highly nonlinear or characterized by noisy objective functions. Furthermore, in several cases, there is no explicit deterministic expression for the optimization problem. Under such conditions, the evaluation of each candidate solution is carried out through the result of an experimental or simulation process. In this context, the metaheuristic methods have been proposed as optimization alternatives.

A metaheuristic approach is a generic search strategy used to solve optimization problems. It employs a cost function in an abstract way, considering only its evaluations in particular positions without considering its mathematical properties. Metaheuristic methods do not need any hypothesis on the optimization problem nor any kind of prior knowledge on the objective function. They consider the optimization formulation as "black boxes" [5]. This property is the most prominent and attractive characteristic of metaheuristic computation.

Metaheuristic approaches collect the necessary knowledge about the structure of an optimization problem by using the information provided by all solutions (i.e., candidate solutions) assessed during the optimization process. Then, this knowledge is employed to build new candidate solutions. It is expected that these new solutions present better quality than the previous ones.

Currently, different metaheuristic approaches have been introduced in the literature with good results. These methods consider modeling our scientific knowledge of biological, natural, or social systems, which, under some perspective, can be understood as optimization problems [6].

These schemes involve the cooperative behavior of bee colonies such as the Artificial Bee Colony (ABC) technique [7], the social behavior of bird flocking and fish schooling such as the Particle Swarm Optimization (PSO) algorithm [8], the emulation of the bat behavior such as the Bat Algorithm (BA) method [9], the improvisation process that occurs when a musician searches for a better state of harmony such as the Harmony Search (HS) [10], the social-spider behavior such as the Social Spider Optimization (SSO) [11], the mating behavior of firefly insects such as the Firefly (FF) method [12], the emulation of immunological systems as the clonal selection algorithm (CSA) [13], the simulation of the animal behavior in a group such as the Collective Animal Behavior [14], the emulation of the differential and conventional evolution in species such as the Differential Evolution (DE) [15], the simulation of the electromagnetism phenomenon as the electromagnetism-Like algorithm [16] and Genetic Algorithms (GA) [17], respectively.

1.3.1 Generic Structure of a Metaheuristic Method

In general terms, a metaheuristic scheme refers to a search strategy that emulates under a particular point of view a specific biological, natural or social system. A generic metaheuristic method involves the following characteristics:

1. Maintain a population of candidate solutions.
2. This population is dynamically modified through the production of new solutions.
3. A cost function associates the capacity of a solution to survive and reproduce similar elements.

4. Different operations are defined in order to explore an appropriately exploit the space of solutions through the production of new promising solutions.

Under the metaheuristic methodology, it is expected that, on average, candidate solutions enhance their quality during the evolution process (i.e., their ability to solve the optimization formulation). In the operation of the metaheuristic scheme, the operators defined in its structure will produce new solutions. The quality of such solutions will be improved as the number of iterations increases. Since the quality of each solution is associated with its capacity to solve the optimization problem, the metaheuristic method will guide the population towards the optimal global solution. This powerful mechanism has allowed the use of metaheuristic schemes to several complex engineering problems in different domains [18–21].

Most of the metaheuristic schemes have been devised to solve the problem of finding a global solution of a nonlinear optimization problem with box constraints in the following form:

$$\begin{array}{ll} \text{Maximize/Minimize } f(\mathbf{x}), & \mathbf{x} = (x_1, \ldots, x_d) \in \Re^d \\ \text{subject to} & \mathbf{x} \in \mathbf{X} \end{array} \tag{1.7}$$

where $f : \Re^d \rightarrow \Re$ represents a nonlinear function whereas $\mathbf{X} = \left\{\mathbf{x} \in \Re^d | l_i \leq x_i \leq u_i, i = 1, \ldots, d\right\}$ corresponds to the feasible search space, restriced by the lower (l_i) and upper (u_i) bounds.

With the objective of solving the problem of Eq. 1.6, under the metaheuristic computation methodology, a group (population) $\mathbf{P}^k(\{\mathbf{p}_1^k, \mathbf{p}_2^k, \ldots, \mathbf{p}_N^k\})$ of N possible solutions (individuals) is modified from a start point ($k = 0$) to a total *gen* number iterations ($k = gen$). In the beginning, the scheme starts initializing the set of N candidate solutions with random values uniformly distributed between the pre-specified lower (l_i) and upper (u_i) limits. At each generation, a group operations are used over the current population $\mathbf{P}^k$ to generate a new set of individuals $\mathbf{P}^{k+1}$. Each possible solution $\mathbf{p}_i^k$ ($i \in [1, \ldots, N]$) corresponds to a d-dimensional vector $\left\{p_{i,1}^k, p_{i,2}^k, \ldots, p_{i,d}^k\right\}$ where each element represents a decision variable of the optimization problem to be solved. The capacity of each possible solution $\mathbf{p}_i^k$ to solve the optimization problem is assessed by considering a cost function $f\left(\mathbf{p}_i^k\right)$ whose delivered value symbolizes the fitness value of $\mathbf{p}_i^k$. As the evolution process progresses, the best solution $\mathbf{g}$ $(g_1, g_2, \ldots g_d)$ seen so-far is maintained since it is the best available solution. Figure 1.4 shows an illustration of the generic procedure of a metaheuristic method.

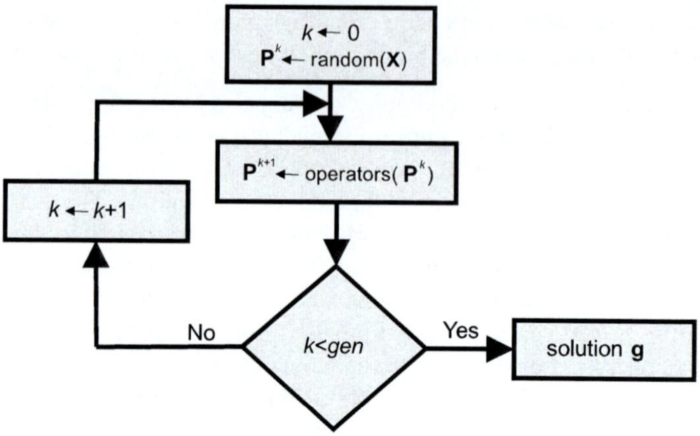

Fig. 1.4 Generic procedure of a metaheuristic scheme

References

1. Akay B, Karaboga D (2015) A survey on the applications of artificial bee colony in signal, image, and video processing. SIViP 9(4):967–990
2. Yang X-S (2010) Engineering optimization. Wiley Inc.
3. Treiber MA (2013) Optimization for computer vision an introduction to core concepts and methods. Springer, Berlin
4. Simon D (2013) Evolutionary optimization algorithms. Wiley
5. Blum C, Roli A (2003) Metaheuristics in combinatorial optimization: overview and conceptual comparison. ACM Comput Surv (CSUR) 35(3):268–308. https://doi.org/10.1145/937503.937505
6. Nanda SJ, Panda G (2014) A survey on nature inspired metaheuristic algorithms for partitional clustering. Swarm Evol Comput 16:1–18
7. Karaboga D (2005) An idea based on honey bee swarm for numerical optimization. TechnicalReport-TR06. Engineering Faculty, Computer Engineering Department, Erciyes University
8. Kennedy J, Eberhart R (1995) Particle swarm optimization. In: Proceedings of the 1995 IEEE international conference on neural networks, vol 4, pp 1942–1948
9. Yang XS (2010) A new metaheuristic bat-inspired algorithm. In: Cruz C, González J, Krasnogor GTN, Pelta DA (eds) Nature inspired cooperative strategies for optimization (NISCO 2010), studies in computational intelligence, vol 284. Springer, Berlin, pp 65–74
10. Geem ZW, Kim JH, Loganathan GV (2001) A new heuristic optimization algorithm: harmony search. Simulations 76:60–68
11. Cuevas E, Cienfuegos M, Zaldívar D, Pérez-Cisneros M (2013) A swarm optimization algorithm inspired in the behavior of the social-spider. Expert Syst Appl 40(16):6374–6384
12. Yang XS (2009) Firefly algorithms for multimodal optimization. In: Stochastic algorithms: foundations and applications, SAGA 2009, lecture notes in computer sciences, vol 5792, pp 169–178
13. de Castro LN, von Zuben FJ (2002) Learning and optimization using the clonal selection principle. IEEE Trans Evol Comput 6(3):239–251
14. Cuevas E, González M, Zaldivar D, Pérez-Cisneros M, García G (2012) An algorithm for global optimization inspired by collective animal behaviour. In: Discrete dynamics in nature and society, art. no. 638275

15. Storn R, Price K (1995) Differential evolution—a simple and efficient adaptive scheme for global optimisation over continuous spaces. TechnicalReportTR-95-012, ICSI, Berkeley, CA
16. Birbil ŞI, Fang SC (2003) An electromagnetism-like mechanism for global optimization. J Glob Optim 25(1):263–282
17. Goldberg DE (1989) Genetic algorithm in search optimization and machine learning. Addison-Wesley
18. Cuevas E (2013) Block-matching algorithm based on harmony search optimization for motion estimation. Appl Intel 39(1):165–183
19. Díaz-Cortés M-A, Ortega-Sánchez N, Hinojosa S, Cuevas E, Rojas R, Demin A (2018) A multi-level thresholding method for breast thermograms analysis using Dragonfly algorithm. Infrared Phys Technol 93:346–361
20. Díaz P, Pérez-Cisneros M, Cuevas E, Hinojosa S, Zaldivar D (2018) An improved crow search algorithm applied to energy problems. Energies 11(3):571
21. Cuevas E, Gálvez J, Hinojosa S, Zaldívar D, Pérez-Cisneros M (2014) A comparison of evolutionary computation techniques for IIR model identification. J Appl Math 827206

Chapter 2
A Metaheuristic Scheme Based on the Hunting Model of Yellow Saddle Goatfish

Animals have demonstrated clever hunting strategies when they collaborate with each other. These techniques involve swarm intelligence and have been applied to develop bio-inspired algorithms, a research field for solving optimization problems. In the last years, several methods arise from inspiration in nature to model collective animal behavior. In this chapter, a biologically-inspired algorithm called Yellow Saddle Goatfish Algorithm (YSGA) is presented. The YSGA is based on collaborative hunting techniques manifested by the yellow saddle goatfish *Parupeneus cyclostomus*. Adult goatfish may form groups and play two different roles to hunt prey in corals: chaser and blocker. Any member of the goatfish shoal can become a chaser by either starting or continuing a hunt while the remaining individuals surround the corals to block escape routes of prey. Depending on the role, different evolutionary operators are applied to each individual to emulate these behaviors. The performance of this approach is tested by comparing optimization results to other related metaheuristic methods. Several standard benchmark functions commonly in the literature were considered to obtain optimal results. Furthermore, YSGA is applied to solve specific engineering optimization problems.

2.1 Introduction

Most real-world optimization problems are highly complex to solve. Fortunately, nature has been a comprehensive source of inspiration for designing metaheuristic algorithms based on biological mechanisms that can solve them. Several bio-inspired metaheuristics use the intelligent collective behavior of many species of animals as a guideline to develop methods that model it. These techniques have demonstrated that the smart mechanisms of nature for solving problems can be applied for optimization.

© The Author(s), under exclusive license to Springer Nature Switzerland AG 2021

E. Cuevas et al., *Recent Metaheuristic Computation Schemes in Engineering*,
Studies in Computational Intelligence 948,
https://doi.org/10.1007/978-3-030-66007-9_2

Cooperative behavior in animals has been a research field commonly referred to as swarm intelligence. There is a broad assortment of optimization techniques in the literature of evolutionary algorithms based on swarm intelligence. Some of the most popular swarm-based methods are the Particle Swarm Optimization (PSO), which mimics the social behavior of birds or fishes [1]; the Synchronous-Asynchronous Particle Swarm Optimization (SA-PSO), that overcomes the PSO and is based on the time in which the particles are updated [2]; the Artificial Bee Colony (ABC), that relies on the collaborative lifestyle of bee colonies [3]; the Crow Search Algorithm (CSA), which is based on the intelligent behavior of crows [4]; the Gray Wolf Optimization (GWO), which emulates the hunting strategies and the leadership of gray wolves [5]; the Natural Aggregation Algorithm (NAA), based on collective decisions of group-living animals [6]; the Whale Optimization Algorithm (WOA), which is inspired by the social behavior of humpback whales [7]; the Selfish Herd Optimizer (SHO), that relies on selfish behavior in herds subjected to a predation risk [8]; or even the Moth-Flame Optimization Algorithm (MFO), based on the navigation of moths in nature [9].

However, other algorithms that are not swarm-based have been reported in the literature. A classification of metaheuristic techniques according to its nature, is given by three main categories: physics-based, evolution-based, and swarm-based. The last type has already been discussed. The universe and physical laws inspire Physics-based methods. Some of the most popular physics-based algorithms are the Gravitational Search Algorithm (GSA), which is founded on the law of gravity and mass interactions [10]; the Gravitational Local Search (GLSA), that relies on natural principles of gravitation [11]; the Big Bang-Big Crunch (BB-BC), inspired by a theory of the universe´s evolution [12]; and the Black Hole (BH), which is based on the black hole phenomenon and how it swallows stars [13]. Evolution-based methods imitate natural evolution by combining the best individuals to generate new ones. Some of the most representative evolution-based algorithms are the Genetic Algorithm (GA), which relies on the Darwinian theory for adaptation and survival in a changing environment [14]; the Differential Evolution (DE), that uses crossover and mutation as operators according to differential variations [15]; and the Evolution Strategy (ES), also based on the evolutionary process [16].

Despite the differences highlighted by the classification of metaheuristics techniques [54], all those algorithms have characteristics in common since they were created to improve the process for solving optimization problems and achieve higher accuracy and efficiency. The improvement implies the design of operators that reach an appropriate balance among the exploration and exploitation process [17, 18, 52, 53]. At the beginning of the search strategy, there must be more exploration for covering all the search space looking for possible solutions. As the iterations progress, the exploitation must be intensified to refine the quality of the solutions. Guiding a successful search strategy involves accomplishing that balance. However, reaching a trade-off between exploration and exploitation is not an easy task. Therefore, the development of new metaheuristic techniques arises to look for enhancement, which is worthwhile work.

In this chapter, a swarm-based algorithm for solving optimization problems called Yellow Saddle Goatfish Algorithm (YSGA) is presented. YSGA is a bio-inspired algorithm founded on the recent discovery of the peculiar hunting behavior of the Yellow Saddle Goatfish *Parupeneus cyclostomus*, which tends to form relatively stable hunting groups and employs sophisticated and highly coordinated actions.

The algorithm considers two different kinds of search agents: blockers and chasers. Based on the natural behavior of the yellow saddle goatfish that has been reported, different operators are applied to each search agent for guiding and controlling the search process.

2.2 Yellow Saddle Goatfish Shoal Behavior

On the reefs of the Indo-Pacific, especially on the Red Sea, researchers have observed the behavior and characteristics of the goatfish *Parupeneus cyclostomus,* which is a very particular kind of fish of the family Mullidae and is often called either yellow saddle goatfish or goldsaddle goatfish due to the distinctive color of the females [19]. Goatfishes has sensory barbels designed to detect hidden prey in sedimentary substrates. Its diet is based on various invertebrates such as polychaetes, crustaceans, sipunculids, and small bivalves [20]. Besides, it has been reported that adults eat vertebrates as well, and they feed mainly on fishes [21, 22]. In addition to these characteristics, specialists have detected that goatfishes manifest cooperative hunting strategies never seen before in fish shoals. This is an unusual behavior that has attracted attention [23]. Collaboration has been observed so far only in a few species, mostly on mammals such as lions, chimpanzees, dolphins, and orcas [24–28].

Adults may live either solitary or in groups. Lonely individuals hunt hidden immobile prey in sandy areas while individuals that form groups hunt mobile prey in corals. Generally, these groups are composed of goatfishes of similar size and not family-related. Besides, groups are not exclusive or permanent. Individuals may join a group and then switch to another. Furthermore, no territorial clashes have been observed; groups usually ignore each other [23].

Goatfish *Parupeneus cyclostomus* play two different roles to hunt prey in corals: the chaser and the blocker. Any member of the goatfish shoal can become a chaser by either starting or continue a hunt while the remaining individuals will become blockers. Quick acceleration indicates that the hunting has begun. Therefore, the prey will try to escape among corals and hide into cracks. The blockers strategically surround the corals to block escape routes of mobile prey, while the chaser attempts to force it out by inserting its barbels into crevices. Goldsaddle goatfish do not share the prey. This behavior causes a problematic situation for chaser fish. The insertion of the barbels by the chaser may direct the prey toward a blocker partner, turning it into the new chaser. Thus, the individual who inserts its barbels takes the risk of not hunting the prey. The chaser goatfish must face this dilemma. If a chaser succeeds, then it swallows the whole prey without sharing it. Roles are flexible and interchangeable. Individuals may act as chasers on some occasions, and as blockers in others. The exchange of roles happens even in the same hunt [23].

2.3 Yellow Saddle Goatfish Algorithm (YSGA)

The collective hunting behavior of yellow saddle goatfish has been a guideline to develop a bio-inspired optimization algorithm. Although goatfish *Parupeneus cyclostomus* may live either solitary or in groups, this method is based only on the group behavior of this species.

The algorithm considers the coral reefs as the search space, goatfishes as search agents, and preys as solutions. Only one prey represents the global best solution. The remaining are local solutions. Every group of goatfish is deemed as a cluster. Two types of search agents are modeled in every cluster: chaser fish and blocker fish. To emulate these distinctive behaviors, different evolutionary operators are applied to each search agent. Every particle within the search space is a candidate solution represented by the position of a search agent. A fitness value related to a successful hunt is assigned to each particle.

2.3.1 Initial Population

A population $\boldsymbol{P}$ of m goatfishes $\{\boldsymbol{p}_1, \boldsymbol{p}_2, \ldots, \boldsymbol{p}_m\}$ is randomly generated and uniformly distributed within the boundaries b^{high} and b^{low} of the n-dimensional search space, where m is the population size and $\boldsymbol{p}_i \in \boldsymbol{P}$ is the vector of decision variables defined as $\boldsymbol{p}_i = \{p_i^1, p_i^2, \ldots, p_i^n\}$. The initialization is formulated as follows:

$$p_i^j = rand \cdot \left(b_j^{high} - b_j^{low}\right) + b_j^{low}$$
$$i = 1, 2, \ldots, m;\ j = 1, 2, \ldots, n \qquad (2.1)$$

where $rand$ is a random number among [0, 1].

The yellow saddle goatfish tend to form groups for collaborative hunting. Under this assertion, the entire population $\boldsymbol{P}$ is divided into clusters to model this behavior. Figure 2.1 illustrates the partition of the population into k clusters where every cluster $\boldsymbol{c}_k$ has a chaser fish $\boldsymbol{\Phi}_l$ at the center and the blocker fish $\boldsymbol{\varphi}_g$ swims around it.

Clustering algorithms have been classified into hierarchical and partitional [29, 30]. Several hierarchical [31] and partitional [32] clustering methods have been proposed in the literature [33]. However, YSGA uses the K-means algorithm [34] for clustering due to its simplicity, efficiency, and easy implementation [33].

K-means is one of the most popular partitional algorithms and has been widely used for clustering [30]. This method partitions a data set into k number of groups

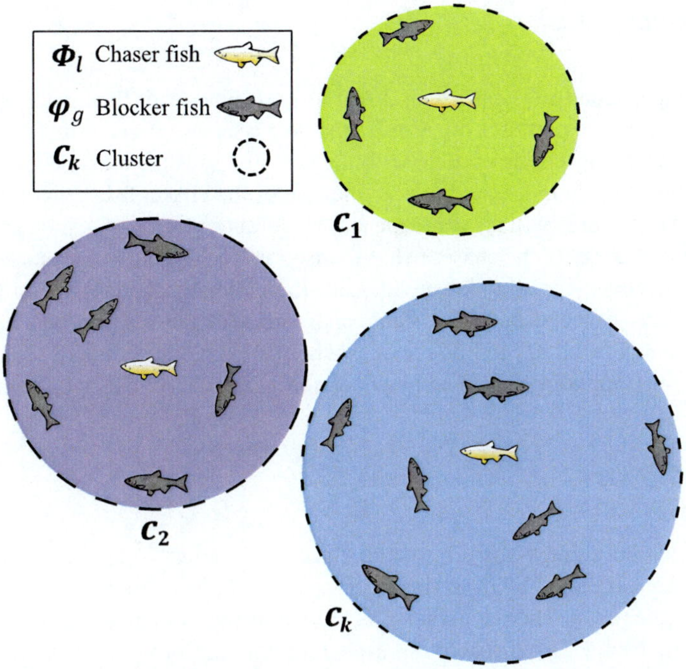

Fig. 2.1 Role identification and partition of the goatfish population P into k clusters

$\{\boldsymbol{c}_1, \boldsymbol{c}_2, \ldots, \boldsymbol{c}_k\}$ called clusters and calculates the mean μ_l of each cluster $\boldsymbol{c}_l$ such that the sum of the squared error of every data point to its nearest cluster's mean is minimized [35–37]. Generally, the Euclidean metric is used to calculate the distance between data points and the mean of the cluster [38]. Considering the population of goatfishes $\boldsymbol{P}$ as the data set; the squared error between μ_l and the set of data points $\{\boldsymbol{p}_1, \boldsymbol{p}_2, \ldots, \boldsymbol{p}_h\}$ in cluster $\boldsymbol{c}_l$ is defined as

$$e(\boldsymbol{c}_l) = \sum_{\boldsymbol{p}_g \in \boldsymbol{c}_l} \left\| \boldsymbol{p}_g - \mu_l \right\|^2$$
$$g = 1, 2, \ldots, h; l = 1, 2, \ldots, k \quad (2.2)$$

In this case, h is determined according to the k-means algorithm, and its value can be different for every cluster $\boldsymbol{c}_l$. The purpose of k-means is to minimize the objective function given by the sum of the squared error over all k clusters. This function is denoted as

$$E(\boldsymbol{C}) = \sum_{l=1}^{k} e(\boldsymbol{c}_l) \quad (2.3)$$

2.3.2 Chaser Fish

Every group of goatfishes has a chaser fish $\boldsymbol{\Phi}_l \in \boldsymbol{P}$ who leads the hunt, see Fig. 2.1. The selection of the particle who represents the chaser fish is according to its fitness value. For each cluster, the particle with the best fitness value must be the closest to the solution. Thus, it is selected to play the role of the chaser fish.

In a chase, the prey will try to escape among corals and attempt to hide in crevices. The hunting strategy of the chaser fish is to insert its barbels into these cracks to force prey out. This behavior is modeled with Lévy flight, which is a class of non-Gaussian random processes that uses a Lévy stable distribution to generate random walks [39].

The chaser fish will try to find crevices where prey may be hiding by changing its position with a random walk. The new location of the chaser fish is calculated as

$$\begin{aligned} \boldsymbol{\Phi}_l^{t+1} &= \boldsymbol{\Phi}_l^t + \alpha \oplus L\grave{e}vy(\beta) \\ 0 &< \beta \leq 2 \end{aligned} \tag{2.4}$$

The new and current position of the chaser fish are $\boldsymbol{\Phi}_l^{t+1}$ and $\boldsymbol{\Phi}_l^t$ respectively. The step size is defined by α, in this approach $\alpha = 1$. The product $\oplus$ means entry-wise multiplications. The parameter β is known as the Lévy index and controls the shape of the probability distribution, especially the tail. For $\beta = 1$ the probability distribution function takes the form of the Cauchy distribution. On the other hand, for $\beta = 2$ the probability distribution corresponds to a Gaussian distribution [40]. In YSGA β increases linearly from 1.99 to 2, the objective is to decrease the length of the steps through the generations, this mechanism leads to a Gaussian distribution which favors the exploitation as iterations increase. The value of β is calculated as

$$\beta = 1.99 + \frac{0.001t}{t_{max}/10} \tag{2.5}$$

where t is the current generation and t_{max} is the maximum number of iterations.

Essentially, Lévy flight draws random walks by generating random steps from a Lévy distribution [41, 42] as follows

$$S = \alpha \oplus levy(\beta) \sim \alpha\left(\frac{u}{|v|^{1/\beta}}\right)\left(\boldsymbol{\Phi}_l^t - \boldsymbol{\Phi}_{best}^t\right) \tag{2.6}$$

Here, S is the random step and $\boldsymbol{\Phi}_{best}^t$ is the best chaser fish so far among all the clusters. From the normal distribution, u and v are defined as

$$\begin{aligned} u &\sim N\left(0, \sigma_u^2\right) \\ v &\sim N\left(0, \sigma_v^2\right) \end{aligned} \tag{2.7}$$

where σ_u and σ_v are denoted as follows, consider Γ as the Gamma function.

$$\sigma_u = \left\{ \frac{\Gamma(1+\beta)\sin\frac{\pi\beta}{2}}{\Gamma\left(\frac{1+\beta}{2}\right)\beta 2^{(\beta-1)/2}} \right\}^{1/\beta}, \sigma_v = 1 \tag{2.8}$$

Under these assertions, the new position of the chaser fish given in Eq. (2.4) can be rewritten as

$$\boldsymbol{\Phi}_l^{t+1} = \boldsymbol{\Phi}_l^t + S \tag{2.9}$$

Equation (2.9) is valid for every chaser fish $\boldsymbol{\Phi}_l$ corresponding to every cluster $\boldsymbol{c}_l$ except for the global best $\boldsymbol{\Phi}_{best}$. If the new position of the best chaser fish is calculated with Eq. (2.9), the subtraction $\left(\boldsymbol{\Phi}_i^t - \boldsymbol{\Phi}_{best}^t\right)$ shown in Eq. (2.6) will be zero. Hence, S will be zero, and the position of the best chaser fish will not change. Thus, the new position of the best chaser fish is calculated as

$$\boldsymbol{\Phi}_{best}^{t+1} = \boldsymbol{\Phi}_{best}^t + S' \tag{2.10}$$

where S' is defined as follows.

$$S' = \alpha\left(\frac{u}{|v|^{1/\beta}}\right) \tag{2.11}$$

2.3.3 Blocker Fish

Once the chaser fish has been chosen for each cluster, the remaining goatfishes become blockers, see Fig. 2.1. The hunting strategy of the blocker fish $\boldsymbol{\varphi}_g \in \boldsymbol{P}$ is to surround the corals to block escape routes of the prey while the chaser fish try to force it out by inserting its barbels into crevices. To model this behavior, a logarithmic spiral is applied to the movement of the blocker fish. Since the chaser fish is the closest to the prey, the behavior of the blocker fish is to draw a spiral path around the chaser goatfish. Figure 2.2 shows the spiral movement of the blocker goatfish. Notice that each blocker fish has its own spiral path, which is changing throughout iterations.

The new position of the blocker fish $\boldsymbol{\varphi}_g^{t+1}$ is calculated according to a logarithmic spiral defined as follows

$$\boldsymbol{\varphi}_g^{t+1} = D_g \cdot e^{b\rho} \cdot \cos 2\pi\rho + \boldsymbol{\Phi}_l \tag{2.12}$$

where ρ is a random number among $[a, 1]$ which defines how close is the blocker to the chaser fish considering the spiral path. To intensify exploitation, a is linearly decreased from -1 to -2 as iterations increase. The position of the chaser fish is

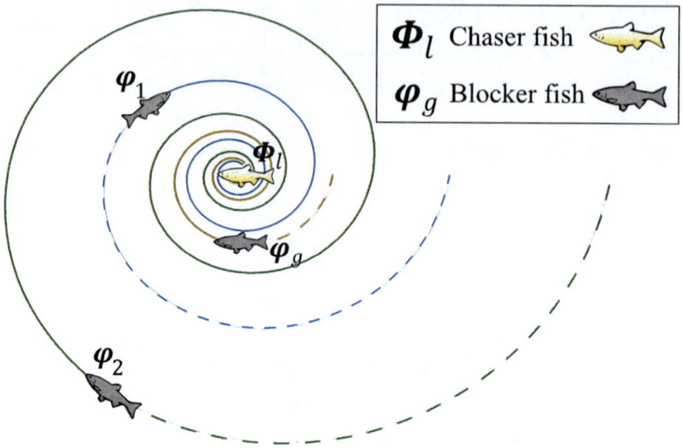

Fig. 2.2 Spiral path of blocker fish

defined when $\rho = -2$. Hence, the closest position to the chaser fish occurs when $\rho \approx -2$ and the farthest when $\rho = 1$, see Fig. 2.3.

Since the range tends to $[-2, 1]$ over iterations, ρ increase the possibility of approaching -2 throughout generations with a minimum probability of reaching it

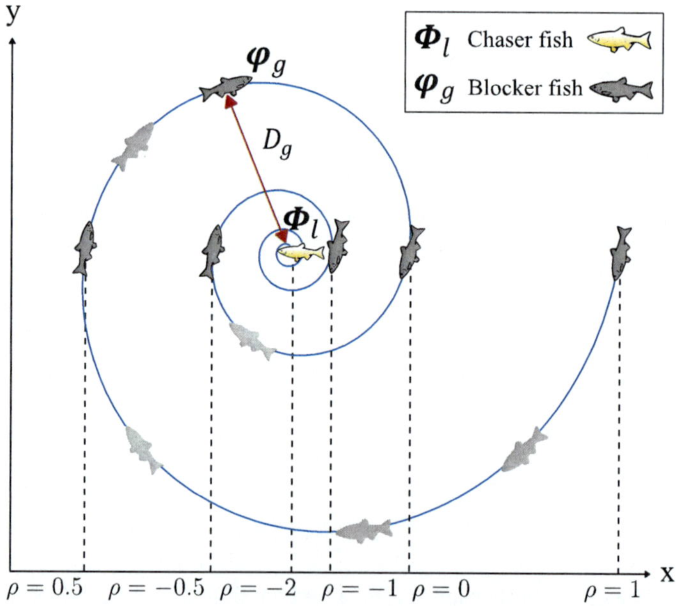

Fig. 2.3 Some of the possible positions reached by a blocker fish depending on the value of ρ

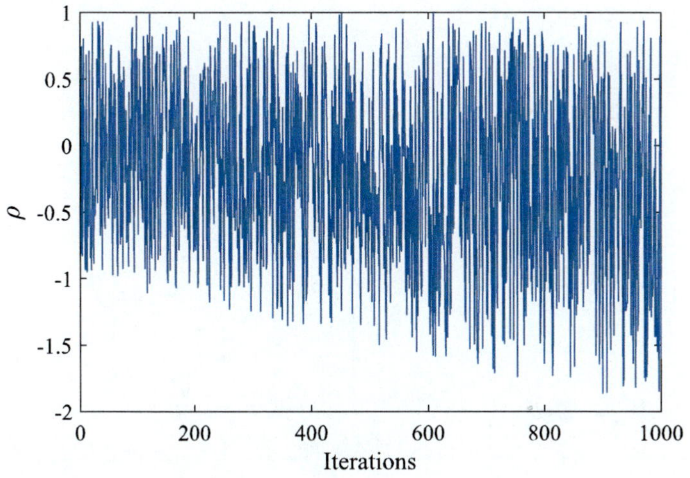

Fig. 2.4 Different values of ρ throughout iterations

(due to randomness), which is just as expected, see Fig. 2.4. Under this schema, blocker goatfish tend to exploit more, proportionally to the number of iterations.

The parameter b in Eq. (2.12) is a constant which determines the shape and the direction of the spiral, in this approach $b = 1$. The term D_g is the distance between the current position of the blocker fish $\boldsymbol{\varphi}_g^t$ and the chaser fish $\boldsymbol{\Phi}_l$ in the cluster $\boldsymbol{c}_l$ described as

$$D_g = \left| r \cdot \boldsymbol{\Phi}_l - \boldsymbol{\varphi}_g^t \right|$$
$$\left\{ \boldsymbol{\Phi}_l, \boldsymbol{\varphi}_g^t \right\} \in \boldsymbol{c}_l \tag{2.13}$$

where r is a random number among $[-1, 1]$ which disturbs the distance D_g avoiding over-exploitation and promoting diversity in the trajectories that define the movement of blocker goatfish.

2.3.4 Exchange of Roles

During the hunting process, blocker goatfish will be waiting for the prey to escape from the chaser fish through some of the cracks. In this case, if the prey manages to escape from the chaser, the blocker fish closest to the prey´s escape route will lead the hunt to become the new chaser fish, and the chaser fish will become a blocker fish. In Fig. 2.5, a role exchange is illustrated. In this example, it is assumed that the blocker fish $\boldsymbol{\varphi}_2$ has achieved better fitness value than the other goatfish in the current iteration t. This means that $\boldsymbol{\varphi}_2$ has moved closer to the prey, becoming a chaser fish. Hence, the role exchange is carried out in iteration $t + 1$.

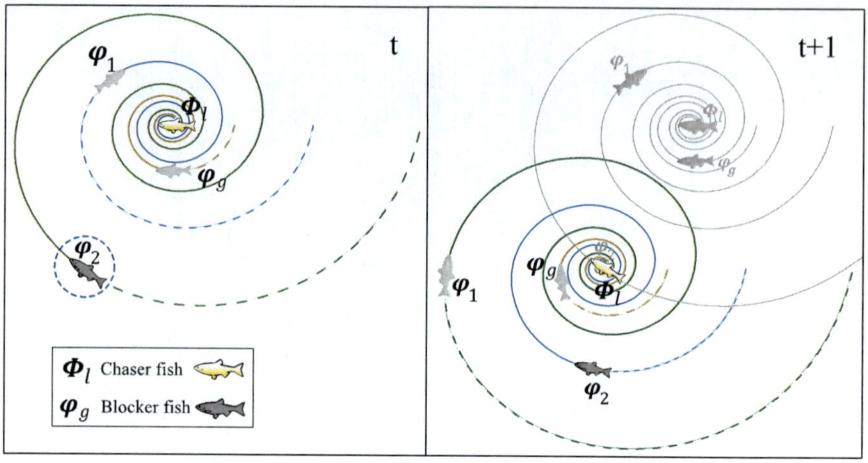

Fig. 2.5 Exchange of roles

YSGA verifies possible role exchanges based on the performance of the search agents. The proximity of a goatfish to the prey determines its role. The goatfish closest to the prey must be the chaser. Closeness is defined according to the fitness value of every goatfish. At the end of each generation, a role update is executed if necessary. For each cluster, if a blocker fish has better fitness value than the chaser, it means that it has found a better solution, then a role exchange is performed.

2.3.5 *Change of Zone*

To avoid being trapped in local optima, a successful hunt is simulated, and consequently, a change of zone is implemented, searching for new prey. For every cluster, if a predefined number of iterations λ has passed without finding a better solution, then it is considered as a successful hunt, so a change of area is performed for all the goatfish in the cluster as follows

$$\boldsymbol{p}_g^{t+1} = \frac{\boldsymbol{\Phi}_{best} + \boldsymbol{p}_g^t}{2} \tag{2.14}$$

where $\boldsymbol{p}_g^{t+1}$ is the new position of a goatfish regardless of whether it is a chaser or blocker. The best solution so far among all the clusters is given by the chaser fish $\boldsymbol{\Phi}_{best}$, $\boldsymbol{p}_g^t$ is the current position of a goatfish (chaser or blocker) member of the cluster. In Fig. 2.6, a change of zone is illustrated. In this example, the chaser $\boldsymbol{\Phi}_1$ has not improved; in that case, the cluster $\boldsymbol{c}_1$ must change to another area considering the position of the global best $\boldsymbol{\Phi}_{best}$.

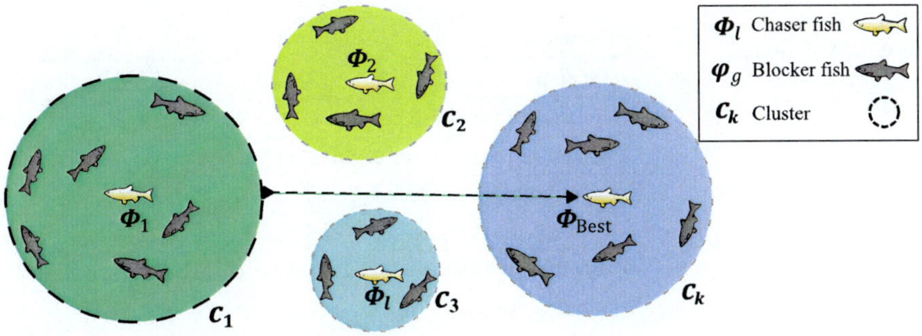

Fig. 2.6 Change of zone of cluster c_1

2.3.6 Computational Procedure

An iterative process is implemented in which a sequence of operations is executed. As input data, the algorithm requires the number of goatfish population m, the number of clusters k, the maximum number of iterations t_{max} and the search space $\mathbf{S}$. The goatfish population $\boldsymbol{P}$ is initialized and uniformly distributed over the search space $\mathbf{S}$ within the predefined limits b^{high} and b^{low} to represent the initial population $\boldsymbol{P}(t)$ where $t = 1$ for the first iteration, also known as the first generation.

After initialization, the fitness value of every goatfish is calculated, the global best $\boldsymbol{\Phi}_{best}$ is identified. Then, the k-means algorithm is applied to partition the goatfish population into k clusters $\{\boldsymbol{c}_1, \boldsymbol{c}_2, \ldots, \boldsymbol{c}_k\}$. For every cluster $\boldsymbol{c}_l$, the goatfish with better fitness value will be identified as the chaser fish $\boldsymbol{\Phi}_l$ of that cluster and the remaining will be identified as blocker fish $\boldsymbol{\varphi}_g$.

Later, different operators are applied to the search agents: a hunting routine for chaser fish and a blocker routine for blocker goatfish. Each particle moves according to these operators. The hunting routine changes the position of the chaser goatfish by generating random steps from a Lévy distribution. In contrast, the blocker routine changes the position of the blocker fish by creating a logarithmic spiral path around the chaser.

Subsequently, the fitness value of every goatfish is calculated. In each cluster, if a blocker has better fitness value than the chaser, then a change of role is made, the current chaser will become a blocker, and the blocker will become the new chaser fish. On the other hand, if any of the chaser goatfish (considering all chasers in all clusters) outperforms the global best, then the global best is updated.

For each iteration and every cluster, if the chaser goatfish has not improved, then a counter q that regulates stagnation is increased until a predefined limit λ is reached. If q becomes bigger than λ, then a change of zone is made, this means that there is no more prey to hunt in that area, so it is necessary to search for the prey in a different zone.

The whole process is repeated until the maximum number of iterations is reached. The YSGA algorithm summary is explained in the pseudo-code of Table 2.1.

Table 2.1 Pseudo-code of the YSGA algorithm

Pseudo-code
Input: m, k, t_{max}, **S**
Initialize the goatfish population $\boldsymbol{P} = \{\boldsymbol{p}_1, \boldsymbol{p}_2, \ldots, \boldsymbol{p}_m\}$
Calculate the fitness value of each particle
Identify the global best $\boldsymbol{\Phi}_{best}$
Partition the population $\boldsymbol{P}$ into k clusters $\{\boldsymbol{c}_1, \boldsymbol{c}_2, \ldots, \boldsymbol{c}_k\}$
Identify the chaser $\boldsymbol{\Phi}_l$ and the blocker $\boldsymbol{\varphi}_g$ fish for every cluster
While ($t < t_{max}$)
For every cluster $\boldsymbol{c}_l$
Execute hunting routine for chaser fish
Execute blocking routine for blocker fish
Calculate the fitness value of each goatfish
If $\boldsymbol{\varphi}_g$ has better fitness value than $\boldsymbol{\Phi}_l$
exchange roles by updating $\boldsymbol{\Phi}_l$
End If
If $\boldsymbol{\Phi}_l$ has better fitness value than $\boldsymbol{\Phi}_{best}$
update $\boldsymbol{\Phi}_{best}$
End If
If the fitness value of $\boldsymbol{\Phi}_l$ has not improved
$q \leftarrow q + 1$
End If
If $q > \lambda$
execute a routine for changing zone
$q \leftarrow 0$
End If
End For
$t \leftarrow t + 1$
End While
Output $\boldsymbol{\Phi}_{best}$

2.4 Experimental Results

To prove the effectiveness, the presented methodology is tested on a set of 27 benchmark functions ($f_1 - f_{27}$) described in detail in Tables 2.23, 2.24, and 2.25. These functions are classified into three main categories according to their characteristics: unimodal, multimodal, and composite. Table 2.23 contains the unimodal test functions, while Table 2.24 the multimodal and Table 2.25 the composite functions.

YSGA is evaluated by comparing the results against other related optimization techniques. In the experiments, the following approaches have been considered:

Table 2.2 Parameter configuration, according to the reported references

Settings configuration	
YSGA	Number of clusters $k = 4$
[5] GWO	Parameters configured according to the reference
[43] PSO	$c_1 = 2$, $c_2 = 2$, velocity clamp = 2, minW = 0.2, maxW = 0.9
[2] SA-PSO	$c_1 = 2$, $c_2 = 2$, velocity clamp = 2, minW = 0.2, maxW = 0.9, number of groups $c = 5$, group size = 10, initial distance 50% search space
[3] ABC	Onlooker 50%, employees 50%, acceleration coefficient upper bound = 1, abandonment limit L = round (0.6 * dimensions * population)
[4] CSA	$p_a = 0.25$
[6] NAA	Number of shelters $N^s = 8$, shelter capacity $Cp^s = 6$, the probability of local search $Cr_{global} = 0.9$, scaling factor $\delta = 1$, generalized movement factor $\alpha = 1.2$

The Gray Wolf Optimization (GWO), the Particle Swarm Optimization (PSO), the Synchronous-Asynchronous Particle Swarm Optimization (SA-PSO), the Artificial Bee Colony (ABC), the Crow Search Algorithm (CSA) and the Natural Aggregation Algorithm (NAA). All these techniques represent some of the most popular swarm-based methods used in the last few years. The set of 27 benchmark functions has been solved using these algorithms, including YSGA, and the results have been compared.

In the experiments, the total population m has been set to 50 individuals, and the maximum number of iterations t_{max} has been set to 1000 for all the algorithms involved. The test functions have been implemented in $n = 30$, $n = 50$, and $n = 100$ dimensions. Each experiment is repeated, considering 30 independent executions to verify the consistency of the results.

The parameters used in each method have been configured according to the reported values in which their best performance is achieved. Table 2.2 shows the configuration of these settings.

For comparisons, three different indicators are considered: The Average Best-so-far (AB) solution, the Median Best-so-far (MB) solution, and the Standard Deviation (SD) of the best-so-far solutions. Every benchmark function is tested on each algorithm 30 times, generating 30 results representing the best-found solutions. The mean of these 30 outcomes is calculated to obtain AB. Similarly, the median of these 30 results is determined to generate MB, and the standard deviation of the 30 results is estimated to acquire SD of the best-so-far solutions. Indicators AB and MB evaluate the accuracy of the solutions, while SD the dispersion and, therefore, the robustness of the algorithm.

Since the benchmark functions were classified into three different categories (unimodal, multimodal, and composite), the experiments performed are presented according to this classification.

2.4.1 Results of Unimodal Test Functions

In Table 2.23, functions from f_1 to f_5 are unimodal. These kinds of functions have a single optimum. The experimental results are listed in Tables 2.3, 2.4, and 2.5, considering the performance of each algorithm for each test function in 30, 50, and 100 dimensions, respectively. Indicators AB, MB, and SD illustrate the accuracy of the results. The best outcomes are highlighted in boldface.

From Table 2.3, it can be seen how the YSGA algorithm has better performance than the other methods. Indicators AB, MB, and SD demonstrate that the presented approach stands out from the rest of the techniques for all the test functions. Although the GWO algorithm has the same SD value as the YSGA in function f_3, the AB and MB values are better in YSGA, which means that both have the same dispersion, but YSGA has better quality in the solution. From Table 2.3, it is also observed that PSO has the worst performance while the other methods achieve different quality values.

To test the scalability of the YSGA algorithm, the experiments have been repeated but now considering 50 dimensions instead of 30. The results of this test are exposed in Table 2.4.

From Table 2.4, it can be appreciated that the YSGA algorithm outperforms the other methods. The best solutions are registered in indicators AB and MB corresponding to the presented approach, while the dispersion SD is also better in the YSGA algorithm than the other techniques. On the other hand, the GWO algorithm reaches equivalent results to the YSGA in the test function f_3.

Additionally, the experiments have been repeated but now considering 100 dimensions, which suppose a more significant challenge. The results of this test are reported in Table 2.5.

From Table 2.5, it can be observed that the YSGA algorithm has better performance than the other methods in all the benchmark functions. Indicators AB, MB, and SD prove that the presented approach records the best results. Although the GWO algorithm has the same SD value as the YSGA for function f_3, the AB and MB values are better in YSGA, which means that both have the same dispersion, but YSGA has better quality in the solution. From Table 2.5, it is also detected that the contending techniques GWO, PSO, SA-PSO, and ABC get worst results as the number of iterations increases. From Tables 2.3, 2.4, and 2.5, it can be concluded that the presented method has a good performance in high-dimensional problems.

The results obtained in Tables 2.3, 2.4, and 2.5 are examined for validation in a statistic context by applying a non-parametric analysis identified as the Wilcoxon test [44, 45]. The examination measures the difference between the two methods that are compared. For both algorithms in comparison, a 5% (0.05) of the significance level for the Average Best-so-far (AB) is considered by the Wilcoxon test. Under this schema, the presented method is compared against the other techniques one by one.

For comparison, the following group configuration is set: YSGA versus GWO, YSGA versus PSO, YSGA versus SA-PSO, YSGA versus ABC, YSGA versus CSA, and YSGA versus NAA. In the Wilcoxon test, a null hypothesis is established in which the discrepancy between the two approaches is not perceptible. In contrast,

Table 2.3 Optimal results of the unimodal test functions considering 30 dimensions

Function		YSGA	GWO	PSO	SA-PSO	ABC	CSA	NAA
$f_1(x)$	**AB**	**1.40E−182**	1.53E−72	8.74E−01	3.84E−07	1.29E−03	3.98E−05	6.36E−58
	MB	**2.60E−188**	1.15E−73	1.60E−09	4.18E−07	1.21E−03	3.92E−05	1.95E−60
	SD	**0.00E+00**	3.53E−72	4.79E+00	1.13E−07	5.74E−04	2.12E−05	2.11E−57
$f_2(x)$	**AB**	**1.80E−172**	5.69E−71	7.80E+02	8.06E−05	4.91E−02	3.68E−01	1.15E−55
	MB	**6.40E−185**	7.00E−72	6.00E+02	3.87E−05	4.55E−02	3.39E−01	1.81E−59
	SD	**0.00E+00**	2.30E−70	7.27E+02	8.04E−05	2.32E−02	2.24E−01	5.48E−55
$f_3(x)$	**AB**	**0.00E+00**	1.06E−235	6.86E−18	1.32E−11	1.73E−03	3.64E−09	5.57E−50
	MB	**0.00E+00**	1.49E−241	1.60E−20	1.64E−12	1.21E−03	1.36E−09	5.41E−63
	SD	**0.00E+00**	0.00E+00	3.57E−17	2.14E−11	1.55E−03	5.61E−09	3.05E−49
$f_4(x)$	**AB**	**2.20E−170**	3.20E−68	1.68E+06	8.02E−03	4.33E+01	1.68E+03	6.93E−52
	MB	**3.00E−182**	7.49E−69	1.51E+06	2.88E−03	3.52E+01	1.38E+03	4.00E−55
	SD	**0.00E+00**	7.80E−68	1.16E+06	1.11E−02	2.54E+01	1.15E+03	3.74E−51
$f_5(x)$	**AB**	**7.50E−174**	5.11E−69	3.02E+04	9.26E−05	2.38E+00	1.39E+01	4.43E−56
	MB	**7.00E−185**	3.23E−70	2.58E+04	5.92E−05	1.98E+00	1.01E+01	2.83E−57
	SD	**0.00E+00**	1.34E−68	2.40E+04	9.28E−05	1.24E+00	9.93E+00	1.40E−55

Table 2.4 Optimal results of the unimodal test functions considering 50 dimensions

Function		YSGA	GWO	PSO	SA-PSO	ABC	CSA	NAA
$f_1(x)$	**AB**	**4.31E−172**	3.96E−87	2.80E+01	3.88E−04	4.20E+00	1.20E−04	1.29E−57
	MB	**3.03E−185**	2.29E−88	2.62E+01	3.97E−04	4.23E+00	1.16E−04	2.75E−62
	SD	**0.00E+00**	1.82E−86	1.94E+01	1.25E−04	9.17E−01	4.78E−05	7.04E−57
$f_2(x)$	**AB**	**3.05E−162**	6.17E−86	3.46E+03	3.49E−02	1.76E+02	1.29E+00	2.05E−54
	MB	**8.58E−184**	2.85E−86	2.90E+03	1.58E−02	1.77E+02	8.99E−01	3.08E−60
	SD	**1.68E−161**	7.57E−86	2.32E+03	4.03E−02	4.25E+01	9.36E−01	1.12E−53
$f_3(x)$	**AB**	**0.00E+00**	**0.00E+00**	1.01E−13	7.29E+09	1.89E−01	8.76E−10	6.84E−47
	MB	**0.00E+00**	**0.00E+00**	1.52E−14	8.11E+05	1.71E−01	6.40E−10	7.52E−53
	SD	**0.00E+00**	**0.00E+00**	2.07E−13	1.26E+10	1.11E−01	9.00E−10	2.74E−46
$f_4(x)$	**AB**	**6.74E−167**	3.10E−82	1.61E+07	6.14E−01	1.58E+05	7.83E+03	2.10E−54
	MB	**2.05E−178**	8.24E−83	1.64E+07	6.16E−01	1.54E+05	7.35E+03	1.33E−56
	SD	**0.00E+00**	6.07E−82	7.26E+06	4.03E−01	3.09E+04	4.22E+03	9.13E−54
$f_5(x)$	**AB**	**2.70E−170**	4.33E−84	1.87E+05	1.53E−02	8.12E+03	5.35E+01	1.08E−54
	MB	**7.86E−181**	8.16E−85	1.68E+05	5.06E−03	7.73E+03	4.98E+01	7.85E−58
	SD	**0.00E+00**	7.30E−84	9.33E+04	1.84E−02	2.18E+03	2.79E+01	3.16E−54

Table 2.5 Optimal results of the unimodal test functions considering 100 dimensions

Function		YSGA	GWO	PSO	SA-PSO	ABC	CSA	NAA
$f_1(x)$	**AB**	**2.55E−162**	5.13E−37	1.23E+02	7.09E−01	2.04E+02	8.83E−05	2.36E−122
	MB	**7.28E−176**	2.44E−37	1.31E+02	1.91E−01	2.06E+02	8.62E−05	5.82E−125
	SD	**1.28E−161**	8.64E−37	4.31E+01	4.31E+01	1.52E+01	1.49E−05	8.19E−122
$f_2(x)$	**AB**	**1.62E−162**	9.32E−35	3.12E+04	8.60E+00	2.52E+04	2.95E+00	1.52E−117
	MB	**1.88E−172**	5.84E−35	2.99E+04	9.17E+00	2.52E+04	2.58E+00	1.94E−123
	SD	**8.89E−162**	9.30E−35	7.94E+03	7.94E+03	2.07E+03	1.18E+00	8.30E−117
$f_3(x)$	**AB**	**3.13E−233**	9.17E−179	2.18E−02	6.18E+46	1.16E+00	1.33E−10	1.42E−82
	MB	**0.00E+00**	1.85E−201	5.48E−07	6.86E+42	1.23E+00	7.55E−11	5.34E−97
	SD	**0.00E+00**	**0.00E+00**	9.31E−02	9.31E−02	2.93E−01	2.16E−10	7.71E−82
$f_4(x)$	**AB**	**6.56E−158**	2.96E−31	1.88E+08	3.59E+02	9.06E+07	3.65E+04	6.40E−115
	MB	**1.76E−172**	2.59E−31	1.97E+08	2.02E+02	9.16E+07	3.50E+04	2.44E−120
	SD	**3.30E−157**	2.11E−31	4.70E+07	4.70E+07	6.79E+06	1.30E+04	3.41E−114
$f_5(x)$	**AB**	**2.61E−159**	4.71E−33	1.20E+06	2.24E+01	1.10E+06	1.01E+02	1.02E−115
	MB	**3.16E−173**	2.75E−33	1.25E+06	2.24E+01	1.12E+06	9.48E+01	1.63E−121
	SD	**1.43E−158**	6.04E−33	3.86E+05	3.86E+05	9.76E+04	4.12E+01	5.61E−115

an alternative hypothesis is also accepted in which the difference between the two methods is significant.

However, the possibility of falling into an error Type I [46] increases as the number of comparisons involved in the Wilcoxon test is incremented. To avoid such error, the significance level must be reconsidered using the Bonferroni correction [47]. Hence, the significance level is recalculated by Bonferroni correction, generating a value of 3.3E−04.

The p-values resulting from the Wilcoxon test, considering the Bonferroni correction, are visualized in Tables 2.6, 2.7, and 2.8. After applying the Wilcoxon test over the AB data from Table 2.3, the p-values generated were reported in Table 2.6. Similarly, the p-values generated by the Wilcoxon test over the AB data from Tables 2.4 and 2.5 were listed in Tables 2.7 and 2.8, respectively. In the test, the null hypothesis is rejected for all the p-values that are lower than 3.3E−04, which means that both methods in comparison differ significantly in their results. To simplify the analysis, the symbols ▲, ▼, and ► are used to identify the outcomes of the Wilcoxon test. The symbol ▲ means that the YSGA approach has better performance than the compared method. The symbol ▼ indicates that the YSGA method performs worse

Table 2.6 p-values of Wilcoxon-Bonferroni test comparing YSGA versus GWO, YSGA versus PSO, YSGA versus SA-PSO, YSGA versus ABC, YSGA versus CSA, and YSGA versus NAA over the AB values from Table 2.3, $n = 30$

Function	YSGA versus GWO	YSGA versus PSO	YSGA versus SA-PSO	YSGA versus ABC	YSGA versus CSA	YSGA versus NAA
$f_1(x)$	3.0E−11▲	3.0E−11▲	3.0E−11▲	3.0E−11▲	3.0E−11▲	3.0E−11▲
$f_2(x)$	3.0E−11▲	3.0E−11▲	3.0E−11▲	3.0E−11▲	3.0E−11▲	3.0E−11▲
$f_3(x)$	1.2E−12▲	1.2E−12▲	1.2E−12▲	1.2E−12▲	1.2E−12▲	1.2E−12▲
$f_4(x)$	3.0E−11▲	3.0E−11▲	3.0E−11▲	3.0E−11▲	3.0E−11▲	3.0E−11▲
$f_5(x)$	3.0E−11▲	3.0E−11▲	3.0E−11▲	3.0E−11▲	3.0E−11▲	3.0E−11▲

Table 2.7 p-values of Wilcoxon-Bonferroni test comparing YSGA versus GWO, YSGA versus PSO, YSGA versus SA-PSO, YSGA versus ABC, YSGA versus CSA, and YSGA versus NAA over the AB values from Table 2.4, $n = 50$

Function	YSGA versus GWO	YSGA versus PSO	YSGA versus SA-PSO	YSGA versus ABC	YSGA versus CSA	YSGA versus NAA
$f_1(x)$	3.0E−11▲	3.0E−11▲	3.0E−11▲	3.0E−11▲	3.0E−11▲	3.0E−11▲
$f_2(x)$	3.0E−11▲	3.0E−11▲	3.0E−11▲	3.0E−11▲	3.0E−11▲	3.0E−11▲
$f_3(x)$	1.2E−12▲	3.3E−11▲	1.2E−12▲	1.2E−12▲	1.2E−12▲	1.2E−12▲
$f_4(x)$	3.0E−11▲	3.0E−11▲	3.0E−11▲	3.0E−11▲	3.0E−11▲	3.0E−11▲
$f_5(x)$	1.2E−12▲	3.0E−11▲	3.0E−11▲	3.0E−11▲	3.0E−11▲	3.0E−11▲

Table 2.8. p-values of Wilcoxon-Bonferroni test comparing YSGA versus GWO, YSGA versus PSO, YSGA versus SA-PSO, YSGA versus ABC, YSGA versus CSA, and YSGA versus NAA over the AB values from Table 2.5, $n = 100$

Function	YSGA versus GWO	YSGA versus PSO	YSGA versus SA-PSO	YSGA versus ABC	YSGA versus CSA	YSGA versus NAA
$f_1(x)$	8.6E−07▲	4.4E−13▲	1.4E−08▲	2.7E−12▲	8.4E−10▲	1.9E−12▲
$f_2(x)$	7.5E−10▲	1.3E−11▲	6.5E−12▲	4.6E−10▲	9.1E−12▲	7.6E−11▲
$f_3(x)$	1.0E−10▲	7.5E−13▲	4.1E−12▲	3.0E−12▲	5.5E−13▲	2.4E−11▲
$f_4(x)$	2.8E−06▲	5.4E−11▲	1.9E−12▲	8.8E−11▲	1.0E−13▲	2.9E−12▲
$f_5(x)$	1.7E−06▲	4.8E−12▲	4.9E−12▲	1.9E−12▲	2.8E−14▲	4.9E−11▲

than its competitor. Finally, the symbol ▶ implies that the Wilcoxon analysis could not identify any difference between the YSGA algorithm and its contender.

From Table 2.6, it can be observed that the column of the YSGA versus GWO has a p-value (equal to 3.0E−11) in the first row (corresponding to the test function f_1), which is lower than the significance value of 3.3E−04 calculated with the Bonferroni correction. This means that the null hypothesis is rejected. Hence, there is a difference between both algorithms, which supports the results obtained in Table 2.3 for that function. Under this assumption, the symbol ▲ is placed on the right side of the p-value, indicating that the YSGA has better performance than the GWO. Examining Table 2.6, it is clear that all the p-values in all the columns show evidence against the null hypothesis. After the analysis, it can be concluded that the YSGA algorithm performs better than the other methods for all the unimodal test functions in 30 dimensions.

The Wilcoxon test for the unimodal functions considering 50 dimensions is displayed in Table 2.7. In this case, it is visible that all the p-values in the columns YSGA versus PSO, YSGA versus SA-PSO, YSGA versus ABC, YSGA versus CSA, and YSGA versus NAA validate that the presented approach outperforms its competitors for all the functions. Besides, despite the tie in function f_3 from Table 2.4 for the YSGA and GWO algorithms, the Wilcoxon analysis could determine a discrepancy between the YSGA and the GWO. Thus, the symbol ▲ is placed on the right side of the p-value, indicating that the YSGA has better performance than the GWO.

The results of the Wilcoxon test for the unimodal functions in 100 dimensions are presented in Table 2.8. From these results, it is evident that all the p-values in the columns YSGA versus PSO, YSGA versus SA-PSO, YSGA versus ABC, YSGA versus CSA, and YSGA versus NAA confirm that the presented approach performs better than its contenders for all the test functions.

From Tables 2.6, 2.7, and 2.8, it can be resolved that the statistical test exhibits evidence supporting the results registered in Tables 2.3, 2.4, and 2.5, exposing that the YSGA algorithm outperforms its opponents.

After the Wilcoxon analysis of the presented approach against the other methods and its performance over the unimodal benchmark functions considering 30, 50, and 100 dimensions, it can be concluded that the YSGA algorithm is functional in high-dimensional optimization problems and achieve better performance than its opponents.

2.4.2 Results of Multimodal Test Functions

Table 2.24 exhibits from f_6 to f_{23} the multimodal functions. These functions are characterized by having multiple local optima that increase exponentially as the number of dimensions gets bigger. Therefore, this kind of function is more challenging to optimize, so this test challenges the algorithms to demonstrate their ability to find the global optimum under these conditions. The experimental results are shown in Tables 2.9, 2.10, and 2.11, considering the performance of each method for each test function in 30, 50, and 100 dimensions, respectively. Indicators AB, MB, and SD illustrate the accuracy of the results. The best results are highlighted in boldface.

From Table 2.9, it can be appreciated that the YSGA algorithm achieves better performance than its competitors in functions f_7, f_8, f_9, f_{12}, f_{13}, f_{14},f_{15}, f_{16} and f_{22}. On the other hand, the NAA method reaches better indicators for function f_6 while the SA-PSO obtains the best results in function f_{10}. In the case of function f_{17}, although the SA-PSO approach achieved the best SD, the best AB and MB indicators were obtained by the GWO algorithm, which means that the SA-PSO only has the best dispersion of the data. However, the GWO accomplishes the best results in terms of quality solution, and also its distribution has the best second value. Consequently, it can be concluded that for function f_{17}, the best performance was obtained by the GWO algorithm. Regarding function f_{11}, the YSGA and the GWO reach the same values in AB and MB, but the YSGA achieves better dispersion than the GWO, which means that the YSGA is a more consistent algorithm in terms of the generated results. Similarly, in function f_{21} the YSGA, the GWO, and the NAA obtain the same values in AB and MB, but the YSGA achieves better dispersion. Finally, in functions f_{18} and f_{19} the YSGA and the ABC reach the same values in AB, MB, and SD indicators while the YSGA and the PSO tie in f_{20} as well as the YSGA and the GWO in f_{23}.

To test the adaptability of the YSGA algorithm, the experiments have been repeated but now considering 50 dimensions. The results of this test are informed in Table 2.10.

From Table 2.10, it can be detected that the YSGA algorithm achieves better performance than its competitors in functions f_7, f_8, f_{10}, f_{12}, f_{13}, f_{14},f_{17}, f_{22} and f_{23}. By contrast, the NAA and the GWO outperform the YSGA in functions f_6 and f_9 respectively while the SA-PSO reaches the best results in functions f_{15} and f_{16}. In function f_{11}, the YSGA, the GWO, and the NAA reach the same values in AB, MB, and SD indicators while the YSGA and the PSO tie in f_{20}. Regarding functions f_{18} and f_{19} the YSGA and the ABC reach the same values in AB and MB, but the YSGA achieves better SD value than the ABC, which means that the YSGA is more consistent. Similarly, the YSGA and the NAA reach the same values AB and MB in function f_{21} but the YSGA achieves better dispersion than the NAA.

Table 2.9 Optimal results of the multimodal test functions considering 30 dimensions

Function		YSGA	GWO	PSO	SA-PSO	ABC	CSA	NAA
$f_6(x)$	**AB**	2.49E−15	1.17E−14	1.86E+01	1.49E−03	8.75E−01	3.11E+00	**1.18E−16**
	MB	3.55E−15	1.42E−14	1.99E+01	1.32E−03	7.90E−01	3.13E+00	**0.00E+00**
	SD	1.66E−15	3.25E−15	5.06E+00	9.52E−04	3.81E−01	8.95E−01	**6.49E−16**
$f_7(x)$	**AB**	**5.23E−08**	4.08E−06	2.02E+03	5.35E−02	3.83E+02	6.33E−01	3.23E−02
	MB	**6.62E−15**	3.18E−06	1.93E+03	2.23E−02	3.82E+02	6.34E−01	1.81E−02
	SD	**2.85E−07**	3.85E−06	1.72E+03	5.57E−02	8.26E+01	3.06E−01	4.65E−02
$f_8(x)$	**AB**	**2.43E+01**	2.67E+01	2.11E+05	7.95E+01	6.90E+02	8.82E+01	6.22E+01
	MB	**2.43E+01**	2.62E+01	1.70E+05	8.22E+01	6.92E+02	7.78E+01	4.51E+01
	SD	**2.24E−01**	8.32E−01	1.42E+05	6.55E+00	1.76E+02	5.84E+01	5.05E+01
$f_9(x)$	**AB**	**2.35E−99**	8.96E−40	9.04E+02	2.07E−03	9.10E+10	1.64E+02	6.77E−46
	MB	**8.64E−106**	5.94E−40	8.50E+02	2.29E−03	1.64E+06	1.69E+02	5.15E−47
	SD	**1.20E−98**	9.45E−40	3.35E+02	1.11E−03	4.86E+11	3.97E+01	2.25E−45
$f_{10}(x)$	**AB**	8.38E+01	1.17E+02	1.25E+38	**1.44E+01**	8.21E+78	4.61E+02	2.06E+02
	MB	3.10E+01	3.90E+01	6.05E+19	**1.30E+01**	4.12E+77	2.87E+02	6.98E+01
	SD	1.40E+02	1.69E+02	6.85E+38	**1.51E+01**	2.64E+79	4.25E+02	2.99E+02
$f_{11}(x)$	**AB**	**6.67E−01**	**6.67E−01**	2.85E+04	7.87E−01	1.48E+02	9.13E−01	1.64E+00
	MB	**6.67E−01**	**6.67E−01**	9.58E+01	8.27E−01	1.46E+02	8.48E−01	6.96E−01
	SD	**1.98E−07**	4.38E−07	5.68E+04	1.05E−01	6.04E+01	1.85E−01	1.60E+00
$f_{12}(x)$	**AB**	**−1.13E+03**	−9.42E+02	−1.07E+03	−1.01E+03	−8.28E+02	−1.01E+03	−1.06E+03
	MB	**−1.13E+03**	−9.44E+02	−1.08E+03	−9.77E+02	−8.15E+02	−9.91E+02	−1.06E+03
	SD	**2.86E+01**	6.84E+01	3.25E+01	4.90E+01	4.72E+01	4.12E+01	3.25E+01

(continued)

Table 2.9 (continued)

Function		YSGA	GWO	PSO	SA-PSO	ABC	CSA	NAA
$f_{13}(x)$	**AB**	**8.74E+00**	9.64E+00	1.46E+01	1.62E+01	1.19E+01	8.83E+00	9.91E+00
	MB	**8.78E+00**	8.79E+00	1.35E+01	1.69E+01	1.20E+01	8.83E+00	9.96E+00
	SD	**3.03E−01**	4.09E−01	4.25E+00	1.71E+00	6.37E−01	4.48E−01	7.39E−01
$f_{14}(x)$	**AB**	**1.03E−01**	1.50E−01	1.31E+00	9.00E−01	2.20E+00	8.03E−01	2.27E−01
	MB	**9.99E−02**	1.50E−01	7.00E−01	9.00E−01	2.24E+00	8.50E−01	2.00E−01
	SD	**1.83E−02**	5.09E−02	2.43E+00	1.00E−01	2.42E−01	1.38E−01	6.92E−02
$f_{15}(x)$	**AB**	**7.16E+01**	1.37E+06	1.05E+02	8.12E+01	1.57E+07	8.83E+01	8.29E+01
	MB	**7.10E+01**	1.00E+06	8.48E+01	8.42E+01	1.63E+07	8.55E+01	8.26E+01
	SD	**1.38E+00**	9.43E+05	5.06E+01	5.32E+00	3.68E+06	1.31E+01	3.38E+00
$f_{16}(x)$	**AB**	**1.25E+01**	1.25E+05	1.13E+02	8.62E+01	1.56E+06	1.15E+02	1.03E+02
	MB	**1.25E+01**	1.25E+05	1.08E+02	8.63E+01	1.59E+06	1.14E+02	1.02E+02
	SD	**3.35E+00**	8.53E+04	1.27E+01	4.26E+00	3.71E+05	5.74E+00	6.99E+00
$f_{17}(x)$	**AB**	1.44E+01	**6.96E−01**	1.40E+02	5.55E+00	2.20E+02	2.05E+01	1.16E+01
	MB	3.51E+00	**4.10E−01**	1.44E+02	5.18E+00	2.21E+02	1.84E+01	1.04E+01
	SD	2.68E+01	2.44E+00	3.33E+01	**1.52E+00**	1.25E+01	7.80E+00	9.34E+00
$f_{18}(x)$	**AB**	**2.00E+00**	3.03E+00	1.17E+02	2.82E+01	**2.00E+00**	7.82E+07	2.07E+00
	MB	**2.00E+00**	2.06E+00	9.00E+00	2.82E+01	**2.00E+00**	1.03E+07	2.57E+00
	SD	**0.00E+00**	2.53E+00	2.32E+02	1.02E+01	**0.00E+00**	1.47E+08	1.81E−01
$f_{19}(x)$	**AB**	**2.00E+00**	9.73E+00	1.36E+02	2.82E+01	**2.00E+00**	8.47E+07	2.05E+00
	MB	**2.00E+00**	2.49E+00	4.35E+01	2.82E+01	**2.00E+00**	2.36E+07	2.01E+00
	SD	**0.00E+00**	3.12E+01	2.26E+02	1.03E+01	**0.00E+00**	1.95E+08	1.04E−01

(continued)

Table 2.9 (continued)

Function		YSGA	GWO	PSO	SA-PSO	ABC	CSA	NAA
$f_{20}(x)$	**AB**	**0.00E+00**	1.57E−11	**0.00E+00**	1.86E−17	9.29E−05	5.69E−12	4.39E−13
	MB	**0.00E+00**	1.49E−11	**0.00E+00**	4.92E−18	7.19E−05	1.72E−12	1.25E−18
	SD	**0.00E+00**	1.04E−11	**0.00E+00**	2.41E−17	8.87E−05	9.92E−12	1.63E−12
$f_{21}(x)$	**AB**	**−3.00E+01**	**−3.00E+01**	−2.82E+01	−4.92E+14	−1.91E+01	−2.98E+01	**−3.00E+01**
	MB	**−3.00E+01**	**−3.00E+01**	−2.81E+01	−5.06E+14	−1.92E+01	−2.99E+01	**−3.00E+01**
	SD	**6.33E−04**	9.57E−04	1.73E+00	4.58E+13	1.31E+00	4.16E−01	3.15E−03
$f_{22}(x)$	**AB**	**2.29E−287**	1.88E−153	4.49E−13	5.07E−12	2.24E−03	3.17E−12	5.03E−36
	MB	**0.00E+00**	4.37E−158	3.51E−15	4.38E−12	1.91E−03	1.51E−14	4.30E−44
	SD	**0.00E+00**	1.00E−152	1.98E−12	4.94E−12	1.68E−03	1.35E−11	2.65E−35
$f_{23}(x)$	**AB**	**0.00E+00**	**0.00E+00**	2.10E+01	9.04E−03	7.95E−01	9.54E−02	4.92E−03
	MB	**0.00E+00**	**0.00E+00**	8.63E−03	9.87E−03	8.05E−01	9.53E−02	1.32E−15
	SD	**0.00E+00**	**0.00E+00**	4.55E+01	8.65E−03	9.20E−02	3.65E−02	1.09E−02

Table 2.10 Optimal results of the multimodal test functions considering 50 dimensions

Function		YSGA	GWO	PSO	SA-PSO	ABC	CSA	NAA
$f_6(x)$	**AB**	2.49E−15	1.58E−14	1.99E+01	1.40E−02	8.57E+00	3.67E+00	**0.00E+00**
	MB	3.55E−15	1.42E−14	1.99E+01	1.57E−02	8.60E+00	3.77E+00	**0.00E+00**
	SD	2.12E−15	3.81E−15	2.44E−01	3.18E−03	5.83E−01	8.38E−01	**0.00E+00**
$f_7(x)$	**AB**	**1.05E−14**	1.95E−06	8.25E+03	4.67E+00	2.84E+03	1.99E+00	6.62E−01
	MB	**1.48E−42**	1.04E−06	7.82E+03	5.78E+00	2.93E+03	1.93E+00	2.49E−01
	SD	**5.76E−14**	2.35E−06	3.69E+03	2.24E+00	3.94E+02	6.17E−01	1.04E+00
$f_8(x)$	**AB**	**5.04E−95**	1.44E−48	7.91E+16	2.24E−01	8.58E+32	3.27E+02	5.33E−60
	MB	**2.09E−101**	6.27E−49	2.31E+03	1.15E−01	1.07E+30	3.18E+02	1.52E−61
	SD	**2.63E−94**	2.47E−48	4.33E+17	2.26E−01	3.26E+33	4.27E+01	1.78E−59
$f_9(x)$	**AB**	2.10E+130	**3.26E+02**	1.09E+114	3.81E+02	2.15E+146	3.04E+16	9.34E+07
	MB	8.07E+03	**1.28E+02**	3.98E+102	3.80E+02	6.48E+144	2.14E+05	4.50E+02
	SD	1.15E+131	**4.09E+02**	5.85E+114	3.73E+02	4.69E+146	1.67E+17	5.11E+08
$f_{10}(x)$	**AB**	**6.67E−01**	7.67E−01	3.22E+05	4.21E+00	1.53E+05	2.20E+00	1.75E+01
	MB	**6.67E−01**	7.67E−01	1.88E+05	5.27E+00	1.57E+05	1.85E+00	1.20E+01
	SD	**1.48E−08**	1.70E−07	3.18E+05	3.06E+00	3.40E+04	9.01E−01	1.88E+01
$f_{11}(x)$	**AB**	**0.00E+00**	**0.00E+00**	8.00E+03	2.07E+01	1.54E+03	3.79E+01	**0.00E+00**
	MB	**0.00E+00**	**0.00E+00**	1.00E+04	1.60E+01	1.57E+03	3.75E+01	**0.00E+00**
	SD	**0.00E+00**	**0.00E+00**	8.47E+03	1.08E+01	2.47E+02	1.07E+01	**0.00E+00**
$f_{12}(x)$	**AB**	**−1.82E+03**	−1.41E+03	−1.74E+03	−1.64E+03	−1.11E+03	−1.65E+03	−1.72E+03
	MB	**−1.89E+03**	−1.42E+03	−1.73E+03	−1.66E+03	−1.09E+03	−1.66E+03	−1.72E+03
	SD	**4.18E+01**	7.93E+01	5.41E+01	5.35E+01	7.25E+01	4.53E+01	4.75E+01

(continued)

Table 2.10 (continued)

Function		YSGA	GWO	PSO	SA-PSO	ABC	CSA	NAA
$f_{13}(x)$	**AB**	**1.19E+01**	1.69E+01	4.35E+01	4.99E+01	3.45E+01	1.73E+01	2.00E+01
	MB	**1.40E+01**	1.70E+01	3.88E+01	5.17E+01	3.39E+01	1.74E+01	2.00E+01
	SD	**2.06E−01**	4.72E−01	1.65E+01	1.84E+01	3.52E+00	5.84E−01	1.60E+00
$f_{14}(x)$	**AB**	**9.99E−02**	1.80E−01	7.52E+00	1.73E+00	8.49E+00	1.19E+00	3.47E−01
	MB	**9.99E−02**	2.00E−01	1.03E+01	1.70E+00	8.39E+00	1.20E+00	4.00E−01
	SD	**3.36E−12**	4.07E−02	5.61E+00	2.52E−01	6.35E−01	1.58E−01	8.19E−02
$f_{15}(x)$	**AB**	4.76E+03	7.79E+06	6.16E+03	**1.41E+02**	1.16E+08	1.62E+02	1.42E+02
	MB	6.00E+02	7.00E+06	9.88E+02	**1.40E+02**	1.14E+08	1.57E+02	1.43E+02
	SD	6.77E+03	2.36E+06	1.92E+04	**2.24E+00**	2.12E+07	1.68E+01	4.76E+00
$f_{16}(x)$	**AB**	2.16E+02	5.17E+05	1.62E+03	**1.52E+02**	1.09E+08	2.08E+02	1.97E+02
	MB	2.16E+02	5.00E+05	4.87E+02	**1.53E+02**	1.07E+08	2.07E+02	1.79E+02
	SD	3.53E+00	1.52E+05	2.61E+03	**2.50E+00**	2.86E+07	1.00E+01	7.05E+01
$f_{17}(x)$	**AB**	**0.00E+00**	1.79E+00	3.33E+02	4.49E+01	4.73E+02	3.36E+01	4.58E+01
	MB	**0.00E+00**	0.00E+00	3.27E+02	1.83E+01	4.71E+02	3.14E+01	4.34E+01
	SD	**0.00E+00**	6.02E+00	5.69E+01	4.88E+01	1.97E+01	1.15E+01	3.10E+01
$f_{18}(x)$	**AB**	**2.00E+00**	3.41E+03	1.81E+09	1.41E+14	**2.00E+00**	5.46E+18	5.52E+00
	MB	**2.00E+00**	5.84E+01	2.10E+06	1.41E+14	**2.00E+00**	1.07E+18	4.24E+00
	SD	**0.00E+00**	1.79E+04	6.57E+09	5.44E−16	7.99E−14	1.04E+19	4.73E+00
$f_{19}(x)$	**AB**	**2.00E+00**	1.26E+03	4.62E+09	1.41E+14	**2.00E+00**	7.69E+19	7.10E+00
	MB	**2.00E+00**	6.81E+01	4.87E+05	1.41E+14	**2.00E+00**	1.21E+17	5.08E+00
	SD	**0.00E+00**	4.86E+03	2.50E+10	5.44E−16	2.22E−13	2.64E+20	6.12E+00

(continued)

Table 2.10 (continued)

Function		YSGA	GWO	PSO	SA-PSO	ABC	CSA	NAA
$f_{20}(x)$	**AB**	**0.00E+00**	2.15E−11	**0.00E+00**	1.85E−12	8.86E−03	3.65E−11	5.09E−07
	MB	**0.00E+00**	3.49E−12	**0.00E+00**	5.58E−13	8.86E−03	2.19E−11	1.10E−08
	SD	**0.00E+00**	3.73E−11	**0.00E+00**	2.60E−12	3.33E−03	4.61E−11	1.85E−06
$f_{21}(x)$	**AB**	**−5.00E+01**	−4.89E+01	−4.48E+01	−7.78E+14	−2.61E+01	−4.90E+01	**−5.00E+01**
	MB	**−5.00E+01**	−4.00E+01	−4.62E+01	−7.71E+14	−2.50E+01	−4.95E+01	**−5.00E+01**
	SD	**2.24E−03**	5.90E+00	3.45E+00	7.73E+13	3.22E+00	1.02E+00	5.40E−03
$f_{22}(x)$	**AB**	**2.69E−288**	6.12E−178	2.50E−08	1.43E−05	4.22E−01	7.62E−08	1.07E−18
	MB	**0.00E+00**	2.29E−182	9.05E−09	2.01E−06	4.42E−01	9.56E−09	5.23E−25
	SD	**0.00E+00**	**0.00E+00**	3.72E−08	2.24E−05	1.18E−01	1.46E−07	5.37E−18
$f_{23}(x)$	**AB**	**0.00E+00**	3.50E−04	9.03E+01	5.76E−03	1.49E+01	1.74E−01	2.63E−03
	MB	**0.00E+00**	**0.00E+00**	9.02E+01	2.21E−05	1.45E+01	1.72E−01	**0.00E+00**
	SD	**0.00E+00**	1.92E−03	8.22E+01	9.95E−03	3.84E+00	3.75E−02	5.37E−03

Table 2.11 Optimal results of the multimodal test functions considering 100 dimensions

Function		YSGA	GWO	PSO	SA-PSO	ABC	CSA	NAA
$f_6(x)$	**AB**	**3.20E−15**	6.94E−14	1.99E+01	3.17E−01	1.88E+01	4.32E+00	2.41E−01
	MB	**3.55E−15**	6.75E−14	2.00E+01	1.86E−01	1.89E+01	4.29E+00	0.00E+00
	SD	**1.43E−15**	5.81E−15	2.26E−01	2.26E−01	3.25E−01	7.39E−01	9.16E−01
$f_7(x)$	**AB**	**1.11E−24**	1.05E−05	3.39E+04	1.13E+02	2.88E+04	4.06E+00	2.13E+00
	MB	**5.17E−103**	5.04E−06	3.37E+04	1.09E+02	2.90E+04	4.08E+00	7.90E−02
	SD	**6.08E−24**	1.47E−05	8.04E+03	8.04E+03	3.63E+03	6.58E−01	1.08E+01
$f_8(x)$	**AB**	7.53E−93	9.59E−20	1.01E+75	1.42E+00	7.13E+103	2.93E+17	**2.32E−123**
	MB	5.37E−100	8.36E−20	4.91E+03	1.33E+00	6.34E+95	6.81E+02	**1.81E−125**
	SD	2.66E−92	4.52E−20	5.50E+75	5.50E+75	2.00E+104	1.60E+18	**6.47E−123**
$f_9(x)$	**AB**	**6.67E−01**	**6.67E−01**	4.05E+06	2.80E+02	7.01E+06	5.51E+00	1.22E+01
	MB	**6.67E−01**	**6.67E−01**	3.60E+06	2.80E+02	6.99E+06	4.73E+00	1.38E+01
	SD	**5.72E−07**	**5.72E−07**	1.93E+06	1.93E+06	9.25E+05	3.22E+00	8.80E+00
$f_{10}(x)$	**AB**	**1.30E+03**	2.42E+03	3.31E+03	3.36E+03	1.53E+03	3.29E+03	1.72E+03
	MB	**1.33E+03**	2.39E+03	3.31E+03	3.36E+03	1.55E+03	3.27E+03	1.72E+03
	SD	**3.51E+01**	1.30E+02	1.31E+02	1.31E+02	1.77E+02	5.92E+01	4.16E+01
$f_{11}(x)$	**AB**	**0.00E+00**	**0.00E+00**	5.70E+04	2.90E+01	7.54E+04	2.19E+02	**0.00E+00**
	MB	**0.00E+00**	**0.00E+00**	5.50E+04	2.90E+01	7.58E+04	2.10E+02	**0.00E+00**
	SD	**0.00E+00**	**0.00E+00**	1.97E+04	1.97E+04	5.15E+03	4.72E+01	**0.00E+00**
$f_{12}(x)$	**AB**	**−3.86E+03**	−3.91E+02	−3.05E+02	−5.17E+02	−5.11E+02	−3.96E+01	−1.97E+01
	MB	**−3.87E+03**	−3.92E+02	−3.03E+02	−5.17E+02	−5.22E+02	−3.99E+01	−1.95E+01
	SD	**7.77E−01**	7.98E−01	1.17E+02	1.17E+02	6.99E+01	9.06E−01	1.67E+00

(continued)

Table 2.11 (continued)

Function		YSGA	GWO	PSO	SA-PSO	ABC	CSA	NAA
$f_{13}(x)$	**AB**	**1.03E−01**	2.57E−01	2.19E+01	4.40E+00	2.99E+01	2.00E+00	2.67E−01
	MB	**9.99E−02**	3.00E−01	2.33E+01	4.40E+00	3.00E+01	2.00E+00	3.00E−01
	SD	**1.83E−02**	5.68E−02	4.23E+00	4.23E+00	1.36E+00	1.67E−01	7.11E−02
$f_{14}(x)$	**AB**	2.27E+05	3.41E+07	9.44E+07	3.59E+02	1.22E+09	3.86E+02	**1.44E+02**
	MB	1.59E+05	3.46E+07	5.95E+05	3.59E+02	1.22E+09	3.48E+02	**1.40E+02**
	SD	2.35E+05	5.07E+06	1.42E+08	1.42E+08	2.10E+08	1.07E+02	**1.86E+01**
$f_{15}(x)$	**AB**	2.56E+03	2.21E+06	3.56E+08	3.75E+02	1.84E+09	4.31E+02	**1.79E+02**
	MB	2.00E+03	2.25E+06	4.10E+08	3.75E+02	1.82E+09	4.27E+02	**1.76E+02**
	SD	2.03E+03	2.64E+05	4.27E+08	4.27E+08	3.63E+08	2.23E+01	**1.38E+01**
$f_{16}(x)$	**AB**	**0.00E+00**	7.36E−01	7.83E+02	1.38E+02	1.21E+03	5.60E+01	3.72E+01
	MB	**0.00E+00**	1.14E−13	7.87E+02	1.38E+02	1.22E+03	5.28E+01	3.88E+01
	SD	**0.00E+00**	2.67E+00	7.37E+01	7.37E+01	4.21E+01	1.47E+01	2.46E+01
$f_{17}(x)$	**AB**	**0.00E+00**	1.71E−12	**0.00E+00**	1.84E−07	1.37E−01	5.05E−11	1.67E−07
	MB	**0.00E+00**	1.81E−14	**0.00E+00**	1.84E−07	1.36E−01	1.21E−11	8.31E−11
	SD	**0.00E+00**	5.08E−12	**0.00E+00**	2.40E−01	2.74E−02	1.02E−10	4.81E−07
$f_{18}(x)$	**AB**	**2.00E+00**	1.16E+24	3.95E+37	−7.04E+13	**2.00E+00**	6.98E+55	5.77E+00
	MB	**2.00E+00**	1.18E+15	1.55E+28	−7.04E+13	**2.00E+00**	6.09E+51	3.18E+00
	SD	**0.00E+00**	5.07E+24	2.16E+38	2.16E+38	2.01E−08	2.87E+56	9.05E+00
$f_{19}(x)$	**AB**	**2.00E+00**	1.81E+22	4.71E+36	−7.04E+13	**2.00E+00**	1.33E+56	4.64E+00
	MB	**2.00E+00**	4.09E+17	1.55E+28	−7.04E+13	**2.00E+00**	1.91E+53	3.19E+00
	SD	**0.00E+00**	9.06E+22	2.53E+37	2.53E+37	2.06E−08	4.94E+56	3.75E+00

(continued)

Table 2.11 (continued)

Function		YSGA	GWO	PSO	SA-PSO	ABC	CSA	NAA
$f_{20}(x)$	**AB**	**1.07E−03**	2.83E−01	2.48E+01	1.39E+15	1.77E+01	8.50E+01	5.07E−01
	MB	**9.99E−02**	9.99E−01	8.08E+01	1.39E+15	4.29E+01	9.78E+01	5.00E−01
	SD	**2.09E−03**	2.84E−02	5.59E+00	5.59E+00	4.48E+00	3.02E+00	1.81E−02
$f_{21}(x)$	**AB**	**−1.00E+02**	−2.58E+01	−1.04E+00	−7.40E+01	−7.96E+01	−7.56E+00	1.98E+01
	MB	**−1.00E+02**	−2.35E+01	−1.16E+00	−7.40E+01	−8.31E+01	−6.75E+00	9.22E+00
	SD	**0.00E+00**	1.35E−73	8.79E−01	8.79E−01	1.85E+00	5.62E−07	1.08E−22
$f_{22}(x)$	**AB**	**0.00E+00**	6.76E−04	4.18E+02	5.65E−02	6.88E+02	1.15E−01	**0.00E+00**
	MB	**0.00E+00**	0.00E+00	4.06E+02	5.65E−02	6.92E+02	1.15E−01	**0.00E+00**
	SD	**0.00E+00**	3.71E−03	1.63E+02	1.63E+02	5.02E+01	1.45E−02	**0.00E+00**
$f_{23}(x)$	**AB**	**3.20E−15**	6.94E−14	1.99E+01	3.17E−01	1.88E+01	4.32E+00	2.41E−01
	MB	**3.55E−15**	6.75E−14	2.00E+01	1.86E−01	1.89E+01	4.29E+00	0.00E+00
	SD	**1.43E−15**	5.81E−15	2.26E−01	2.26E−01	3.25E−01	7.39E−01	9.16E−01

Additionally, the experiments have been repeated but now considering 100 dimensions. The results of this test are exposed in Table 2.11.

From Table 2.11, it can be visualized that the YSGA algorithm achieves better performance than its competitors in functions f_6, f_7, f_{10}, f_{12}, f_{13}, f_{16}, f_{20}, f_{21} and f_{23}. By contrast, the NAA outperforms the YSGA in functions f_8, f_{14} and f_{15}. In function f_{11}, the YSGA, the GWO, and the NAA reach the same values in AB, MB, and SD indicators while the YSGA and the PSO tie in f_{17}. Similarly, the YSGA achieves the same values than the GWO and NAA in terms of AB, MB, and SD for functions f_9 and f_{22} respectively. Regarding function f_{18} and f_{19} the YSGA and the ABC reach the same values in AB and MB, but the YSGA achieves better SD value than the ABC, which means that the YSGA is more consistent concerning the produced results.

The p-values resulting from the Wilcoxon test, considering the Bonferroni correction, are reported in Tables 2.12, 2.13, and 2.14. After applying the Wilcoxon test over the AB data from Table 2.9, the p-values generated were registered in Table 2.12. Similarly, the p-values generated by the Wilcoxon test over the AB data from Tables 2.10 and 2.11 were reported in Tables 2.13 and 2.14, respectively.

Table 2.12 p-values of Wilcoxon-Bonferroni test comparing YSGA versus GWO, YSGA versus PSO, YSGA versus SA-PSO, YSGA versus ABC, YSGA versus CSA, and YSGA versus NAA over the AB values from Table 2.9, $n = 30$

Function	YSGA versus GWO	YSGA versus PSO	YSGA versus SA-PSO	YSGA versus ABC	YSGA versus CSA	YSGA versus NAA
$f_6(x)$	1.2E−06▲	6.6E−12▲	1.3E−11▲	1.4E−11▲	1.4E−11▲	8.2E−06▼
$f_7(x)$	6.0E−07▲	3.0E−11▲	3.0E−11▲	3.0E−11▲	3.0E−11▲	3.0E−11▲
$f_8(x)$	3.0E−11▲	3.0E−11▲	3.0E−11▲	3.0E−11▲	3.0E−11▲	1.1E−08▲
$f_9(x)$	3.0E−11▲	3.0E−11▲	3.0E−11▲	3.0E−11▲	3.0E−11▲	3.0E−11▲
$f_{10}(x)$	9.5E−05▲	4.5E−07▲	3.0E−11▼	3.0E−11▲	7.7E−06▲	9.1E−06▲
$f_{11}(x)$	1.7E−12▲	8.9E−06▲	3.7E−11▲	3.0E−11▲	3.0E−11▲	2.6E−06▲
$f_{12}(x)$	2.5E−07▲	1.7E−07▲	4.4E−07▲	3.0E−11▲	1.7E−08▲	5.1E−06▲
$f_{13}(x)$	3.0E−11▲	9.6E−05▲	3.0E−11▲	3.0E−11▲	5.0E−08▲	3.4E−08▲
$f_{14}(x)$	3.0E−11▲	6.0E−08▲	3.0E−11▲	3.0E−11▲	3.0E−11▲	3.0E−11▲
$f_{15}(x)$	1.4E−08▲	2.2E−10▲	6.5E−08▲	3.0E−11▲	1.7E−07▲	6.0E−07▲
$f_{16}(x)$	3.0E−11▲	1.1E−06▲	3.3E−06▲	3.0E−11▲	3.6E−06▲	3.3E−11▲
$f_{17}(x)$	8.5E−06▼	2.7E−05▲	1.2E−11▲	1.1E−11▲	5.5E−07▲	6.7E−08▲
$f_{18}(x)$	5.5E−06▲	1.2E−12▲	4.3E−09▲	1.1E−11▲	1.2E−12▲	1.2E−12▲
$f_{19}(x)$	3.3E−07▲	1.2E−12▲	1.4E−11▲	1.1E−11▲	1.2E−12▲	1.2E−12▲
$f_{20}(x)$	2.2E−08▲	4.2E−04►	1.4E−11▲	1.2E−12▲	1.2E−12▲	1.2E−12▲
$f_{21}(x)$	1.2E−12▲	3.0E−11▲	1.9E−06▲	3.0E−11▲	3.0E−11▲	5.6E−10▲
$f_{22}(x)$	4.1E−12▲	4.1E−12▲	4.1E−12▲	4.1E−12▲	4.1E−12▲	4.1E−12▲
$f_{23}(x)$	4.1E−12▲	4.1E−12▲	1.2E−12▲	1.2E−12▲	1.2E−12▲	1.4E−08▲

Table 2.13 p-values of Wilcoxon-Bonferroni test comparing YSGA versus GWO, YSGA versus PSO, YSGA versus SA-PSO, YSGA versus ABC, YSGA versus CSA, and YSGA versus NAA over the AB values from Table 2.10, $n = 50$

Function	YSGA versus GWO	YSGA versus PSO	YSGA versus SA-PSO	YSGA versus ABC	YSGA versus CSA	YSGA versus NAA
$f_6(x)$	1.3E−11▲	4.1E−12▲	1.1E−11▲	1.2E−11▲	1.2E−11▲	1.8E−07▼
$f_7(x)$	7.0E−07▲	3.0E−11▲	3.0E−11▲	3.0E−11▲	3.0E−11▲	3.0E−11▲
$f_8(x)$	3.0E−11▲	2.0E−10▲	3.0E−11▲	3.0E−11▲	1.1E−07▲	3.0E−11▲
$f_9(x)$	3.0E−11▼	4.6E−10▲	7.0E−07▲	3.0E−11▲	2.9E−07▲	3.0E−11▲
$f_{10}(x)$	1.3E−11▲	3.2E−09▲	1.3E−08▲	3.0E−11▲	1.8E−09▲	2.4E−05▲
$f_{11}(x)$	1.1E−06▲	1.1E−10▲	3.0E−11▲	3.0E−11▲	3.0E−11▲	5.6E−10▲
$f_{12}(x)$	6.0E−08▲	3.0E−11▲	3.6E−11▲	3.0E−11▲	2.4E−10▲	1.6E−11▲
$f_{13}(x)$	3.0E−11▲	3.0E−11▲	3.0E−11▲	3.0E−11▲	1.2E−12▲	4.1E−11▲
$f_{14}(x)$	3.0E−11▲	7.0E−07▲	3.0E−11▲	3.0E−11▲	3.0E−11▲	3.0E−11▲
$f_{15}(x)$	5.4E−11▲	3.0E−11▲	3.1E−12▼	3.0E−11▲	6.0E−08▲	8.5E−09▲
$f_{16}(x)$	3.0E−11▲	3.0E−11▲	3.0E−11▼	3.0E−11▲	4.0E−10▲	1.1E−07▲
$f_{17}(x)$	2.0E−11▲	2.3E−05▲	3.2E−12▲	3.2E−12▲	9.4E−12▲	4.0E−10▲
$f_{18}(x)$	2.8E−11▲	1.2E−12▲	1.1E−12▲	1.3E−05▲	1.2E−12▲	1.2E−12▲
$f_{19}(x)$	2.3E−11▲	1.2E−12▲	1.2E−12▲	2.2E−06▲	1.2E−12▲	1.2E−12▲
$f_{20}(x)$	3.7E−11▲	1.2E−12▲	2.0E−10▲	1.2E−12▲	1.2E−12▲	1.2E−12▲
$f_{21}(x)$	1.2E−12▲	1.9E−06▲	1.1E−06▲	3.0E−11▲	3.0E−11▲	2.5E−09▲
$f_{22}(x)$	1.4E−11▲	1.4E−11▲	1.4E−11▲	1.4E−11▲	1.4E−11▲	1.4E−11▲
$f_{23}(x)$	1.3E−11▲	1.4E−11▲	1.2E−12▲	1.2E−12▲	1.2E−12▲	5.6E−08▲

The Wilcoxon test for the multimodal functions considering 30 dimensions is listed in Table 2.12. In this case, it is visible that all the p-values in the columns YSGA versus ABC and YSGA versus CSA validate that the presented approach outperforms its competitors for all the functions. Regardless of the tie in functions f_{18} and f_{19} from Table 2.9 between YSGA and ABC, the Wilcoxon analysis could determine the discrepancy between both algorithms. Hence, the symbol ▲ is placed on the right side of the p-value, indicating that the YSGA has better performance than the ABC.

In the same way, despite the tie in function f_{23} from Table 2.9 between YSGA and GWO, the Wilcoxon analysis could determine the discrepancy between both algorithms. Therefore, the symbol ▲ is placed on the right side of the p-value, indicating that the YSGA has better performance than the GWO. In the case of function f_{20}, in which the YSGA and PSO also tie, the Wilcoxon test shows (in column YSGA vs. PSO) that the p-value is higher than the significance value of 3.3E−04. This result indicates that the statistical test could not determine the discrepancy between the two approaches. Thus, the symbol ► is placed on the right side of the p-value.

Table 2.14 p-values of Wilcoxon-Bonferroni test comparing YSGA versus GWO, YSGA versus PSO, YSGA versus SA-PSO, YSGA versus ABC, YSGA versus CSA, and YSGA versus NAA over the AB values from Table 2.11, $n = 100$

Function	YSGA versus GWO	YSGA versus PSO	YSGA versus SA-PSO	YSGA versus ABC	YSGA versus CSA	YSGA versus NAA
$f_6(x)$	1.0E−07▲	3.4E−11▲	2.7E−11▲	1.3E−10▲	1.6E−10▲	6.0E−08▲
$f_7(x)$	2.3E−10▲	1.4E−12▲	2.3E−11▲	2.3E−11▲	4.6E−12▲	1.9E−11▲
$f_8(x)$	1.6E−06▲	3.7E−10▲	7.7E−12▲	1.1E−11▲	2.5E−11▲	3.0E−09▼
$f_9(x)$	1.3E−06▲	7.8E−06▲	2.7E−03►	1.7E−11▲	4.5E−09▲	1.5E−11▲
$f_{10}(x)$	4.7E−10▲	1.2E−12▲	1.2E−12▲	1.3E−13▲	1.7E−12▲	6.9E−13▲
$f_{11}(x)$	4.6E−10▲	9.6E−12▲	2.7E−11▲	1.6E−12▲	2.4E−12▲	8.1E−09▲
$f_{12}(x)$	2.9E−06▲	2.9E−11▲	9.0E−02►	1.6E−11▲	3.0E−05▲	7.2E−12▲
$f_{13}(x)$	2.9E−06▲	9.8E−10▲	1.7E−11▲	2.4E−11▲	2.2E−10▲	3.7E−12▲
$f_{14}(x)$	1.5E−06▲	1.3E−11▲	4.2E−12▲	2.8E−11▲	2.9E−11▲	1.5E−11▼
$f_{15}(x)$	2.4E−06▲	1.2E−11▲	5.0E−12▲	3.9E−12▲	1.4E−13▲	7.2E−12▼
$f_{16}(x)$	4.3E−07▲	2.3E−11▲	7.8E−12▲	1.7E−11▲	4.3E−11▲	1.3E−11▲
$f_{17}(x)$	1.3E−06▲	3.1E−08▲	1.0E−12▲	5.7E−13▲	9.9E−13▲	9.6E−12▲
$f_{18}(x)$	7.5E−10▲	2.3E−13▲	2.9E−13▲	1.4E−14▲	1.1E−12▲	1.1E−12▲
$f_{19}(x)$	6.5E−10▲	5.9E−13▲	9.5E−13▲	4.1E−13▲	1.0E−13▲	1.1E−12▲
$f_{20}(x)$	7.9E−10▲	5.4E−13▲	2.8E−13▲	2.0E−13▲	4.8E−13▲	5.9E−13▲
$f_{21}(x)$	2.0E−06▲	5.8E−02►	2.8E−11▲	2.4E−11▲	1.9E−11▲	1.5E−11▲
$f_{22}(x)$	9.0E−08▲	8.6E−13▲	4.2E−13▲	3.8E−13▲	9.7E−13▲	4.1E−13▲
$f_{23}(x)$	7.0E−10▲	2.5E−01►	2.4E−13▲	6.4E−13▲	5.2E−13▲	1.1E−12▲

Besides, it can be observed that functions f_6, f_{10} and f_{17} have the symbol ▼ as an indicator that for these functions, the NAA, SA-PSO, and GWO (respectively) reached better performances than the YSGA. For the rest of the functions in all comparisons, the Wilcoxon test supports the results recorded in Table 2.9, exposing that the presented approach outperforms its opponents.

The Wilcoxon test for the multimodal functions considering 50 dimensions is visualized in Table 2.13. In this case, it is visible that all the p-values in the columns YSGA versus PSO, YSGA versus ABC, and YSGA versus CSA validate that the presented approach outperforms its competitors for all the functions. The Wilcoxon analysis could determine the discrepancy between the YSGA versus NAA and YSGA versus GWO in functions f_6 and f_9 respectively, the symbol ▼ indicates that the YSGA has worse performance than its contenders. In the same way, there was no difference detected by the statistical test between YSGA and SA-PSO in functions f_{15} an f_{16}, hence the symbol ▼ is placed as well. For the rest of the functions in all comparisons, the Wilcoxon test supports the results recorded in Table 2.10, exposing that the presented approach outperforms its opponents.

The results of the Wilcoxon test for the multimodal functions in 100 dimensions are presented in Table 2.14. From these results, it is evident that all the p-values in the columns YSGA versus GWO, YSGA versus ABC, and YSGA versus CSA confirm that the presented approach performs better than its contenders for all the test functions. The Wilcoxon analysis could determine the discrepancy between the YSGA versus NAA in functions f_8, f_{14} and f_{15}. Hence, the symbol ▼ is located to indicate that the YSGA has worse performance than its contenders, according to the results from Table 2.11. On the other hand, there was no difference detected by the statistical test in YSGA versus PSO for functions f_{21} and f_{23}, thus the symbol ▶ is placed to indicate it. Similarly, there was no difference detected by the statistical test in YSGA versus SA-PSO in functions f_9 and f_{12}. For the rest of the functions in all comparisons, the Wilcoxon test supports the results registered in Table 2.11, exposing that the presented approach outperforms its adversaries.

From Tables 2.12, 2.13, and 2.14, it can be concluded that the statistical test exhibits evidence that ratifies the results registered in Tables 2.9, 2.10, and 2.11, showing that the presented approach outperforms its challengers. The statistical analysis of the YSGA algorithm against the other methods and its performance over the multimodal benchmark functions considering 30, 50, and 100 dimensions demonstrate that the YSGA method is efficient in high-dimensional optimization problems and reaches better performance than its adversaries.

2.4.3 Results of Composite Test Functions

Functions from f_{24} to f_{27} are included in Table 2.25. These are composite functions that are used to test the effectiveness of the presented approach at a higher level of complexity. This kind of function is generated as a mixture of different multimodal functions. Therefore, it is evident that composite functions represent a greater difficulty in solving them. The experimental results collected from 30 distinct executions are exhibited in Tables 2.15, 2.16, and 2.17. Indicators AB, MB, and SD illustrate the accuracy of the results considering the performance of each algorithm for every test function in 30, 50, and 100 dimensions, respectively. The best results are highlighted in boldface.

The results contained in Table 2.15 exhibit evidence that the YSGA algorithm achieves superior performance than its competitors in functions f_{24}, f_{25} and f_{27}. On the other hand, the GWO method reaches better indicators AB, MB, and SD for function f_{26}.

To examine the adjustability of the YSGA algorithm, the experiments have been repeated but now considering 50 dimensions instead of 30. The results of this test are shown in Table 2.16.

From Table 2.16, it can be detected that the YSGA algorithm achieves better performance than its competitors in functions f_{24}, f_{25} and f_{27}. By contrast, the GWO outperforms the YSGA in function f_{26}.

Table 2.15 Optimal results of the composite test functions considering 30 dimensions

Function		YSGA	GWO	PSO	SA-PSO	ABC	CSA	NAA
$f_{24}(x)$	**AB**	**3.71E−99**	9.19E−41	2.61E+04	7.28E−05	5.28E−01	8.72E−02	7.91E−46
	MB	**4.51E−106**	7.31E−41	2.01E+04	6.72E−05	5.17E−01	8.26E−02	4.76E−47
	SD	**2.03E−98**	7.98E−41	1.70E+04	1.26E−05	1.49E−01	3.28E−02	2.24E−45
$f_{25}(x)$	**AB**	**2.90E+01**	3.90E+01	1.73E+02	8.52E+01	2.12E+02	1.05E+02	6.32E+01
	MB	**2.90E+01**	3.90E+01	9.31E+01	8.37E+01	2.12E+02	1.03E+02	2.90E+01
	SD	**3.63E−06**	5.52E−05	1.57E+02	1.09E+01	1.97E+01	2.34E+01	5.28E+01
$f_{26}(x)$	**AB**	4.11E+01	**3.20E+01**	1.48E+02	3.30E+01	1.89E+05	6.01E+01	4.38E+01
	MB	3.97E+01	**3.20E+01**	1.46E+02	3.30E+01	1.42E+05	6.01E+01	3.91E+01
	SD	9.23E+00	**1.14E−05**	3.63E+01	4.05E−01	1.59E+05	9.86E+00	1.74E+01
$f_{27}(x)$	**AB**	**2.90E+01**	3.00E+01	1.02E+03	1.07E+02	1.36E+02	1.20E+02	3.28E+01
	MB	**2.90E+01**	3.00E+01	9.68E+02	1.05E+02	1.36E+02	1.19E+02	2.90E+01
	SD	**0.00E+00**	9.38E−15	5.24E+02	4.18E+01	3.57E+01	2.83E+01	2.06E+01

Table 2.16 Optimal results of the composite test functions considering 50 dimensions

Function		YSGA	GWO	PSO	SA-PSO	ABC	CSA	NAA
$f_{24}(x)$	**AB**	**1.67E−96**	1.74E−49	7.56E+04	1.56E−02	1.81E+03	2.03E−01	5.37E−58
	MB	**1.29E−101**	8.93E−50	8.03E+04	7.97E−03	1.84E+03	1.88E−01	7.30E−60
	SD	**5.86E−96**	1.94E−49	3.75E+04	1.37E−02	3.50E+02	6.21E−02	1.66E−57
$f_{25}(x)$	**AB**	**4.90E+01**	5.90E+01	6.80E+02	1.57E+02	6.83E+02	1.80E+02	1.81E+02
	MB	**4.90E+01**	5.90E+01	6.84E+02	1.71E+02	6.86E+02	1.76E+02	1.68E+02
	SD	**1.38E−05**	1.77E−04	3.07E+02	5.14E+01	3.08E+01	3.31E+01	1.30E+02
$f_{26}(x)$	**AB**	1.11E+02	**5.40E+01**	7.91E+02	6.54E+01	9.76E+07	1.46E+02	6.72E+02
	MB	1.04E+02	**5.40E+01**	7.85E+02	6.53E+01	9.52E+07	1.46E+02	6.33E+02
	SD	3.87E+01	**2.38E−05**	1.77E+02	7.00E−01	2.52E+07	2.02E+01	1.95E+02
$f_{27}(x)$	**AB**	**4.90E+01**	5.10E+01	2.57E+03	2.04E+02	6.30E+02	2.03E+03	6.68E+01
	MB	**4.90E+01**	5.00E+01	2.86E+03	2.08E+02	6.34E+02	9.40E+02	5.90E+01
	SD	**4.76E−15**	2.21E−14	1.16E+03	4.73E+01	3.28E+01	1.73E+03	5.74E+01

Additionally, the experiments have been repeated but now considering 100 dimensions as an alternative to provide increased complexity to the test. The results are reported in Table 2.17.

From Table 2.17, it can be visualized that the YSGA algorithm achieves better performance than its competitors in functions f_{24}, f_{25} and f_{27}. On the other hand, the GWO outperforms the YSGA in functions f_8, f_{14} and f_{15}.

The p-values resulting from the Wilcoxon test, considering the Bonferroni correction, are stated in Tables 2.18, 2.19, and 2.20. After applying the Wilcoxon test over the AB data from Table 2.15, the p-values generated were listed in Table 2.18. Similarly, the p-values generated by the Wilcoxon test over the AB data from Tables 2.16 and 2.17 were registered in Table 2.19 and 2.20, respectively.

Table 2.17 Optimal results of the composite test functions considering 100 dimensions

Function		YSGA	GWO	PSO	SA-PSO	ABC	CSA	NAA
$f_{24}(x)$	**AB**	**5.32E−116**	2.55E−20	3.22E+05	2.93E+00	2.57E+05	2.26E−01	1.09E−93
	MB	**1.38E−122**	2.17E−20	3.35E+05	2.93E+00	1.46E+05	2.28E−01	3.89E−97
	SD	**2.91E−115**	1.41E−20	6.78E+04	6.78E+04	5.93E+05	3.39E−02	4.58E−93
$f_{25}(x)$	**AB**	**9.90E+01**	1.01E+02	2.17E+03	2.84E+02	3.37E+03	3.38E+02	1.38E+02
	MB	**9.90E+01**	1.01E+02	2.13E+03	2.84E+02	3.36E+03	3.36E+02	4.90E+02
	SD	**1.02E−04**	4.69E−04	5.21E+02	5.21E+02	1.86E+02	4.92E+01	1.09E+02
$f_{26}(x)$	**AB**	4.51E+02	**1.09E+02**	2.87E+08	3.46E+02	1.92E+09	3.72E+02	5.42E+02
	MB	4.32E+02	**1.09E+02**	4.10E+08	3.46E+02	1.95E+09	3.68E+02	4.85E+02
	SD	1.90E+02	**6.09E−05**	2.88E+08	2.88E+08	3.60E+08	4.74E+01	2.74E+02
$f_{27}(x)$	**AB**	**9.90E+01**	1.04E+02	9.00E+06	3.38E+02	6.24E+03	3.94E+02	8.69E+02
	MB	**9.90E+01**	1.04E+02	1.08E+04	3.38E+02	4.63E+03	3.88E+02	4.90E+02
	SD	**0.00E+00**	3.24E−14	4.93E+07	4.93E+07	8.03E+03	5.70E+01	7.03E+01

Table 2.18 p-values of Wilcoxon-Bonferroni test comparing YSGA versus GWO, YSGA versus PSO, YSGA versus SA-PSO, YSGA versus ABC, YSGA versus CSA, and YSGA versus NAA over the AB values from Table 2.15, $n = 30$

Function	YSGA versus GWO	YSGA versus PSO	YSGA versus SA-PSO	YSGA versus ABC	YSGA versus CSA	YSGA versus NAA
$f_{24}(x)$	3.0E−11▲	3.0E−11▲	3.0E−11▲	3.0E−11▲	3.0E−11▲	3.0E−11▲
$f_{25}(x)$	2.9E−11▲	4.6E−06▲	3.0E−11▲	3.0E−11▲	3.0E−11▲	3.8E−06▲
$f_{26}(x)$	1.1E−11▼	3.0E−11▲	3.0E−11▲	3.0E−11▲	2.3E−04▲	1.3E−10▲
$f_{27}(x)$	7.2E−13▲	5.4E−13▲	2.4E−12▲	2.4E−12▲	2.4E−12▲	6.0E−07▲

Table 2.19 p-values of Wilcoxon-Bonferroni test comparing YSGA versus GWO, YSGA versus PSO, YSGA versus SA-PSO, YSGA versus ABC, YSGA versus CSA, and YSGA versus NAA over the AB values from Table 2.16, $n = 50$

Function	YSGA versus GWO	YSGA versus PSO	YSGA versus SA-PSO	YSGA versus ABC	YSGA versus CSA	YSGA versus NAA
$f_{24}(x)$	3.0E−11▲	3.0E−11▲	3.0E−11▲	3.0E−11▲	3.0E−11▲	3.0E−11▲
$f_{25}(x)$	2.9E−11▲	3.0E−11▲	3.0E−11▲	3.0E−11▲	3.0E−11▲	9.2E−05▲
$f_{26}(x)$	1.8E−11▼	3.0E−11▲	3.0E−11▲	3.0E−11▲	4.0E−05▲	3.0E−11▲
$f_{27}(x)$	7.4E−13▲	8.2E−13▲	4.1E−12▲	4.1E−12▲	4.1E−12▲	1.0E−11▲

The Wilcoxon test for the multimodal functions considering 30 dimensions is listed in Table 2.18. In this case, it is visible that all the p-values in the columns YSGA versus PSO, YSGA versus SA-PSO, YSGA versus ABC, and YSGA versus CSA, and YSGA versus NAA validate that the presented approach outperforms its

Table 2.20 *p*-values of Wilcoxon-Bonferroni test comparing YSGA versus GWO, YSGA versus PSO, YSGA versus SA-PSO, YSGA versus ABC, YSGA versus CSA, and YSGA versus NAA over the AB values from Table 2.17, $n = 100$

Function	YSGA versus GWO	YSGA versus PSO	YSGA versus SA-PSO	YSGA versus ABC	YSGA versus CSA	YSGA versus NAA
$f_{24}(x)$	2.8E−06▲	3.0E−12▲	2.8E−12▲	1.8E−12▲	1.0E−11▲	4.1E−12▲
$f_{25}(x)$	3.0E−07▲	2.1E−11▲	1.9E−11▲	1.8E−11▲	5.5E−12▲	7.4E−02▲
$f_{26}(x)$	2.3E−06▼	2.0E−11▲	1.4E−11▲	7.9E−12▲	7.6E−04▲	1.7E−05▲
$f_{27}(x)$	2.2E−06▲	1.8E−13▲	4.3E−13▲	7.9E−13▲	1.8E−13▲	3.6E−05▲

competitors for all the functions. Besides, it can be observed that function f_{26} in column YSGA versus GWO has the symbol ▼ as an indicator that for this function, the GWO reaches better performances than the YSGA. For the rest of the functions in the same column, the Wilcoxon test supports the results recorded in Table 2.15, revealing that the YSGA algorithm outperforms its opponents.

The Wilcoxon test for the multimodal functions considering 50 dimensions is written in Table 2.19. In this case, it is visible that all the *p*-values in the columns YSGA versus PSO, YSGA versus SA-PSO, YSGA versus ABC, and YSGA versus CSA, and YSGA versus NAA validate that the presented approach outperforms its competitors for all the functions. The Wilcoxon analysis could determine the discrepancy between the YSGA versus GWO in function f_{26} therefore the symbol ▼ specifies that the YSGA has worse performance than the GWO. For the rest of the functions in the same column, the Wilcoxon test supports the results registered in Table 2.16, exposing that the presented approach outperforms its opponent.

The results of the Wilcoxon test for the multimodal functions in 100 dimensions are presented in Table 2.20. From these results, it is evident that all the *p*-values in the columns YSGA versus PSO, YSGA versus SA-PSO, YSGA versus ABC, and YSGA versus CSA, and YSGA versus NAA confirm that the presented approach performs better than its contenders for all the test functions. The Wilcoxon analysis could determine the discrepancy between the YSGA versus GWO in function f_{26}, thus the symbol ▼ is placed on the right side of the *p*-value, indicating that the YSGA has worse performance than the GWO according to the results of Table 2.16.

From Tables 2.18, 2.19, and 2.20, it can be concluded that the Wilcoxon test validates the results contained in Tables 2.15, 2.16, and 2.17, showing that the presented approach outperforms its opponents. The statistical study of the YSGA algorithm against the other algorithms and its performance over the composite benchmark functions considering 30, 50, and 100 dimensions prove that the YSGA method is effective in high-dimensional optimization problems and obtains better performance than its challengers.

2.4.4 Convergence Analysis

To visualize the evolution over the iterations of the presented approach, the performance results of each algorithm in a single execution have been plotted to illustrate the speed of convergence. This test examines the evolution of the optimization process for each method on a particular benchmark function.

For the convergence analysis, six distinct benchmark functions were contemplated: f_4, f_5, f_{13}, f_8, f_{24} and f_{25}. The selection of the functions for the examination was made considering two unimodal functions (f_4 and f_5), two multimodal functions (f_{13} and f_8) and two composite functions (f_{24} and f_{25}). The test evaluates the performance of every algorithm when solving each function in 50 dimensions. The graphs with the results of the convergence test are illustrated in Figs. 2.7, 2.8, and 2.9.

In Fig. 2.7, the convergence curves of the unimodal functions f_4 and f_5 are visualized in (a) and (b), respectively. Multimodal functions f_{13} and f_8 can be observed in (c) and (d) respectively from Fig. 2.8, while in Fig. 2.9, the convergence of the composite functions f_{24} and f_{25} are visualized in (e) and (f), respectively. It is clear

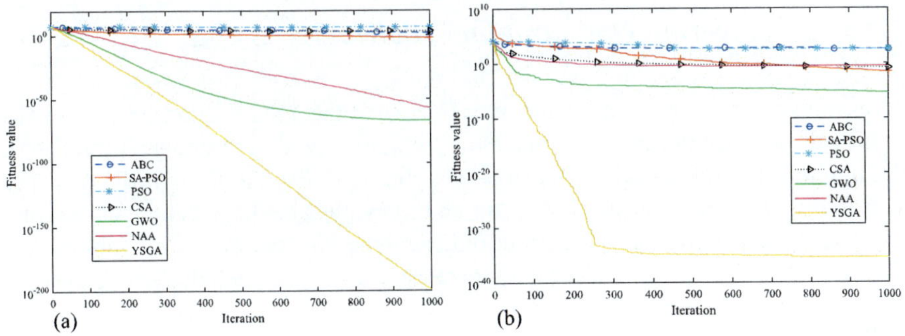

Fig. 2.7 Convergence test for functions **a** f_4 and **b** f_5 in 50 dimensions

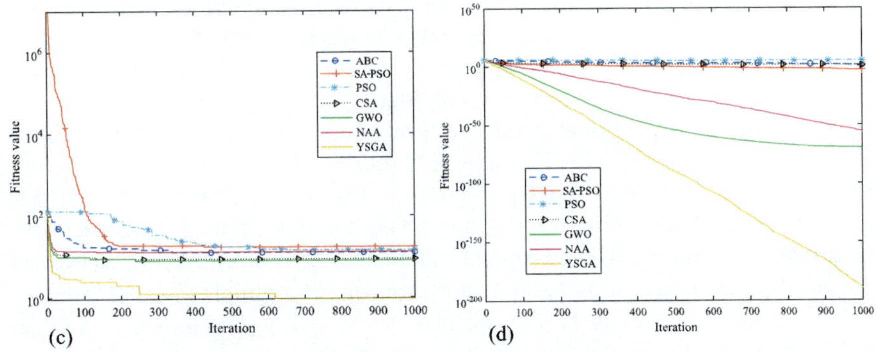

Fig. 2.8 Convergence test for functions **c** f_{13} and **d** f_8 in 50 dimensions

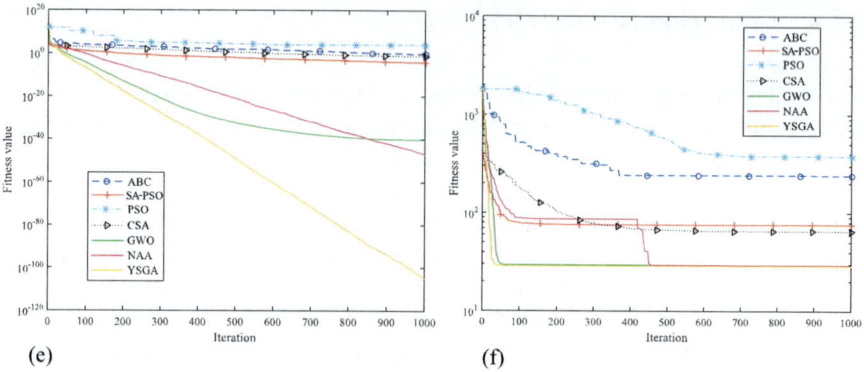

Fig. 2.9 Convergence test for functions **e** f_{24} and **f** f_{25} in 50 dimensions

that the presented approach converges faster than the ABC, SA-PSO, PSO, CSA, GWO, and NAA methods.

2.4.5 Engineering Optimization Problems

As a complementary evaluation, the performance of the YSGA algorithm was tested in the following engineering optimization problems: the design of a gear train [48], the minimization of a spring´s tension–compression [49], the design of a pressure vessel [48], the parameter estimation for an FM synthesizer [50] and the design of a three-bar truss [51]. The cost functions of the engineering optimization problems are f_{B1}, f_{B2}, f_{B3}, f_{B4} and f_{B5} respectively. See Sect. 2.4.7 for a detailed explanation of the engineering problems.

To ensure the uniformity of the experiments, the algorithms implicated were the same as in Sects. 2.4.1, 2.4.2, and 2.4.3. Thus, the experimental results were compared against the performance of the GWO, PSO, SA-PSO, ABC, CSA, and NAA algorithms. The parameters considered for the tests that are described in this section were taken from Table 2.2. The total population m has been set to 50 individuals, and the maximum number of iterations t_{max} has been set to 1000 for all the algorithms involved. Each experiment is repeated, considering 30 independent executions.

The performance of each algorithm for every engineering optimization problem can be analyzed in Table 2.21. The registered results are classified as the *Worst* and *Best* fitness value over the 30 independent runs. Likewise, the average *Avg* and the standard deviation *Std* are listed in the same Table. The *Avg* and the *Std* are calculated considering the results collected from the 30 distinct executions. The best outcomes obtained by each engineering optimization problem are highlighted in boldface.

From Table 2.21, it can be noticed that the YSGA algorithm achieves better performance than its competitors for all the engineering optimization problems.

Table 2.21 Optimal results of the engineering optimization problems

Function		YSGA	GWO	PSO	SA-PSO	ABC	CSA	NAA
f_{B1}	***Worst***	**1.23E−11**	4.32E−10	6.93E−08	8.79E−09	1.90E−09	2.12E−09	1.32E−11
	Best	**1.92E−12**	3.32E−12	2.31E−11	3.24E−12	2.81E−12	6.43E−11	4.14E−12
	Avg	**1.97E−12**	4.46E−12	2.53E−11	7.22E−12	4.87E−12	3.51E−11	4.45E−12
	Std	**4.20E−14**	7.82E−13	3.34E−12	1.68E−13	5.69E−13	1.20E−13	2.06E−12
f_{B2}	***Worst***	**1.20E−02**	8.81E−02	1.32E−02	2.68E−02	2.06E−02	8.42E−01	1.53E−02
	Best	**7.28E−03**	7.39E−03	7.30E−03	1.16E−02	7.39E−03	1.04E−02	7.68E−03
	Avg	**7.77E−03**	7.90E−03	8.36E−03	1.17E−02	7.89E−03	8.20E−03	1.20E−02
	Std	**2.70E−04**	3.58E−04	1.54E−03	1.32E−03	3.93E−04	2.95E−04	2.19E−03
f_{B3}	***Worst***	**6.25E+03**	6.42E+03	7.30E+03	6.88E+03	6.46E+03	6.79E+03	6.71E+03
	Best	**5.88E+03**	5.97E+03	5.89E+03	6.10E+03	5.98E+03	5.89E+03	5.89E+03
	Avg	**6.10E+03**	6.11E+03	6.36E+03	6.11E+03	6.16E+03	6.17E+03	6.12E+03
	Std	**5.30E+01**	1.04E+02	6.70E+02	1.07E+02	1.16E+02	1.89E+02	7.75E+01
f_{B4}	***Worst***	**2.08E+01**	2.51E+01	2.61E+01	2.68E+01	2.13E+01	2.53E+01	2.36E+01
	Best	**1.63E+01**	1.79E+01	1.75E+01	2.50E+01	1.70E+01	1.64E+01	1.68E+01
	Avg	**1.72E+01**	1.89E+01	1.86E+01	2.60E+01	1.97E+01	2.06E+01	1.86E+01
	Std	**1.17E+00**	4.80E+00	5.66E+00	1.63E+00	1.30E+00	4.35E+00	4.96E+00
f_{B5}	***Worst***	**2.65E+02**	2.83E+02	2.83E+02	2.82E+02	2.83E+02	2.89E+02	2.82E+02
	Best	**2.64E+02**	2.80E+02	2.80E+02	2.70E+02	2.80E+02	2.79E+02	2.79E+02
	Avg	**2.64E+02**	2.80E+02	2.81E+02	2.80E+02	2.80E+02	2.80E+02	2.80E+02
	Std	**3.95E−11**	9.51E−01	1.57E+00	4.57E−09	1.79E−03	1.49E−10	9.40E−08

Table 2.22 The best parameter configuration for the design variables obtained by the YSGA for each engineering optimization problem

Function	x_1	x_2	x_3	x_4	x_5	x_6	$f_{Bi}(\mathrm{x})$ for $i = 1, \ldots, 5$
f_{B1}	1.20E+01	2.10E+01	5.90E+01	3.00E+01			1.94E−12
f_{B2}	5.00E−02	4.79E−01	4.07E+00				7.28E−03
f_{B3}	7.78E−01	3.84E−01	4.04E+01	1.99E+02			5.88E+03
f_{B4}	−5.35E−01	4.76E+00	3.36E+00	4.98E+00	3.28E−01	4.47E+00	1.63E+01
f_{B5}	7.88E−01	4.09E−01	7.88E−01				2.64E+02

Table 2.23 Unimodal benchmark functions

Name	Function	S	Dim	Minimum
Sphere	$f_1(\mathbf{x}) = \sum_{i=1}^{n} x_i^2$	$[-5, 5]^n$	$n = 30$ $n = 50$ $n = 100$	$f(\mathbf{x}^*) = 0;$ $\mathbf{x}^* = 0, \ldots, 0$
Sum squares	$f_2(\mathbf{x}) = \sum_{i=1}^{n} i x_i^2$	$[-10, 10]^n$	$n = 30$ $n = 50$ $n = 100$	$f(\mathbf{x}^*) = 0;$ $\mathbf{x}^* = (0, \ldots, 0)$
Sum of different powers	$f_3(\mathbf{x}) = \sum_{i=1}^{n} \lvert x_i \rvert^{i+1}$	$[-1, 1]^n$	$n = 30$ $n = 50$ $n = 100$	$f(\mathbf{x}^*) = 0;$ $\mathbf{x}^* = (0, \ldots, 0)$
Schwefel 2	$f_4(\mathbf{x}) = \sum_{i=1}^{n} \left(\sum_{j=1}^{i} x_i \right)^2$	$[-100, 100]^n$	$n = 30$ $n = 50$ $n = 100$	$f(\mathbf{x}^*) = 0;$ $\mathbf{x}^* = (0, \ldots, 0)$
Rotated hyper-ellipsoid	$f_5(\mathbf{x}) = \sum_{i=1}^{n} \sum_{j=1}^{i} x_j^2$	$[-65.5, 65.5]^n$	$n = 30$ $n = 50$ $n = 100$	$f(\mathbf{x}^*) = 0;$ $\mathbf{x}^* = (0, \ldots, 0)$

Finally, the best set of design variables **x** are listed in Table 2.22. These results define the best parameter configuration for the best solutions founded by the YSGA approach overall 30 independent executions for every engineering optimization problems.

2.4.6 *Benchmark Functions*

The set of benchmark test functions implemented in the experiments is described in Tables 2.23, 2.24, and 2.25. In the tables, $f(\mathbf{x}^*)$ is the optimum value of the function, $\mathbf{x}^*$ the optimum position and $\mathbf{S} \in \mathbb{R}^n$ the search space. The benchmark test functions are classified in unimodal (Table 2.23), multimodal (Table 2.24), and composite (Table 2.25).

Table 2.24 Multimodal benchmark functions

Name	Function	S	Dim	Minimum
Ackley	$f_6(\mathbf{x}) = -20e^{-0.2\sqrt{\frac{1}{n}\sum_{i=1}^{n} x_i^2}} - e^{\frac{1}{n}\sum_{i=1}^{n}\cos(2\pi x_i)} + 20 + e$	$[-30, 30]^n$	$n = 30$ $n = 50$ $n = 100$	$f(\mathbf{x}^*) = 0$; $\mathbf{x}^* = (0, \ldots, 0)$
Powell	$f_{754}(\mathbf{x}) =$ $\sum_{i=1}^{n/4}\left[(x_{4i-3} + 10x_{4i-2})^2 + 5(x_{4i-1} - x_{4i})^2 + (x_{4i-2} - 2x_{4i-1})^4 + 10(x_{4i-3} - x_{4i})^4\right]$	$[-4, 5]^n$	$n = 30$ $n = 50$ $n = 100$	$f(\mathbf{x}^*) = 0$; $\mathbf{x}^* = (0, \ldots, 0)$
Rosenbrock	$f_8(\mathbf{x}) = \sum_{i=1}^{n-1}\left[100\left(x_{i+1} - x_i^2\right)^2 + (x_i - 1)^2\right]$	$[-5, 10]^n$	$n = 30$ $n = 50$ $n = 100$	$f(\mathbf{x}^*) = 0$; $\mathbf{x}^* = (1, \ldots, 1)$
Schwefel 22	$f_9(\mathbf{x}) = \sum_{i=1}^{n}\lvert x_i\rvert + \prod_{i=1}^{n}\lvert x_i\rvert$	$[-100, 100]^n$	$n = 30$ $n = 50$ $n = 100$	$f(\mathbf{x}^*) = 0$; $\mathbf{x}^* = (0, \ldots, 0)$
Perm2	$f_{10}(\mathbf{x}) = \sum_{i=1}^{n}\left[\sum_{j=1}^{n}(j + \beta)\left(x_j^i - \frac{1}{j}\right)\right]^2$	$[-n, n]^n$	$n = 30$ $n = 50$ $n = 100$	$f(\mathbf{x}^*) = 0$; $\mathbf{x}^* = \left(1, \frac{1}{2}, \ldots, \frac{1}{n}\right)$
Dixon Price	$f_{11}(\mathbf{x}) = (x_i - 1)^2 + \sum_{i=2}^{n} i\left(2x_i^2 - x_{i-1}\right)^2$	$[-10, 10]^n$	$n = 30$ $n = 50$ $n = 100$	$f(\mathbf{x}^*) = 0$; $\mathbf{x}^* = 2^{-\frac{2^i-2}{2^i}}$ for$i =$ $1, \ldots, n$

(continued)

Table 2.24 (continued)

Name	Function	S	Dim	Minimum
Styblinski Tang	$f_{12}(\mathbf{x}) = \frac{1}{2}\sum_{i=1}^{n}\left(x_i^4 - 16x_i^2 + 5x_i\right)$	$[-5, 5]^n$	$n = 30$ $n = 50$ $n = 100$	$f(\mathbf{x}^*) = -39.1659n$; $\mathbf{x}^* = (-2.90, \ldots, 2.90)$
Quartic	$f_{13}(\mathbf{x}) = \sum_{i=1}^{n}\left\{(ix_i)^4 + rand[0, 1)\right\}$	$[-1.28, 1.28]^n$	$n = 30$ $n = 50$ $n = 100$	$f(\mathbf{x}^*) = 0$; $\mathbf{x}^* = (0, \ldots, 0)$
Salomon	$f_{14}(\mathbf{x}) = 1 - \cos\left(2\pi\sqrt{\sum_{i=1}^{n} x_i^2}\right) + 0.1\sqrt{\sum_{i=1}^{n} x_i^2}$	$[-100, 100]^n$	$n = 30$ $n = 50$ $n = 100$	$f(\mathbf{x}^*) = 0$; $\mathbf{x}^* = (0, \ldots, 0)$
Penalty1A	$f_{15}(\mathbf{x}) = \frac{\pi}{30}\left\{\begin{array}{c} 10\sin^2(\pi y_1) \\ + \sum_{i=1}^{n-1}(y_i - 1)^2\left[1 + 10\sin^2(\pi y_i + 1)\right] \\ +(y_n - 1)^2 \end{array}\right\} + \sum_{i=1}^{n} u(x_i, 10, 100, 4)$; $y_i = 1 + \frac{x_i + 1}{4}$; $u(x_i, a, k, m) = \begin{cases} k(x_i - a)^m, x_i > a \\ 0, -a \le x_i \le a \\ k(-x_i - a)^m, x_i < -a \end{cases}$	$[-50, 50]^n$	$n = 30$ $n = 50$ $n = 100$	$f(\mathbf{x}^*) = 0$; $\mathbf{x}^* = (-1, \ldots, -1)$

(continued)

Table 2.24 (continued)

Name	Function	S	Dim	Minimum
Penalty2A	$f_{16}(\mathbf{x}) = 0.1\left\{\begin{array}{c}\sin^2(3\pi x_1) \\ +\sum_{i=1}^{n-1}(x_i-1)^2\left[1+\sin^2(3\pi x_{i+1})\right] \\ +(x_n-1)^2\left[1+\sin^2(2\pi x_n)\right]\end{array}\right\} + \sum_{i=1}^{n} u(x_i, 5, 100, 4);$ $u(x_i, a, k, m) = \begin{cases} k(x_i-a)^m, x_i > a \\ 0, -a \le x_i \le a \\ k(-x_i-a)^m, x_i < -a \end{cases}$	$[-50, 50]^n$	$n = 30$ $n = 50$ $n = 100$	$f(\mathbf{x}^*) = 0;$ $\mathbf{x}^* = (1, \ldots, 1)$
Rastrigin A	$f_{17}(\mathbf{x}) = 10n + \sum_{i=1}^{n}\left[x_i^2 - 10\cos(2\pi x_i)\right]$	$[-5.12, 5.12]^n$	$n = 30$ $n = 50$ $n = 100$	$f(\mathbf{x}^*) = 0;$ $\mathbf{x}^* = (0, \ldots, 0)$
Mishra1	$f_{18}(\mathbf{x}) = (1 + x_n)^{x_n}; x_n = n - \sum_{i=1}^{n-1} x_i$	$[0, 1]^n$	$n = 30$ $n = 50$ $n = 100$	$f(\mathbf{x}^*) = 2;$ $\mathbf{x}^* = (1, \ldots, 1)$
Mishra2	$f_{19}(\mathbf{x}) = (1 + x_n)^{x_n}; x_n = n - \sum_{i=1}^{n-1} \frac{(x_i + x_{i+1})}{2}$	$[0, 1]^n$	$n = 30$ $n = 50$ $n = 100$	$f(\mathbf{x}^*) = 2;$ $\mathbf{x}^* = (1, \ldots, 1)$
Mishra11	$f_{20}(\mathbf{x}) = \left[\frac{1}{n}\sum_{i=1}^{n}\lvert x_i\rvert - \left(\prod_{i=1}^{n}\lvert x_i\rvert\right)^{\frac{1}{n}}\right]^2$	$[-10, 10]^n$	$n = 30$ $n = 50$ $n = 100$	$f(\mathbf{x}^*) = 0;$ $\mathbf{x}^* = (0, \ldots, 0)$
Vincent	$f_{21}(\mathbf{x}) = -\sum_{i=1}^{n}\sin(10\log x_i)$	$[0.25, 10]^n$	$n = 30$ $n = 50$ $n = 100$	$f(\mathbf{x}^*) = -n;$ $\mathbf{x}^* =$ $(7.70, \ldots, 7.70)$

(continued)

Table 2.24 (continued)

Name	Function	S	Dim	Minimum
Infinity	$f_{22}(\mathbf{x}) = \sum_{i=1}^{n} x_i^6 \left[\sin\left(\frac{1}{x_i}\right) + 2 \right]$	$[-1, 1]^n$	$n = 30$ $n = 50$ $n = 100$	$f(\mathbf{x}^*) = 0$; $\mathbf{x}^* = (0, \ldots, 0)$
Griewank	$f_{23}(\mathbf{x}) = \frac{1}{4000} \sum_{i=1}^{n} x_i^2 - \prod_{i=1}^{n} \cos\left(\frac{x_i}{\sqrt{i}}\right) + 1$	$[-600, 600]^n$	$n = 30$ $n = 50$ $n = 100$	$f(\mathbf{x}^*) = 0$; $\mathbf{x}^* = (0, \ldots, 0)$

Table 2.25 Composite benchmark functions

Name	Function	S	Dim	Minimum
Fx24	$f_{24}(\mathbf{x}) = f_1(\mathbf{x}) + f_9(\mathbf{x}) + f_{17}(\mathbf{x})$	$[-100, 100]^n$	$n = 30$ $n = 50$ $n = 100$	$f(\mathbf{x}^*) = 0$; $\mathbf{x}^* = (0, \ldots, 0)$
Fx25	$f_{25}(\mathbf{x}) = f_8(\mathbf{x}) + f_{17}(\mathbf{x}) + f_{23}(\mathbf{x})$	$[-100, 100]^n$	$n = 30$ $n = 50$ $n = 100$	$f(\mathbf{x}^*) = n - 1$; $\mathbf{x}^* = (0, \ldots, 0)$
Fx26	$f_{26}(\mathbf{x}) = f_4(\mathbf{x})+f_6(\mathbf{x})+f_8(\mathbf{x})+f_{16}(\mathbf{x})$	$[-100, 100]^n$	$n = 30$ $n = 50$ $n = 100$	$f(\mathbf{x}^*) = (1.1n) - 1$; $\mathbf{x}^* = (0, \ldots, 0)$
Fx27	$f_{27}(\mathbf{x}) = f_6(\mathbf{x}) + f_8(\mathbf{x}) + f_9(\mathbf{x}) + f_{17}(\mathbf{x}) + f_{23}(\mathbf{x})$	$[-100, 100]^n$	$n = 30$ $n = 50$ $n = 100$	$f(\mathbf{x}^*) = n - 1$; $\mathbf{x}^* = (0, \ldots, 0)$

2.4.7 *Description of Engineering Problems*

In this section, the engineering optimization problems considered in the experiments are described in detail.

2.4.7.1 Gear Train Problem

Given a gear train, as shown in Fig. 2.10, it is required to minimize the squared difference between the teeth ratio of the gear and a given scalar value. The decision variables are the number of teeth corresponding to each gear. Labels A, B, D, and F are used to identify the gears. The decision variables are $x_1 = A$, $x_2 = B$, $x_3 = D$ and $x_4 = F$. The scalar value is $1/6.931$. The cost function and constraints are defined as follows:

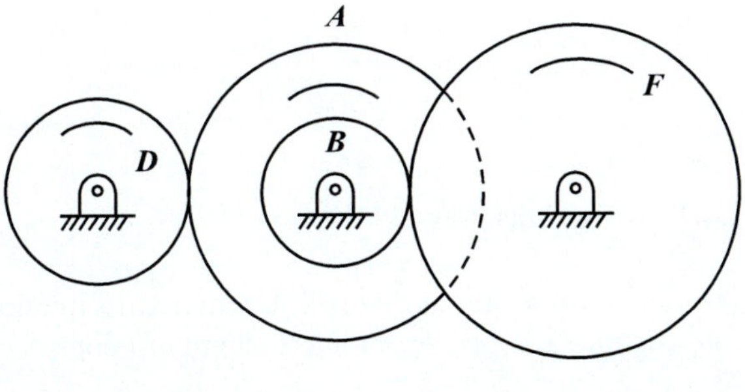

Fig. 2.10 Gear train design

Fig. 2.11 Spring design

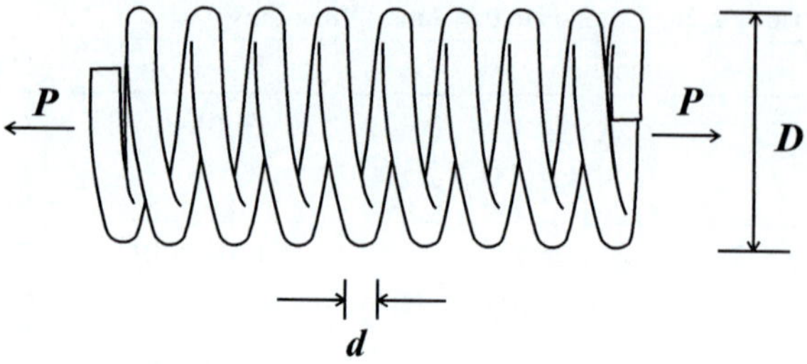

Problem 2.1. Gear train

Minimize:

$f_{B1}(\mathbf{x}) = \left(\frac{1}{6.931} - \frac{x_3 x_2}{x_1 x_4}\right)^2$

Subject to:

$0 \leq x_i \leq 600, i = 1, 2, 3, 4$

2.4.7.2 Spring Problem

The problem is to minimize the tension or compression experienced by a spring when a load P is applied. For optimization, the wire diameter d, the coil diameter D, and the number of active coils n are considered. The decision variables are $x_1 = d$, $x_2 = D$ and $x_3 = n$, see Fig. 2.11. The design problem is formulated as:

Problem 2.2. Spring

Minimize:

$f_{B2}(\mathrm{x}) = (x_3 + 2)x_2 x_1^2$

Subject to:

$g_1(\mathrm{x}) = 1 - \frac{x_2^3 x_3}{71{,}785 x_1^4} \leq 0$

$g_2(\mathrm{x}) = \frac{4x_2^2 - x_1 x_2}{12{,}566(x_2 x_1^3 - x_1^4)} + \frac{1}{5{,}108 x_1^2} - 1 \leq 0$

$g_3(\mathrm{x}) = 1 - \frac{140.45 x_1}{x_2^2 x_3} \leq 0 \; g_4(\mathrm{x}) = \frac{x_1 + x_2}{1.5} \leq 0$

$0.5 \leq x_1 \leq 2 0.25 \leq x_2 \leq 1.3 \; 2 \leq x_3 \leq 15$

2.4.7.3 Pressure Vessel Problem

The design of a pressure vessel is required to minimize the material used for its construction. The optimization problem must consider the thickness of the shell T_s,

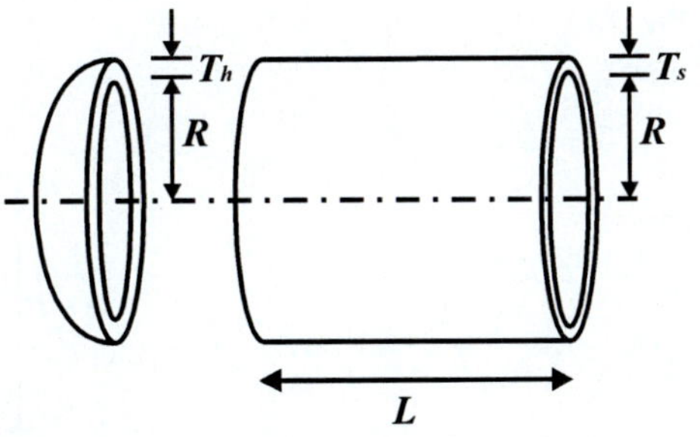

Fig. 2.12 Pressure vessel design

the thickness of the head T_h, the internal radius of the vessel R, and the length of the vessel L, see Fig. 2.12. The decision variables are $\mathbf{x} = [x_1, x_2, x_3, x_4]$ were $x_1 = T_s$, $x_2 = T_h$, $x_3 = R$ and $x_4 = L$. The cost function and constraints are defined as follows:

Problem 2.3. Pressure vessel

Minimize:

$f_{B3}(\mathrm{x}) = 0.6224x_1x_3x_4 + 1.7781x_2x_3^2 + 3.1661x_1^2x_4 + 19.84x_1^2x_3$

Subject to:

$g_1(\mathrm{x}) = -x_1 + 0.0193x_3 \leq 0 g_2(\mathrm{x}) = -x_2 + 0.00954x_3 \leq 0 g_3(\mathrm{x}) =$
$-\pi x_3^2 x_4 - (4/3)\pi x_3^2 + 1,296,000 \leq 0 g_4(\mathrm{x}) = x_4 - 240 \leq 0$
$0 \leq x_i \leq 100, i = 1, 20 \leq x_i \leq 200, i = 3, 4$

2.4.7.4 FM Synthesizer Problem

An FM synthesizer generates a signal $y(\mathbf{x}, t)$ similar to a target signal $y_0(t)$. To minimize the error between the signal and the target signal, a parameter estimator for the FM synthesizer is designed considering a finite wave amplitude a_i and the frequency ω_i. The decision variables are $\mathbf{x} = [x_1 = a_1, x_2 = \omega_1, x_3 = a_2, x_4 = \omega_2, x_5 = a_3, x_6 = \omega_3]$. The cost function and constraints are defined as follows:

Problem 2.4. FM synthesizer

Minimize:

$$f_{B4}(\mathrm{x}) = \sum_{t=0}^{100} (y(\mathrm{x}, t) - y_0(t))^2$$

$$y(\mathrm{x}, t) = x_1 \sin(x_2\theta t) + x_3 \sin(x_4\theta t) + x_5 \sin(x_6\theta t)$$

Subject to:

$-6.4 \leq x_i \leq 6.35, i = 1, 2, 3, 4, 5, 6$

$\theta = \frac{2\pi}{100}$

Fig. 2.13 Three-bar truss design

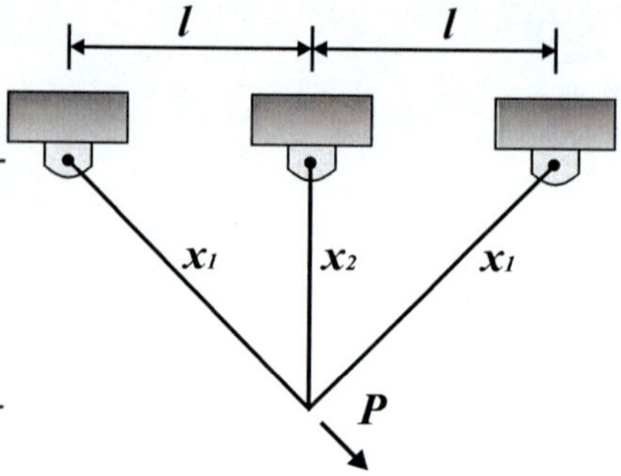

2.4.7.5 Three Bar Truss Problem

The manufacturing cost of a three-bar truss, subject to a load P and a stress σ, must be minimized. The decision variables are the cross-sectional areas x_1, x_2 and x_3 shown in Fig. 2.13, where $x_3 = x_1$ due to the design symmetry. The cost function and constraints are defined as follows:

Problem 2.5. Three bar truss

Minimize:
$f_{B5}(\mathrm{x}) = l\left(2\sqrt{2x_1} + x_2\right)$

Subject to:

$$g_1(\mathrm{x}) = \frac{\sqrt{2x_1+x_2}}{\sqrt{2x_1^2+2x_1x_2}}P - \sigma \leq 0 g_2(\mathrm{x}) = \frac{x_2}{\sqrt{2x_1^2+2x_1x_2}}P - \sigma \leq 0 g_3(\mathrm{x}) = \frac{1}{\sqrt{2x_2^2+x_1}}P - \sigma \leq 0 0 \leq$$

$x_i \leq 1, i = 1, 2$
$l = 100$ cm, $P = 2$ kN/cm^2, $\sigma = 2$ kN/cm^2

2.5 Summary

In this chapter, an approach for solving global optimization problems was presented. The YSGA is a bio-inspired algorithm based on the hunting behavior of the Yellow Saddle Goatfish *Parupeneus cyclostomus*. The described method has implemented operators that mimic these hunting techniques to design a search strategy that overcomes several issues manifested in other swarm-based algorithms, such as premature convergence, stagnation in local optima, lack of balance in the exploration–exploitation process, and the diversity of the possible solutions.

The performance of the YSGA was tested on a set of 27 well-known benchmark functions. The results were compared to other similar algorithms, which are some of the most popular in the literature of the meta-heuristic techniques, including the Gray Wolf Optimization (GWO), the Particle Swarm Optimization (PSO), the Synchronous-Asynchronous Particle Swarm Optimization (SA-PSO), the Artificial

Bee Colony (ABC), the Crow Search Algorithm (CSA) and the Natural Aggregation Algorithm (NAA). Statistical validation of the results has been corroborated by applying a non-parametric framework to ensure the consistency of the algorithm and to guarantee there is no random effect. Additionally, YSGA was also tested over several engineering optimization problems. These results were compared to the performance of other algorithms. The experimental results demonstrate this technique is fast, accurate, and robustness over its competitors.

References

1. Kennedy J, Eberhart R (1995) Particle swarm optimization. In: Proceedings of ICNN'95—international conference on neural networks, vol 4, pp 1942–1948
2. Ab Aziz NA, Mubin M, Mohamad MS, Ab Aziz K (2014) A synchronous-asynchronous particle swarm optimization algorithm. Sci World J 17
3. Karaboga D (2005) An idea based on Honey Bee Swarm for Numerical Optimization. In: Tech. Rep. TR06, Erciyes Univ., no. TR06, p 10
4. Askarzadeh A (2016) A novel metaheuristic method for solving constrained engineering optimization problems: crow search algorithm. Comput Struct 169:1–12
5. Mirjalili S, Mirjalili SM, Lewis A (2014) Grey wolf optimizer. Adv Eng Softw 69:46–61
6. Luo F, Zhao J, Dong ZY (2016) A new metaheuristic algorithm for real-parameter optimization: natural aggregation algorithm. In: 2016 The IEEE congress on evolutionary computation, pp 94–103
7. Mirjalili S, Lewis A (2016) The whale optimization algorithm. Adv Eng Softw 95:51–67
8. Fausto F, Cuevas E, Valdivia A, González A (2017) A global optimization algorithm inspired in the behavior of selfish herds. Biosystems 160:39–55
9. Mirjalili S (2015) Moth-flame optimization algorithm: A novel nature-inspired heuristic paradigm. Knowl Based Syst 89:228–249
10. Rashedi E, Nezamabadi-pour H, Saryazdi S (2009) GSA: a gravitational search algorithm. Inf Sci 179(13):2232–2248
11. Webster B, Bernhard PJ A local search optimization algorithm based on natural principles of gravitation
12. Erol OK, Eksin I (2006) A new optimization method: Big Bang-Big Crunch. Adv Eng Softw 37(2):106–111
13. Hatamlou A (2013) Blackhole: a new heuristic optimization approach for data clustering. Inf Sci 222:175–184
14. Schmitt LM (2001) Theory of genetic algorithms. Theor Comput Sci 259(1):1–61
15. Price JA, Kenneth S, Lampinen RM (2005) Differential evolution. Springer, Berlin/Heidelberg
16. Rechenberg I (1978) Evolutions strategien. Springer, Berlin, pp 83–114
17. Olorunda O, Engelbrecht AP (2008) Measuring exploration/exploitation in particle swarms using swarm diversity. In: 2008 IEEE congress on evolutionary computation (IEEE world congress on computational intelligence), pp 1128–1134
18. Lin L, Gen M (2009) Auto-tuning strategy for evolutionary algorithms: balancing between exploration and exploitation. Soft Comput 13(2):157–168
19. Randall JE (2014) The goatfishes Parupeneus cyclostomus, *P. macronemus* and freeloaders, vol 20, no 2. Aquaprint
20. McCormick MI (1995) Fish feeding on mobile benthic invertebrates: influence of spatial variability in habitat associations. Mar Biol 121(4):627–637
21. Randall JE (1983) Red Sea reef fishes. IMMEL
22. Randall JE (2007) Reef and shore fishes of the Hawaiian Islands. Sea Grant College Program, University of Hawai

23. Strübin C, Steinegger M, Bshary R (2011) On group living and collaborative hunting in the yellow saddle goatfish (*Parupeneus cyclostomus*)1. Ethology 117(11):961–969
24. Arnegard ME, Carlson BA (2005) Electric organ discharge patterns during group hunting by a mormyrid fish. Proc Biol Sci 272(1570):1305–1314
25. Bshary R, Hohner A, Ait-el-Djoudi K, Fricke H (2006) Interspecific communicative and coordinated hunting between groupers and giant moray eels in the red sea. PLoS Biol 4(12):e431
26. Boesch C, Boesch H (1989) Hunting behavior of wild Chimpanzees in the tai' National Park. Am J Phys Anthropol 78547–78573
27. Stander PE (1992) Cooperative hunting in lions: the role of the individual. Source Behav Ecol Sociobiol Behav Ecol Sociobiol 29(6):445–454
28. Biro D, Sasaki T, Portugal SJ (2016) Bringing a time-depth perspective to collective animal behaviour. Trends Ecol Evol 31(7):550–562
29. Kalyani S, Swarup KS (2011) Particle swarm optimization-based K-means clustering approach for security assessment in power systems. Expert Syst Appl 38(9):10839–10846
30. Al-Harbi SH, Rayward-Smith VJ (2006) Adapting k-means for supervised clustering. Appl Intell 24(3):219–226
31. Kaufman L, Rousseeuw PJ (eds) (1990) Finding groups in data. Wiley Inc., Hoboken, NJ, USA
32. MacQueen J (1967) Some methods for classification and analysis of multivariate observations. In: Proceedings of the fifth berkeley symposium on mathematical statistics and probability, vol 1, Statistics, pp 281–297
33. Jain AK (2010) Data clustering: 50 years beyond K-means. Pattern Recognit Lett 31(8):651–666
34. Likas A, Vlassis N, Verbeek JJ (2003) The global k-means clustering algorithm. Pattern Recognit 36(2):451–461
35. Hartigan JA, Wong MA (1979) A K-means clustering algorithm. Source J R Stat Soc Ser C (Appl Stat) 28(1):100–108
36. Forgy E (1965) Cluster analysis of multivariate data: efficiency versus interpretability of classification. Biometrics 21(3):768–769
37. Lloyd S (1982) Least squares quantization in PCM. IEEE Trans Inf Theory 28(2):129–137
38. Redmond SJ, Heneghan C (2007) A method for initializing the K-means clustering algorithm using kd-trees. Pattern Recognit Lett 28(8):965–973
39. Chechkin A, Metzler R, Klafter J, Gonchar V (2008) Introduction to the theory of lévy flights. In: Anomalous transport: foundations and applications, pp 129–162
40. Haklı H, Uğuz H (2014) A novel particle swarm optimization algorithm with Levy flight. Appl Soft Comput 23:333–345
41. Yang X-S, Deb S (2013) Multiobjective cuckoo search for design optimization. Comput Oper Res 40(6):1616–1624
42. Yang X-S (2010) Engineering optimization: an introduction with metaheuristic applications, 1st edn. Wiley
43. Marini F, Walczak B (2015) Particle swarm optimization (PSO). A tutorial. Chemom Intell Lab Syst 149:153–165
44. García S, Molina D, Lozano M, Herrera F (2009) A study on the use of non-parametric tests for analyzing the evolutionary algorithms' behaviour: a case study on the CEC'2005 special session on real parameter optimization. J Heuristics 15(6):617–644
45. Wilcoxon F (1945) Individual comparisons by ranking methods. Biometrics Bull 1(6):80–83
46. Hochberg Y (1988) A sharper Bonferroni procedure for multiple tests of significance. Biometrika
47. Armstrong RA (2014) When to use the Bonferroni correction. Ophthalmic Physiol Opt 34(5):502–508
48. Sandgren E (1990) Nonlinear integer and discrete programming in mechanical design optimization. J Mech Des 112(2):223
49. Arora JS (2012) Chapter 12—numerical methods for constrained optimum design. In: Introduction to optimum design, pp 491–531

50. Das S, Suganthan P (2018) Problem definitions and evaluation criteria for CEC 2011 competition on testing evolutionary algorithms on real world optimization problems
51. Koski J (1985) Defectiveness of weighting method in multi-criterion optimization of structures. Commun Appl Numer Methods 1(6):333–337
52. Cuevas E (2013) Block-matching algorithm based on harmony search optimization for motion estimation. Appl Intell 39(1):165–183
53. Díaz P, Pérez-Cisneros M, Cuevas E, Hinojosa S, Zaldivar D (2018) An improved crow search algorithm applied to energy problems. Energies 11(3):571
54. Cuevas E, Gálvez J, Hinojosa S, Zaldívar D, Pérez-Cisneros M (2014) A comparison of evolutionary computation techniques for IIR model identification. J Appl Math 827206

Chapter 3
Metaheuristic Algorithm Based on Hybridization of Invasive Weed Optimization asnd Estimation Distribution Methods

Hybrid metaheuristic methods combine approaches extracted from different techniques to build a single optimization method. The design of such systems represents a current trend in the metaheuristic optimization literature. In hybrid algorithms, the objective is to extend the potential advantages of the integrated approaches and eliminates their main drawbacks. In this chapter, a hybrid method for solving optimization problems is presented. The presented approach combines (1) the explorative characteristics of the invasive weed optimization (IWO) method, (2) the probabilistic models of the estimation distribution algorithms (EDA), and (3) the dispersion capacities of a mixed Gaussian-Cauchy distribution to produce its own search strategy. With these mechanisms, the method conducts an optimization strategy over search areas that deserve a special interest according to a probabilistic model and the fitness value of the existent solutions. In the presented method, each individual of the population generates new elements around its own location, dispersed according to the mixed distribution. The number of new elements depends on the relative fitness value of the individual regarding the complete population. After this process, a group of promising solutions is selected from the set compound by the (1) new elements and the (2) original individuals. Based on the selected solutions, a probabilistic model is built from which a certain number of members (3) is sampled. Then, all the individuals of the sets (1), (2), and (3) are joined in a single group and ranked in terms of their fitness values. Finally, the best elements of the group are selected to replace the original population. This process is repeated until a termination criterion has been reached. To test the performance of the method, several comparisons to other well-known metaheuristic methods has been conducted. The comparison consists of analyzing the optimization results over different standard benchmark functions within a statistical framework. Conclusions based on the comparisons exhibit the accuracy, efficiency, and robustness of the presented approach.

© The Author(s), under exclusive license to Springer Nature Switzerland AG 2021

E. Cuevas et al., *Recent Metaheuristic Computation Schemes in Engineering*,
Studies in Computational Intelligence 948,
https://doi.org/10.1007/978-3-030-66007-9_3

3.1 Introduction

The main objective of an optimization method is to obtain the optimal solution from a set of possible alternatives which formulate a minimization/maximization problem [51]. Optimization processes emerge in diverse fields such as engineering, economics, biology, and others, where a solution must be explored to find the best possible performance of a particular system [40]. Currently, different optimization methods have been introduced in the literature. In general, they can be divided into gradient-based approaches and Evolutionary Computation (EC) methods. Gradient-based algorithms efficiently find the optimum solution of optimization problems formulated through unimodal objective functions. However, practical optimization formulations are prone to generate multimodal objective functions [10, 11]. In this scenario, conventional gradient-based techniques tend to find sub-optimal solutions as a consequence of their incapacity to escape local minima. On the contrary, EC methods are non-gradient algorithms which are specially appropriated to optimize multimodal formulations due to its stochastic properties [25].

In the last years, different optimization methods based on evolutionary principles have been introduced with interesting results [13]. These methods emulate biological or social systems, which, according to a certain point of view, can be conceived as optimization processes. Some examples of these approachesconsider the collaborative behavior of bird flocking and fish schooling in the Particle Swarm Optimization (PSO) method [27], the collective behavior of bee colonies in the Artificial Bee Colony (ABC) method [26], the composition process that takes place when a musician seeks for a better level of harmony in the Harmony Search (HS), the gravitational forces that takes place among particles in the Gravitational Search Algorithm (GSA) [41], the mating process of firefly insects in the Firefly (FF) method [50], the social response of spiders in the Social Spider Optimization (SSO) [9], the interaction behavior of an animal group in the Collective Animal Behavior (CAB) [12] and the mimicry of the conventional and differential evolution in species in the Genetic Algorithms (GA) [22] and the Differential Evolution (DE) [44], respectively.

Another interesting evolutionary algorithm is the Invasive Weed Optimization (IWO) [33] method, which emulates the proliferation of weed in a plantation. In IWO, each new individual simulates a weed that grows around a certain area determined by a previous weed. The number of new weeds and its position depends on the quality of the previous week (in terms of its fitness value) and Gaussian distribution, respectively. Different from other EC methods, IWO maintains excellent explorative capacities over the entire search space [3]. This feature has motivated its use to solve a wide variety of engineering problems such as the optimal localization of piezoelectric actuators [34], antenna design [32], and decoding DNA sequences [56],to name a few.

Estimation Distribution Algorithms (EDA) [36] are a new paradigm of generic EC methods that have received great attention in the last years. EDA produces promising solutions that have higher probabilities of reaching a global optimum according to a probabilistic model. In its operation, EDA selects a subset of promising solutions from the current population to build a parent group. Considering the information provided by the parent group, a probabilistic model is generated. Finally, by using the model, new individuals are produced to replace some or all individuals from the original population. Strong evidence of the EDA capacities is its extensive application in several practical problems such as protein synthesis [43], antenna design [54], economic dispatch [8], forest management [14], portfolio management [31], and others. According to its operation, EDA can significantly augment the solution quality of the population in the first generations. However, its diversity decreases quickly as the search process evolves, producing its premature convergence [38].

Gaussian and Cauchy distributions have been frequently used to produce new individuals in several EC methods [2, 42]. The lingering tails of the Cauchy distribution maintain a higher probability of making long jumps, which contribute to escape from local optima [52]. Under this distribution, the exploration of the optimization strategy is promoted. Different from Cauchy approaches, Gaussian distributions generate regulated dispersions that produce a faster local convergence. Therefore, the production of individuals through Gaussian distributions privilege the exploitation of a search strategy [52]. On the other hand, the linear combination of both distributions merges their desirable properties yielding a better exploration–exploitation rate than the use of each distribution individually [7].

The fact that all EC techniques produce similar results when they are compared to all possible objective functions has been extensively reported in the literature [49]. From these works, it has been reported that if Algorithm A surpasses algorithm B on some optimization problems. Then, it is expected that Algorithm B outperforms A in many other problems. Currently, it is widely accepted that the performance of an EC depends on its faculty to obtain an appropriate exploitation-exploration rate of the search space [45]. Exploration represents the inspection of new promising solutions, while the process of exploitation considers the operation of refining previously visited locations with the intention of improving their fitness quality. Using primarily exploration deteriorates the accuracy of the produced solutions but enhances the potential to locate new promising solutions [1]. Contrariwise, if the exploitation operation is mainly conducted, then available solutions are essentially refined. Nevertheless, this fact unfavorably produces that the search strategy gets trapped in local minima due to the ineffective exploration of the search space [39]. On the other hand, each optimization problem demands a specific balance between exploration and exploitation [10, 11]. Consequently, each EC algorithm suggests a particular exploration–exploitation rate using the integration of deterministic rules and stochastic elements. Under such conditions, it is complicated to design a generic optimization method that could competitively solve all formulated problems.

Researchers have proposed the integration of diverse EC methods for reaching new performance results when it is apparent that single EC algorithms had attained their limits. In the last years, several approaches have been introduced that do not fully match the structure of an EC method. Alternatively, they merge several computational components, derived from other evolutionary or metaheuristic approaches. These kinds of algorithms are designated as hybrid EC methods [6]. The central objective of hybridization is to combine the best characteristics of different EC algorithms together, to build a new EC method that is expected to have a better performance than the original approaches. With the integration, the intention is to eliminate the weaknesses of every single method and to reach a synergetic impact through their union [5, 20]. Several hybrid EC methods have been reported in the literature. They have demonstrated to be successful for many particular problems where single EC methods have obtained only marginal results [15]. Some examples of such methods include the combinations of PSO-GA [17], GA-DE [46], PSO-ABC [29], HS-DE [55], GSA-PSO [4], to name a few.

In this chapter, a hybrid method for solving optimization problems is presented. The approach combines (1) the explorative characteristics of the IWO method, (2) the probabilistic models of EDA, and (3) the dispersion capacities of a mixed Gaussian-Cauchy distribution to produce its own search strategy. With these mechanisms, the presented method conducts an optimization strategy over search areas that deserve a special interest according to a probabilistic model and the fitness value of the existent solutions. In the presented method, each individual of the population generates new elements around its own location, dispersed according to the mixed distribution. The number of new elements depends on the relative fitness value of the individual regarding the complete population. After this process, a group of promising solutions is selected from the set compound by the (1) new elements and the (2) original individuals. Based on the selected solutions, a probabilistic model is built from which a certain number of members (3) is sampled. Then, all the individuals of the sets (1), (2), and (3) are joined in a single group and ranked in terms of their fitness values. Finally, the best elements of the group are selected to replace the original population. This process is repeated until a termination criterion has been reached. To test the performance of the method, several comparisons to other well-known metaheuristic methods has been conducted. The comparison consists of analyzing the optimization results over different standard benchmark functions within a statistical framework. Conclusions based on the comparisons exhibit the accuracy, efficiency, and robustness of this approach.

3.2 The Invasive Weed Optimization (IWO) Algorithm

IWO is a metaheuristic method based on the population that emulates the proliferation of weed in a plantation. The central feature of a weed is that it produces new individuals in a specified region, which can be large or small. The operation of IWO can be divided into four steps: Initialization, reproduction, spatial localization, and competitive exclusion.

3.2.1 Initialization

In the initial step, a population $\mathbf{P}$ of m weeds $\{\mathbf{p}_1, \mathbf{p}_2, \ldots, \mathbf{p}_m\}$ is randomly distributed over the complete n-dimensional search space where $\mathbf{p}_i = \{p_i^1, p_i^2, \ldots, p_i^n\}$ $(i \in \{1, \ldots, m\})$.

3.2.2 Reproduction

Each individual $\mathbf{p}_i$ of $\mathbf{P}$ produces a group of E_i elements within a given area around its own position. The number of elements E_i generated by $\mathbf{p}_i$ is determined in terms of its relative fitness value with regard to the best and worst fitness values of the population. Under these conditions, the amount of generated elements depends on two parameters: E^{worst} and E^{best}. The parameter E^{worst} corresponds to the number of elements generated by the individual $\mathbf{p}^{worst}$ with the worst fitness value. On the other hand, E^{best} represents the number of seeds produced by the individual $\mathbf{p}^{best}$ with the best fitness quality. Any other individual $\mathbf{p}_i$ yields a number of seeds E_i which linearly varies from E^{worst} to E^{best} in terms of its respective fitness value.

3.2.3 Spatial Localization

The position of each element from E_i is randomly calculated around $\mathbf{p}_i$ according to a Gaussian distribution with zero mean and standard deviation $\sigma(N(0, \sigma))$. In IWO, during the evolution process, the value of σ is gradually modified so that the exploration of the newly produced elements is reduced as the number of generations increases. Therefore, the algorithm adjusts the standard deviation σ in terms of the generation number k, according to the following model:

$$\sigma(k) = \sigma_{\min} + \left(\frac{k_{\max} - k}{k_{\max}}\right)^{\gamma} \cdot (\sigma_{\max} - \sigma_{\min}) \tag{3.1}$$

where $\sigma_{\max}$ and $\sigma_{\min}$ represent the maximum and minimum standard deviations that limit the maximal and minimal exploration of the method. $k_{\max}$ symbolizes the maximum number of generations considered in the optimization process. γ is a calibration parameter that nonlinearly adjusts the modification of σ.

3.2.4 Competitive Exclusion

During the optimization process, several new individuals are produced by using the reproduction procedure. All of them are included in the population until the number of elements s in **P**reaches a limit $m_{\max}(m_{\max} \gg m)$. Once the population size surpasses its limit, an elimination mechanism is used to delete the rest of the individuals. In this process, the entire set of elements is ranked according to their fitness values. Then, those individuals with better fitness values persist for reproduction while the excess is discarded. All IWO operations are summarized in the form of pseudo-code in Algorithm 1.

Algorithm 3.1. Summarized operations for the IWO method

Algorithm 3.1. Pseudo-code for IWO method
Input: m, $\sigma_{\min}$, $\sigma_{\max}$, $k_{\max}$, γ, E^{worst}, E^{best},$m_{\max}$
P ←**Initialize**(m);
for ($k = 1$; $k < = k_{\max}$;$k + +$)
$\mathbf{p}^{worst}$ ←**SelectWorstElement**(**P**);
$\mathbf{p}^{best}$ ←**SelectBestElement**(**P**);
s ←**Size**(**P**);
for ($i = 1$; $i< = s$; $i + +$)
E_i ←**Reproduction**(**X**,E^{worst}, E^{best},$\mathbf{p}^{best}$,$\mathbf{p}^{worst}$);
$\sigma(k) \leftarrow \sigma_{\min} + \left(\frac{k_{\max}-k}{k_{\max}}\right)^{\gamma} \cdot (\sigma_{\max} - \sigma_{\min})$
P ←**SpatialLocalization**($\sigma(k)$,E_i);
end for
s ←**Size**(**P**);
If (s > *max_pop*) {**P** ←**CompetitiveExclusion** (**P**)}
end for
Output:$\mathbf{p}^{best}$

3.3 Estimation Distribution Algorithms (EDA)

Estimation Distribution Algorithms (EDA) [36] are a kind of generic evolutionary method which estimates a probability distribution model to produce new candidate solutions. This model is generated from a set of solutions chosen in terms of their quality. The main objective of the probabilistic model is to assign a higher probability of selection to areas of the search space that maintains better fitness values. With the use of this probabilistic model, EDA is able to utilize the knowledge obtained during the optimization process to conduct the search strategy. An EDA method includes four operations: Initialization, selection, model construction, individual production, and truncation.

3.3.1 *Initialization*

The method begins defining an initial population $\mathbf{Y}$ of p elements $\{\mathbf{y}_1, \mathbf{y}_2, \ldots, \mathbf{y}_p\}$ that are randomly generated in the n-dimensional search space where $\mathbf{y}_j = \{y_j^1, y_j^2, \ldots, y_j^n\}$ $(j \in \{1, \ldots, p\})$.

3.3.2 *Selection*

The operation of selection is used to choose a set $\mathbf{M} = \{\mathbf{m}_1, \mathbf{m}_2, \ldots, \mathbf{m}_g\}$ of g elements in the current population that will be considered to build the probabilistic model ($g \ll p$). EDAs hold three different selection techniques: proportional, tournament, and truncation.

In the proportional method [23], each individual from the g elements is selected according to a probability assigned in terms of its relative fitness value in the population. Under the tournament method [19], the best element is chosen from a set of individuals (generally two) randomly selected. This process is repeated g times for completing the set $\mathbf{M}$. The truncation technique [37] is the most common selection approach used in EDA algorithms. In this method, all solutions of the population $\mathbf{Y}$ are ranked according to their fitness values. Then, the first g elements are selected to build a new group $\mathbf{M}$ that contains the best g elements of $\mathbf{Y}$ while the rest is not considered.

3.3.3 *Model Construction*

In each generation, a probability distribution model is produced. This model is generated from the individuals of $\mathbf{M}$ so that it integrates their main features. The model consists of n Gaussian distributions, where each one of them captures the data behavior of a decision variable. Therefore, the d-th Gaussian distribution $N_d(\mu_d, \sigma_d)$ $(d \in \{1, \ldots, n\})$ is estimated through the data of $\mathbf{M}$ in the following form:

$$\mu_d = \frac{\sum_{j=1}^{p} m_j^d}{p}, \sigma_d^2 = \frac{\sum_{j=1}^{p} (m_j^d - \mu_d)^2}{p-1} \quad (d = 1, \ldots, n;\ j = 1, \ldots, p), \tag{3.2}$$

where m_j^d represents the d-th decision variable of the element $\mathbf{m}_j$.

3.3.4 Individual Production

In this stage, a set $\mathbf{Z} = \{\mathbf{z}_1, \mathbf{z}_2, \ldots, \mathbf{z}_w\}$ of w new individuals is produced by using the Gaussian models. Therefore, each decision variable v of the element $\mathbf{z}_u$ is computed as follows:

$$z_u^v = N(\mu_v, \sigma_v), \tag{3.3}$$

Once calculated the w new elements, the original population $\mathbf{Y}$ and the produced $\mathbf{Z}$ are merged in a temporary population $\mathbf{T} = \mathbf{Y} \cup \mathbf{Z}$. Under such conditions, the temporary population $\mathbf{T}$ would contain $p + w$ elements.

3.3.5 Truncation

Finally, the population **T**is ranked according to their fitness values. Then, the best p individuals are selected to build the new population $\mathbf{Y}$ to be evolved (the other w elements are discarded). Therefore, the operations selection, model construction, individual production, and truncation are repeatedly processed until the maximum number of generations has been reached. All EDA operations are exposed in the form of pseudo-code in Algorithm 2.

Algorithm 3.2. Summarized operations for the EDA method

Algorithm 2. Pseudo-code for EDA method
Input: p, g, w
Y ←**Initialize**(p);
for ($k = 1; k < = k_{\max}; k++$)
Y ←**Rank(Y)**;
M ←**Truncate(Y**,g);
$N(\mu, \sigma)$ ←**ModelConstruction(M)**;
Z ←**IndividualProduction**($N(\mu, \sigma)$,w);
T ← **Y** ∪ **Z**
T ←**Rank(T)**;
Y ←**Truncate(T**,p);
$\mathbf{y}^{best}$ ←**SelectBestElement(Y)**;
end for
Output:$\mathbf{y}^{best}$

3.4 Mixed Gaussian-Cauchy Distribution

3.4.1 Gaussian Distribution

The normal or Gaussian distribution is the most extensively used distribution in statistics. Gaussian distributions are representative approaches of stable processes with a finite second moment. This fact confers its characteristic dispersion (scale) and behavior around the central value. The formulation of the normal distribution with a mean value of 0 and variance σ is established as follows:

$$N(0, \sigma) = \frac{1}{\sigma\sqrt{2\pi}} e^{-\frac{-x^2}{2\sigma^2}} \tag{3.4}$$

Gaussian distributions have been extensively utilized in several optimization approaches. They are considered to produce new individuals as a result of the Gaussian perturbation (mutation) of existent elements [2]. Under a Gaussian perturbation, a new individual a_G^N is produced from the existent a^E considering the following model:

$$a_G^N = a^E + N(0, \sigma) \tag{3.5}$$

In order to make easy the computation of the random numbers, the Gaussian perturbation is calculated by a normal distribution with zero mean and variance one $N(0, 1)$. Therefore, the magnitude of the perturbation is yielded by the product $\sigma \cdot N(0, 1)$. Gaussian distributions generate regulated perturbations that produce exploitation of a search space around a central location [52].

3.4.2 Cauchy Distribution

The Cauchy density function resembles a Gaussian distribution. However, due to its long flat tails, the Cauchy distribution decreases its probability so slowly that its variance is not defined. Under this condition, the Cauchy distribution maintains a high probability of producing a random value far away from its central position. The Cauchy distribution $C\ (t, s)$ is defined through two parameters: location parameter (t) and scale (s). The location parameter defines its central position while the scale its dispersion. The Cauchy density function with $t = 0$ and $s = 1$ can be formulated as follows:

$$C(0, 1) = \frac{1}{\pi(x^2 + 1)} \tag{3.6}$$

Even though the use of Cauchy distribution as a mutation operator is less frequent than the Gaussian operator, it has been used in several EC approaches [53] as a search strategy. Under a Cauchy perturbation, a new individual a_C^N is produced from the existent a^E considering the following model:

$$a_C^N = a^E + \sigma \cdot C(0, \sigma), \tag{3.7}$$

where the value of σ modifies the extent of the perturbation. Since the Cauchy distribution has a high probability of producing a random value far away from its central position, it is mainly used to explore the search space.

3.4.3 Mixed Distribution

Cauchy perturbations allow the exploration of the search space and provide a mechanism to escape from local minima. On the other hand, Gaussian mutations promote not only fast convergence but also the exploitation of existent solutions. In order to fuse both paradigms, the linear combination of both distributions has been considered to produce a single perturbation that integrates their desirable properties. This approach has been used as the basis of some EC approaches [7] to produce new individuals. The mixed distribution $M(\alpha)$ is defined by the following formulation:

$$M(\alpha) = (1 - \alpha) \cdot N(0, 1) + \alpha \cdot C(0, 1), \tag{3.8}$$

where α ($\alpha \in (0, 1)$) weights the relative importance of each distribution in the final mixture. Under the Mixed perturbation, a new individual a_M^N is produced from the existent a^E considering the following model:

$$a_M^N = a^E + \sigma \cdot M(\alpha), \tag{3.9}$$

where σ represents the extent of the perturbation.

The comparison among the Cauchy, Gaussian, and mixed distributions are shown in Fig. 3.1. The marked differences among the density functions can be easily appreciated in Fig. 3.1a, considering the one-dimensional case. On the other hand, Fig. 3.1b–d exhibit the differences among the distributions considering point distributions. Figure 3.1b shows the exploitation properties of the Gaussian distribution. Figure 3.1c presents the explorative capacities of the Cauchy distribution. Finally, Fig. 3.1d exhibits the fusion of both behaviors in a single distribution, balancing the exploration and exploitation characteristics from the Cauchy and Gaussian distributions, respectively.

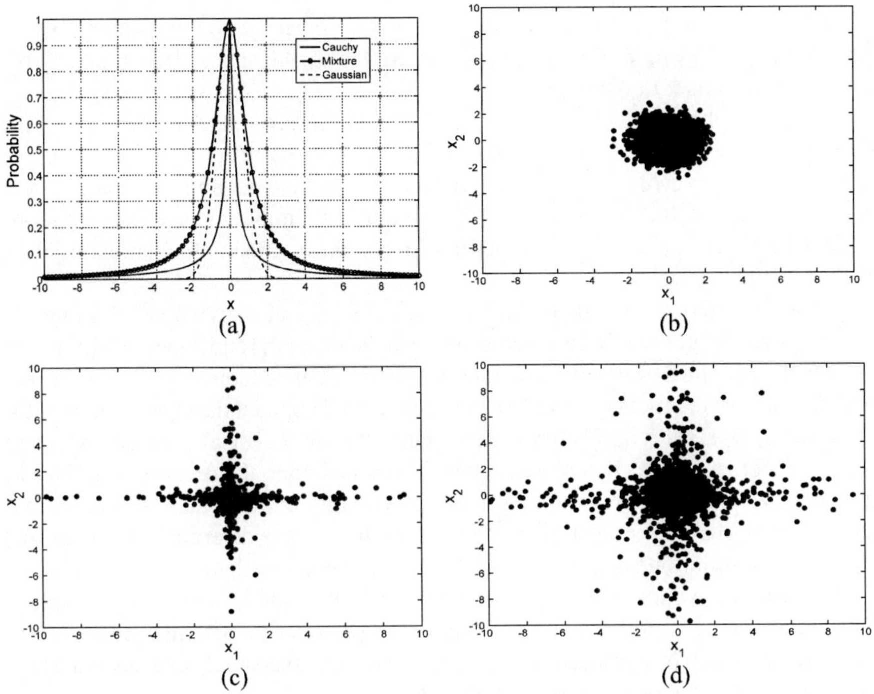

Fig. 3.1 Comparison among Gaussian, Cauchy, and Mixed distributions considering 3000 sampled points. **a** Density functions, **b** Gaussian point distribution, **c** Cauchy point distribution, and **d** mixed point distribution with $\alpha = 0.5$

3.5 Hybrid Algorithm

In this section, the optimization process of the hybrid method is presented. The approach combines (1) the explorative characteristics of the IWO method, (2) the probabilistic models of EDA, and (3) the dispersion capacities of a mixed Gaussian-Cauchy distribution to produce its own search strategy. With these mechanisms, the method conducts an optimization strategy over search areas that deserve a special interest according to a probabilistic model and the fitness value of the existent solutions.

The algorithm has been conceived to solve the problem of obtaining a global solution to a nonlinear optimization formulation under a defined search space. The structure of this optimization problem, considering the maximization case, can be summarized as follows:

$$\begin{aligned} &\text{Maximize } f(\boldsymbol{x}),\ \boldsymbol{x} = \left(x^1, \ldots, x^n\right) \in \mathbb{R}^n \\ &\text{Subject } to\, \boldsymbol{x} \in \boldsymbol{D} \end{aligned} \tag{3.10}$$

with $f : \mathbb{R}^n \rightarrow \mathbb{R}$ represents a nonlinear function while $\boldsymbol{D} = \left\{x \in \mathbb{R}^n \middle| l_i \leq x^i \leq u_i, i = 1, \ldots, n\right\}$ is a restricted search space, determined by the lower (l_i) and upper (u_i) bounds.

To obtain the best solution of the formulation presented in Eq. 3.10, the hybrid method uses a set $\mathbf{S}(k)(\{\mathbf{s}_1(k), \mathbf{s}_2(k), \ldots, \mathbf{s}_N(k)\})$ of N candidate solutions (elements), which evolves from an initial state ($k = 0$) to a maximum number $k_{\max}$ of cycles. ($k = k_{\max}$). It is important to point out that the number of candidate solutions N must be at least equal to the number of dimensions n of the optimization problem to be solved.

In the first step, the method generates the initial set of individuals with random positions that are uniformly distributed between the lower (l_i) and upper (u_i) limits. At each cycle, a defined group of evolutionary operations are conducted over the current population $\mathbf{S}(k)$ to produce the new population $\mathbf{S}(k+1)$. Under the algorithm, an individual $\mathbf{s}_i(k)$ ($i \in [1, \ldots, N]$) represents a current n-dimensional candidate solution $\left\{s_i^1(k), s_i^2(k), \ldots, s_i^n(k)\right\}$ whose dimensions correspond to the decision variables of the optimization formulation to be solved. The quality of an element $\mathbf{s}_i(k)$ is evaluated in terms of an objective function $f(\mathbf{s}_i(k))$ whose value determines how optimal the candidate solution $\mathbf{s}_i(k)$ is to solve the optimization problem.

After randomly initialize the population $\mathbf{S}(k)$ in the search space, the set of candidate solutions are conducted considering three operations: clustering, intra-cluster, and extra-cluster. The application of these operations is repeatedly executed until the number $k_{\max}$ of generations has been attained.

3.5.1 *Reproduction*

Similar to the IWO algorithm, in the hybrid approach, each individual $\mathbf{s}_i(k)$ produces a group of E_i elements which linearly varies in terms of its relative fitness value with regard to the best (E^{best}) and worst (E^{worst}) fitness values of the population.

3.5.2 *Spatial Localization*

In this process, the position of each element from E_i is randomly calculated around $\mathbf{s}_i(k)$ according to the mixed distribution $M(\alpha)$(Eq. 3.9). In the hybrid algorithm, during the evolution process, the values of $\sigma(k)$ and $\alpha(k)$ are gradually modified as the number of generations k increases. The parameter $\sigma(k)$ is varied from $\sigma_{\max}$ to $\sigma_{\min}$.

Different from the original IWO, in the hybrid approach, the values of $\sigma_{\max}$ and $\sigma_{\min}$ are computed in terms of the search space $\mathbf{D}$, which involves the optimization problem to be solved. Under such conditions, it is selected the lower (l_i) and upper (u_i) limits of the decision variable i with the smallest interval. Therefore, the values of $\sigma_{\max}$ and $\sigma_{\min}$ are calculated as follows:

$$\sigma_{max} = (u_i - l_i) \cdot 0.01, \sigma_{min} = \frac{\sigma_{max}}{10} \tag{3.11}$$

On the other hand, the parameter $\alpha(k)$ is adjusted from 0 to 1. In the first generations, the initial values of $\sigma(k)$ and $\alpha(k)$ allow the exploration of the search space **D** (long perturbations and almost a Cauchy distribution). As the number of generations advances, the process of exploration progressively changes to exploitation refining the existing solutions (small perturbations and almost a Gaussian distribution). The parameter $\sigma(k)$ is adapted considering the same model as the IWO algorithm shown in Eq. 3.1. Meanwhile, the factor $\alpha(k)$ is nonlinearly adapted according to the following model:

$$\alpha(k) = 1 - e^{-\left(\frac{1}{0.15} \cdot \frac{k}{k_{max}}\right)}, \tag{3.12}$$

Figure 3.2 presents the effect of $\sigma(k)$ in the perturbation of a two-dimensional individual distribution. From the figure, it is clear that long perturbations are tolerated in the first cycles, but as the generations increase, they are gradually reduced. It is important to point out that in the hybrid algorithm, the value of γ is fixed to 0.5.

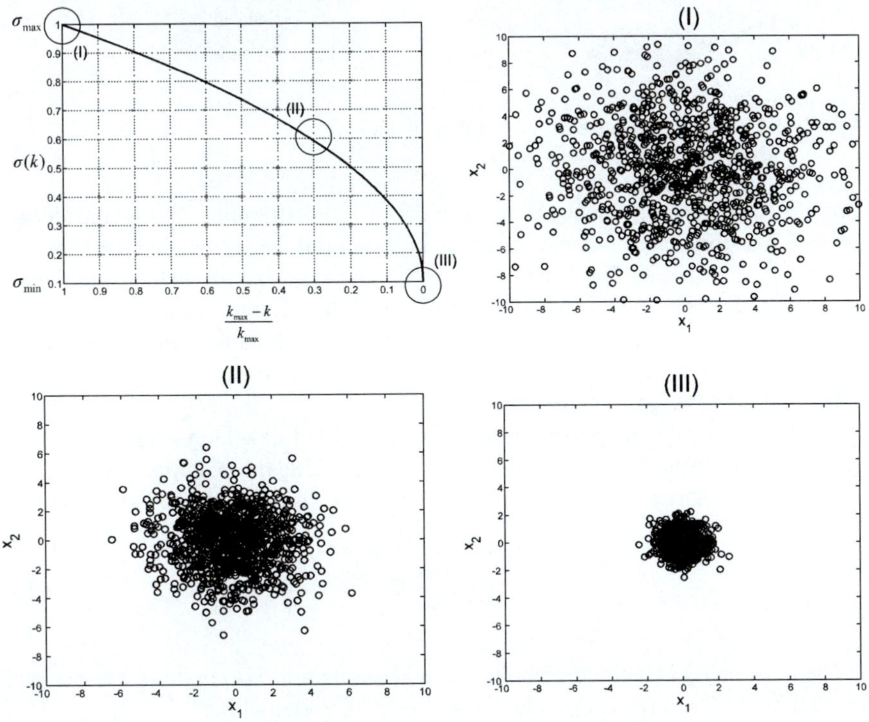

Fig. 3.2 Effect of $\sigma(k)$ in the perturbation of an individual distribution considering $\sigma_{max} = 1$ $\sigma_{min} = 0.1$, and $\gamma = 0.5$

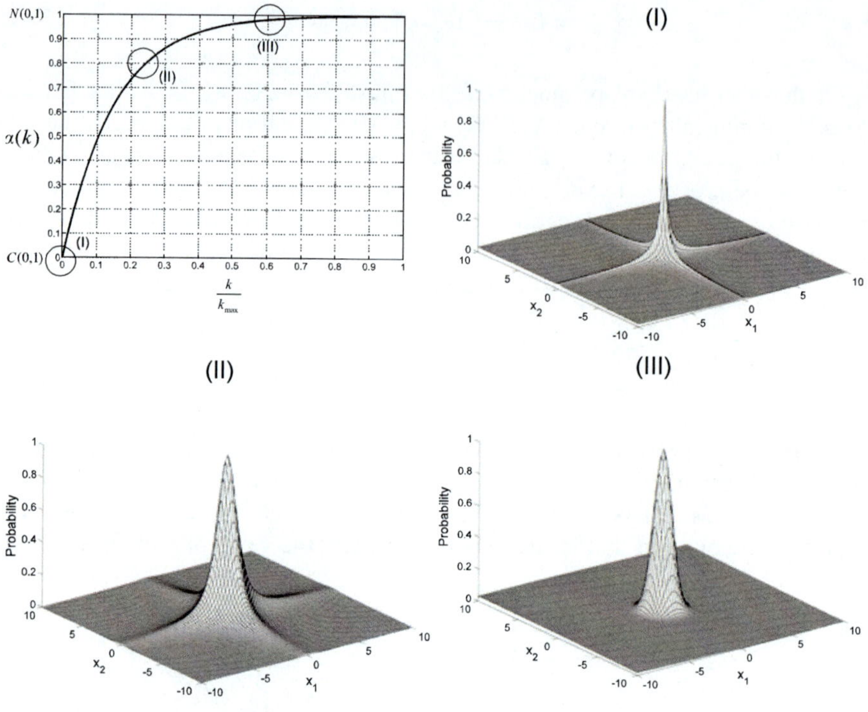

Fig. 3.3 Influence of $\alpha(k)$ in the mixed distribution $M(\alpha)$

Figure 3.3 shows the influence of $\alpha(k)$ in the mixed distribution $M(\alpha)$ considering a two-dimensional case. As can be seen, in the initial stage, the characteristics of the Cauchy distribution dominate the mixture (pure exploration). In the central area, the characteristics of both distributions are fused in the mixture, producing a balance between exploration and exploitation of the search space. Finally, in the last section of the optimization process, the characteristics of the Gaussian density function prevail in the final distribution (pure exploitation).

The spatial localization process generates a set **Q** of new individuals by considering mechanisms that modify their characteristics during the evolution process. The number of elements produced in this procedure is q, that is calculated as follows:

$$q = \sum_{i=1}^{N} E_i, \tag{3.13}$$

Then, the set **Q** is merged with the original population $\mathbf{S}(k)$ to build a temporary population **T** ($\mathbf{T} = \mathbf{S}(k) \cup \mathbf{Q}$), which contains $N + q$ elements.

3.5.3 *Model Construction*

At each generation k, a probability distribution scheme is modeled. The application of this probabilistic model provides a more effective method for generating new individuals with the information extracted during the search process. The process of model construction is divided into two steps: Data preprocessing and parameter calculation.

3.5.3.1 Data Preprocessing

In this step, the data used to build the probabilistic model are obtained. Under this operation, the elements of the temporary population $\mathbf{T}$ are ranked according to their fitness values. Then the best n values are separated in a set $\mathbf{A}^{I}$, where n corresponds to the number of dimensions determined by the optimization problem. The data from $\mathbf{A}^{I}$ are re-sampled by the proportional selection method (roulette) in order to choose n elements in a new group $\mathbf{A}^{II}$. The selection is conducted with replacement. Under such conditions, the new set $\mathbf{A}^{II}$ would contain multiple copies from the best elements $\mathbf{A}^{I}$ while the worst elements of $\mathbf{A}^{I}$ would be eliminated from $\mathbf{A}^{II}$.

3.5.3.2 Parameter Calculation

With the information of the n individuals from $\mathbf{A}^{II} = \{\mathbf{a}_1^{II}, \ldots, \mathbf{a}_n^{II}\}$, the probabilistic model is produced. The model consists of n Gaussian distributions, where each one of them captures the data behavior of a decision variable. Therefore, each i-th Gaussian distribution $N(\mu_i, \sigma_i)(i \in 1 \ldots n)$ is estimated in the following form:

$$\mu_i = \frac{\sum_{j=1}^{n} a_j^i}{n}, \quad \sigma_i^2 = \frac{\sum_{j=1}^{n} (a_j^i - \mu_i)^2}{n-1} \quad (i = 1, \ldots, n;\ j = 1, \ldots, n), \quad (3.14)$$

where a_j^i represents the i-th decision variable of the element $\mathbf{a}_j^{II}$.

3.5.4 *Individual Generation*

In this step, a set $\mathbf{L} = \{\mathbf{l}_1, \mathbf{l}_2, \ldots, \mathbf{l}_r\}$ of r new individuals is produced by using the Gaussian models. So, each decision variable v of the element $\mathbf{l}_u$ is computed as follows:

$$l_u^v = N(\mu_v, \sigma_v), \quad (3.15)$$

3.5.5 Selection of the New Population

In this stage, all the individuals produced during the algorithm process, in the sets **Q** and **L,** are merged along with the original population **S**(k) to produce a final population **F** (**F** = **S**(k) ∪ **Q** ∪ **L**). Then, the elements of F are sorted according to their fitness quality. Finally, the best N individuals are selected to generate the new population **S**($k + 1$). The complete process is repeated until the maximum number of generations $k_{\max}$ has been reached.

3.5.6 Computational Procedure

The hybrid algorithm is designed as an iterative process in which different operations are processed. These operations can be summarized in the form of pseudo-code in Algorithm 3. The algorithm requires as input data the number of candidate solutions (N), the maximum number of generations ($k_{\max}$), the search space **D**, and the tuning parameters E^{best} and E^{worst}. The search space **D** cannot be considered as an additional element since it is known in advance as a part of the optimization formulation. Therefore, from **D**, the values of $\sigma_{\max}$ and $\sigma_{\min}$ are firstly calculated. Identical to other EC algorithms, the method generates the first N candidate solutions with values that are uniformly distributed within the pre-specified limits. These elements represent the first population **S**(k) ($k = 1$). After initialization, the best $\mathbf{s}^{best}(k)$ and the worst $\mathbf{s}^{worst}(k)$ elements are found. Then, the number of elements E_i to be generated for each individual $\mathbf{s}_i(k)$ is calculated. Afterward, the parameters of the mixed distribution $M(\alpha)$ for the k iteration are updated. With these values, the set of E_i new elements are generated for each candidate solution $\mathbf{s}_i(k)$ by using the mixed distribution $M(\alpha)$. Considering the best elements of the produced set **Q** and the original population S(k), the probabilistic model is built. Then, r new candidate solutions are produced though the probabilistic model. Later, all the produced elements are merged in a final population F. Finally, the best elements of F are selected for producing the new population **S**($k + 1$). These operations are repeated until the maximum number $k_{\max}$ of generations is reached.

Algorithm 3.3. Summarized operations for the Hybrid method

Algorithm 3. Pseudo-code for the Hybrid method
Input: N, **D**, E^{best} and E^{worst}
[$\sigma_{\max}$,$\sigma_{\min}$]←**EstimateDesviations(D)**;
S(1)←**Initialize(**p**)**;
for ($k = 1$; $k <= k_{\max}$; $k++$)
$\mathbf{s}^{best}(k)$ ←**SelectBestElement(S(**k**))**;
$\mathbf{s}^{worst}(k)$ ←**SelectWorstElement(S(**k**))**;

(continued)

(continued)

Algorithm 3. Pseudo-code for the Hybrid method	
7	E_i ←**Reproduction(X,**E^{worst}, E^{best},**s**$^{best}(k)$,**s**$^{worst}(k)$);
8	$\sigma(k) \leftarrow \sigma_{\min} + \left(\frac{k_{\max}-k}{k_{\max}}\right)^{0.5} \cdot (\sigma_{\max} - \sigma_{\min})$ $\alpha(k) \leftarrow 1 - e^{-\left(\frac{1}{0.15}\cdot\frac{k}{k_{\max}}\right)}$
9	**Q** ←**SpatialLocalization**($\sigma(k)$,$\alpha(k)$);
10	$N(\mu,\sigma)$ ←**ModelConstruction(Q, S(*k*))**;
11	**L** ←**IndividualGeneration(*r*,**$N(\mu,\sigma)$);
12	**F** ←**Merge(L, Q, S(*k*))**;
15	**S(*k*+ 1)←SelectBestElements(F,***n*);
16	**end for**
17	**Output:s**$^{best}(k)$

3.6 Experimental Study

To evaluate the performance of the hybrid algorithm, a representative collection of 38 functions ($f_1 - f_{38}$) has been considered. This group of functions is extracted from the CEC2016 [30] competition, in the section of single-objective optimization problems. Tables 3.19, 3.20, and 3.21 presents the benchmark functions used during the experiments. According to their characteristics, the functions are classified into three distinct sets: Unimodal (Table 2.19), multimodal (Table 2.20), and composite (Table 2.21) functions. In all these tables, n denotes the number of decision variables with which the function is conducted, **D** symbolizes the feasible search space while $\mathbf{x}^*$ describes the optimal solution and $f(\mathbf{x}^*)$ its fitness value.

In this section, the hybrid algorithm is evaluated in relation to other well-known optimization methods based on evolutionary principles. In the comparison, the following set of algorithms has been contemplated: The original Invasive Weed Optimization (IWO) [33], the Artificial Bee Colony (ABC) method [26], the Particle Swarm Optimization (PSO) technique [27]), the Gravitational Search Algorithm (GSA) [41], the Genetic Algorithms (GA) [22], the Harmony Search (HS) [18] and the Differential Evolution (DE) [44]. Such techniques represent the most popular metaheuristic methods currently in use according to the scientific databases, ScienceDirect, Springer Link, and IEEE-Xplore, over the last ten years.

In the analysis, the hybrid approach is used to solve the complete set of 38 benchmark functions from Appendix A. Then, its results are compared to those produced by the set of methods used in the comparison. In the experiments, the number of individuals N has been set to 50. The results are reported considering the operation of the benchmark functions in 30, 50, and 100 dimensions. To reduce the random effect, each experiment is repeated, considering 30 independent executions. In the study, each optimization algorithm is operated over a determined function until a fixed number *fn* of function evaluations have been reached. Under such conditions, each execution of a test function consists of 50×10^3, 80×10^3 and 160×10^3 function evaluations for 30, 50, and 100 dimensions, respectively. This termination criterion has been selected to maintain affinity with similar works found in the literature [21, 28, 35].

For the experiments, all algorithms have been configured with the parameters, which, according to their reported references, reach their best performance. Such configurations can be summarized as follows:

- **IWO**: $m = 50$, $\sigma_{\max} = 1.2$, $\sigma_{\min} = 0.01$, $\gamma = 0.5$, $E^{best} = 3$, $E^{worst} = 0$ and $m_{\max} = 60$.
- **ABC**: $limit = 50$.
- **PSO**: $c_1 = 2$ and $c_2 = 2$; the weight factor decreases linearly from 0.9 to 0.2.
- **GSA**: $G_t = G_0 e^{-\alpha \frac{t}{T}}$, where $\alpha = 20$, $G_0 = 100$ and $T = 1000$.
- **GA**: The method employs arithmetic crossover, Gaussian mutation, and proportional selection. The crossover and mutation probabilities have been configured to $p_c = 0.8$ and $p_m = 0.1$.
- **HS**: $HCMR = 0.7$ and $PArate = 0.3$.
- **DE**: $CR = 0.5$ and $F = 0.2$.
- **Hybrid**: $E^{best} = 2$ and $E^{worst} = 0$.

In the optimization process, each method finds a solution at each execution. This vector corresponds to the position in the search space **D** with the best possible fitness value produced by a certain function under test. In the experiments, three distinct performance indexes are considered in the comparison: the Average Best-so-far (**AB**) solution, the Median Best-so-far (**MB**) solution, and the Standard Deviation (**SD**) of the best-so-far solutions. The first two indicators **AB** and **MB,** evaluate the accuracy of the solutions while the **SD** index assesses the dispersion or robustness of the produced solutions. The Average Best-so-far (**AB**) solution represents the numerical mean of the best-found solutions, which corresponds to the best values of the certain function ($f_1 - f_{38}$) obtained by each algorithm considering a set of 30 distinct executions. The Median Best-so-far (**MB**) solution corresponds to the solution that divides the set of the 30 best solutions into two sections. The Standard Deviation (**SD**) of the best-so-far solutions is an indicator that expresses the dispersion of the 30 best-found solutions. Therefore, **SD** allows evaluating the robustness of the produced solutions.

Several experiments have been accomplished for analyzing the performance of the hybrid method. The evaluation has been divided into three categories: Unimodal functions (Table 3.19), Multimodal functions (Table 3.20), and composite functions (Table 3.21).

3.6.1 Unimodal Test Functions

In this evaluation, the performance of the hybrid method is compared to IWO, ABC, PSO, GSA, GA, HS, and DE, considering functions containing a single optimum. These functions are characterized by functions from f_1 to f_5 in Table 3.19. In the experiment, each function is tested on 30, 50 and 100 dimensions ($n = 30$, $n = 50$ and $n = 100$). The optimization results collected from 30 distinct executions are shown in Tables 3.1, 3.2, and 3.3 considering 30, 50, and 100 dimensions, respectively. They register the results in terms of the **AB**, **MB**, and **SD** indexes. In all tables, the best results obtained by each function are highlighted in boldface numbers.

From Table 3.1, it can be concluded that the hybrid method performs better than other algorithms in functions f_1, f_2 and f_5 regarding the indexes AB, MD, and SD. In the case of functions f_3 and f_4, the DE technique produces the best solutions. From Table 3.1, it is also observed that the other methods reach different quality levels, being IWO, and HS the methods with the worst performance.

As a complement to the analysis in 30 dimensions, a set of experiments on 50 dimensions have also been conducted to test the scalability of the hybrid method. In the analysis, all the compared optimization techniques have been considered. The results are displayed in Table 3.2, which registers the performance indexes produced in 30 different executions. According to the results, the hybrid method produces better solutions (AB and MD) and less dispersed (SD) than the other methods, in functions f_1, f_2, f_3 and f_5. By contrast, the DE method reaches the best performance in function f_4. After an analysis of Table 3.2, it is clear that the performance of ABC, PSO, GSA, GA, and HS get worse as the number of dimensions increases.

Besides the experiments in 30 and 50 dimensions, the comparison has been extended to consider 100 different dimensions. The results are presented in Table 3.3. An analysis of the results shows that the hybrid method presents the best performance in comparison to IWO, ABC, GSA, GA, HS, and DE. From the results of Tables 3.1, 3.2, and 3.3, it can be concluded that the hybrid method operates in high-dimensional problems without substantially reducing its good performance.

To examine the results of Tables 3.1, 3.2, and 3.3 from a statistic context, a non-parametric analysis identified as the Wilcoxon test [16, 48] has been considered. The analysis permits to estimate the possible differences between two distinct methods that are compared. The test uses as a reference the 5% (0.05) of significance level over the "the Average Best-so-far (**AB**)" elements. In the test, seven different sets are created, one group for each method under analysis. Therefore, the groups are set as follows: Hybrid versus IWO, Hybrid versus ABC, Hybrid versus PSO, Hybrid versus GSA, Hybrid versus GA, Hybrid versus HS, and Hybrid versus DE. Under the Wilcoxon proof, it is contemplated as a null hypothesis that there is not a perceptible distinction between the two algorithms. On the contrary, it is accepted as an alternative hypothesis that there is a significant discrepancy between the two methods. Tables 3.4, 3.5, and 3.6 show the *p*-values produced by the Wilcoxon test after the analysis of the seven groups. Table 3.4 corresponds to the *p*-values generated by the Wilcoxon test over the AB data from Table 3.1 ($n = 30$). Similarly, Tables 3.5 and 3.6 represent

Table 3.1 Results of the unimodal benchmark functions (AI) with 30 dimensions considering 50,000 function evaluations

Function		Hybrid	IWO	ABC	PSO	GSA	GA	HS	DE
$f_1(\boldsymbol{x})$	AB	**7.3E−85**	9.5E+01	1.2E−04	3.1E−09	2.2E−16	6.5E−04	1.4E+01	1.4E−14
	MD	**4.0E−85**	9.7E+01	9.1E−05	8.3E−10	4.5E−16	5.7E−04	1.4E+01	1.2E−14
	SD	**9.2E−85**	1.4E+01	1.5E−04	9.0E−09	1.0E−16	3.1E−04	3.4E+00	5.8E−15
$f_2(\boldsymbol{x})$	AB	**5.8E−72**	7.4E+03	3.7E−03	6.8E+02	4.8E−14	6.5E−02	7.7E+02	5.7E−13
	MD	**2.0E−72**	7.6E+03	2.9E−03	6.0E+02	1.2E−05	5.6E−02	7.8E+02	5.2E−13
	SD	**9.7E−72**	1.0E+03	2.0E−03	5.4E+02	6.6E−05	3.6E−02	1.3E+02	2.2E−13
$f_3(\boldsymbol{x})$	AB	6.0E−09	1.3E−07	2.4E−03	2.1E−19	5.0E−11	8.8E−08	1.9E−03	**1.4E−12**
	MD	5.6E−09	1.3E−07	2.1E−03	8.8E−21	2.7E−08	1.6E−10	1.7E−03	**6.8E−12**
	SD	3.7E−09	4.5E−08	1.6E−03	6.9E−19	4.3E−08	2.0E−07	8.7E−04	**1.8E−12**
$f_4(\boldsymbol{x})$	AB	9.7E−01	1.9E+07	4.2E+00	1.5E+06	1.8E+01	2.0E+02	1.2E+06	**7.9E−10**
	MD	7.0E−52	1.9E+07	3.5E+00	1.5E+06	2.7E+02	1.6E+02	1.1E+06	**7.5E−10**
	SD	3.8E+00	2.2E+06	2.9E+00	9.2E+05	2.5E+02	1.9E+02	2.3E+05	**3.0E−10**
$f_5(\boldsymbol{x})$	AB	**2.8E−70**	4.1E+05	1.7E−01	3.9E+04	4.9E−14	2.6E+00	3.1E+04	5.6E−13
	MD	**7.6E−71**	4.1E+05	1.4E−01	3.0E+04	1.8E−06	2.3E+00	3.2E+04	4.9E−13
	SD	**7.3E−70**	4.8E+04	1.0E−01	2.8E+04	9.9E−06	1.8E+00	6.3E+03	2.3E−13

Bold values represent the main results

Table 3.2 Results of the unimodal benchmark functions (AI) with 50 dimensions considering 80,000 function evaluations

Function		Hybrid	IWO	ABC	PSO	GSA	GA	HS	DE
$f_1(\boldsymbol{x})$	AB	**1.4E–76**	1.6E+02	1.1E+00	2.4E+01	9.9E–16	2.1E–04	2.9E+01	1.9E–13
	MD	**1.0E–76**	1.5E+02	1.0E+00	2.6E+01	9.0E–16	1.9E–04	2.9E+01	1.8E–13
	SD	**1.3E–76**	1.5E+01	4.1E–01	2.0E+01	3.3E–16	1.1E–04	5.0E+00	6.3E–14
$f_2(\boldsymbol{x})$	AB	**9.2E–63**	2.1E+04	6.8E+01	4.3E+03	1.9E–03	2.1E–02	2.4E+03	1.7E–11
	MD	**5.7E–63**	2.1E+04	6.6E+01	4.1E+03	9.1E–09	1.7E–02	2.3E+03	1.6E–11
	SD	**9.4E–63**	2.3E+03	1.9E+01	2.4E+03	9.6E–03	1.4E–02	5.0E+02	7.2E–12
$f_3(\boldsymbol{x})$	AB	**7.2E–49**	7.2E–08	2.7E–01	1.9E–12	4.6E–10	5.3E–08	1.4E–03	5.3E–44
	MD	**6.0E–46**	6.7E–08	2.4E–01	1.2E–14	7.3E–11	4.0E–10	1.2E–03	1.1E–44
	SD	**4.2E–45**	2.1E–08	1.4E–01	7.4E–12	1.5E–09	1.3E–07	7.5E–04	1.7E–43
$f_4(\boldsymbol{x})$	AB	1.1E+01	9.2E+07	1.3E+05	1.6E+07	1.3E+03	7.7E+01	6.4E+06	**3.1E–08**
	MD	**3.2E–43**	9.3E+07	1.3E+05	1.6E+07	1.1E+03	4.8E+01	6.5E+06	3.2E–08
	SD	6.2E+01	1.1E+07	4.6E+04	8.1E+06	9.8E+02	8.7E+01	1.0E+06	**1.1E–08**
$f_5(\boldsymbol{x})$	AB	**2.3E–61**	1.1E–07	1.5E+02	1.8E+05	1.2E–03	7.1E–01	1.0E+05	7.0E–10
	MD	**1.8E–61**	1.1E–07	1.5E+02	1.7E+05	1.2E–06	6.3E–01	1.1E+05	6.1E–10
	SD	**2.5E–61**	3.2E–08	2.5E+01	9.5E+04	3.6E–03	4.2E–01	1.3E+04	2.6E–10

Bold values represent the main results

Table 3.3 Results of the unimodal benchmark functions (AI) with 100 dimensions considering 160,000 function evaluations

Function		Hybrid	IWO	ABC	PSO	GSA	GA	HS	DE
$f_1(\boldsymbol{x})$	AB	**3.5E–51**	2.5E+02	5.7E+02	1.2E+02	6.5E–15	3.4E–04	8.3E+01	5.8E–10
	MD	**2.9E–51**	2.5E+02	5.7E+02	1.3E+02	3.9E–15	3.2E–04	8.2E+01	5.7E–10
	SD	**1.8E–51**	1.9E+01	4.6E+01	4.3E+01	8.9E–15	1.7E–04	6.9E+00	1.4E–10
$f_2(\boldsymbol{x})$	AB	**3.2E–41**	7.9E+04	7.8E+04	3.1E+04	1.7E–02	7.0E–02	1.4E+04	7.6E–08
	MD	**2.7E–41**	7.9E+04	7.9E+04	3.0E+04	1.1E–03	6.5E–02	1.4E+04	7.3E–08
	SD	**2.9E–41**	7.0E+03	6.8E+03	7.9E+03	4.0E–02	3.3E–02	1.2E+03	2.2E–08
$f_3(\boldsymbol{x})$	AB	**7.7E–27**	2.5E–08	1.1E+00	2.2E–02	5.0E–15	1.2E–07	1.1E–03	7.2E–27
	MD	**7.1E–27**	2.4E–08	1.1E+00	5.5E–07	3.4E–16	2.0E–12	1.0E–03	3.1E–27
	SD	**4.0E–28**	5.4E–09	2.5E–01	9.3E–02	1.1E–14	4.0E–07	4.5E–04	1.1E–26
$f_4(\boldsymbol{x})$	AB	**8.4E–04**	8.1E+08	1.1E+00	1.9E+08	8.5E+03	4.5E+02	7.2E+07	2.5E–04
	MD	**4.5E–27**	8.1E+08	1.1E+00	2.0E+08	6.6E+03	4.2E+02	7.0E+07	2.4E–04
	SD	**4.6E–05**	6.0E+07	2.5E–01	4.7E+07	6.5E+03	2.2E+02	7.1E+06	8.8E–05
$f_5(\boldsymbol{x})$	AB	**9.9E–40**	2.6E–08	1.0E+00	1.8E–01	1.9E–01	2.9E+00	5.9E+05	7.4E–08
	MD	**8.5E–40**	2.7E–08	1.1E+00	5.0E–04	2.3E–02	2.8E+00	5.9E+05	7.4E–08
	SD	**6.4E–40**	5.4E–09	3.0E–01	3.6E–01	3.6E–01	1.3E+00	6.1E+04	2.3E–08

Bold values represent the main results

Table 3.4 *p*-values produced by the Wilcoxon test comparing Hybrid versus IWO, Hybrid versus ABC, Hybrid versus PSO, Hybrid versus GSA, Hybrid versus GA, Hybrid versus HS, and Hybrid versus DE over the AB values from Table 3.1 ($n = 30$)

Function	Hybrid versus IWO	Hybrid versus ABC	Hybrid versus PSO	Hybrid versus GSA	Hybrid versus GA	Hybrid versus HS	Hybrid versus DE
$f_1(\mathbf{x})$	5.4E–10▲	5.4E–10▲	5.4E–10▲	5.4E–10▲	5.4E–10▲	5.4E–10▲	5.4E–10▲
$f_2(\mathbf{x})$	3.0E–11▲	3.0E–11▲	3.0E–11▲	3.0E–11▲	3.0E–11▲	3.0E–11▲	3.0E–11▲
$f_3(\mathbf{x})$	3.0E–11▲	3.0E–11▲	1.5E–10▲	1.1E–07▲	3.0E–11▲	3.0E–11▲	0.08211▶
$f_4(\mathbf{x})$	8.5E–09▲	8.5E–09▲	8.5E–09▲	8.5E–09▲	8.5E–09▲	8.5E–09▲	9.3E–09▼
$f_5(\mathbf{x})$	3.0E–11▲	3.0E–11▲	3.0E–11▲	3.0E–11▲	3.0E–11▲	3.0E–11▲	3.0E–11▲

Table 3.5 *p*-values produced by the Wilcoxon test comparing Hybrid versus IWO, Hybrid versus ABC, Hybrid versus PSO, Hybrid versus GSA, Hybrid versus GA, Hybrid versus HS, and Hybrid versus DE over the AB values from Table 3.2 ($n = 50$)

Function	Hybrid versus IWO	Hybrid versus ABC	Hybrid versus PSO	Hybrid versus GSA	Hybrid versus GA	Hybrid versus HS	Hybrid versus DE
$f_1(\mathbf{x})$	5.4E–10▲	5.4E–10▲	5.4E–10▲	5.4E–10▲	5.4E–10▲	5.4E–10▲	5.4E–10▲
$f_2(\mathbf{x})$	3.0E–11▲	3.0E–11▲	3.0E–11▲	3.0E–11▲	3.0E–11▲	3.0E–11▲	3.0E–11▲
$f_3(\mathbf{x})$	3.0E–11▲	3.0E–11▲	3.0E–11▲	7.2E–01▲	3.0E–11▲	3.0E–11▲	3.0E–11▲
$f_4(\mathbf{x})$	5.6E–10▲	5.6E–10▲	5.6E–10▲	5.6E–10▲	5.6E–10▲	5.6E–10▲	4.2E–09▼
$f_5(\mathbf{x})$	3.0E–11▲	3.0E–11▲	3.0E–11▲	3.0E–11▲	3.0E–11▲	3.0E–11▲	3.0E–11▲

Table 3.6 *p*-values produced by the Wilcoxon test comparing Hybrid versus IWO, Hybrid versus ABC, Hybrid versus PSO, Hybrid versus GSA, Hybrid versus GA, Hybrid versus HS, and Hybrid versus DE over the AB values from Table 3.3 ($n = 100$)

Function	Hybrid versus IWO	Hybrid versus ABC	Hybrid versus PSO	Hybrid versus GSA	Hybrid versus GA	Hybrid versus HS	Hybrid versus DE
$f_1(\mathbf{x})$	5.4E–10▲	5.4E–10▲	5.4E–10▲	5.4E–10▲	5.4E–10▲	5.4E–10▲	5.4E–10▲
$f_2(\mathbf{x})$	3.0E–11▲	3.0E–11▲	3.0E–11▲	3.0E–11▲	3.0E–11▲	3.0E–11▲	3.0E–11▲
$f_3(\mathbf{x})$	3.0E–11▲	3.0E–11▲	3.0E–11▲	3.0E–11▲	3.0E–11▲	3.0E–11▲	3.0E–11▲
$f_4(\mathbf{x})$	5.6E–10▲	5.6E–10▲	5.6E–10▲	5.6E–10▲	5.6E–10▲	5.6E–10▲	2.2E–01▲
$f_5(\mathbf{x})$	3.0E–11▲	3.0E–11▲	3.0E–11▲	3.0E–11▲	3.0E–11▲	3.0E–11▲	3.0E–11▲

the results of the Wilcoxon study for the AB data from Tables 3.2 ($n = 50$) and 3.3 ($n = 100$), respectively. All the p-measures that maintain a value under 0.05 (5% significance level) confirm evidence against the null hypothesis. Under such conditions, it is supposed that both methods present a significant difference in their results. To qualitatively identify the results from Tables 3.4, 3.5, and 3.6, the symbols ▲, ▼, and ▶ have been selected. ▲ means that the Hybrid method performs better than the algorithm under comparison on the particular function. ▼ expresses that the hybrid algorithm performs worse than its competitor, and ► implies that the Wilcoxon analysis cannot identify any difference between the Hybrid method and the compared algorithm.

In Table 3.4, all p-values in the columns, Hybrid versus IWO, Hybrid versus ABC, Hybrid versus PSO, Hybrid versus GSA, Hybrid versus GA, and Hybrid versus HS, maintain a value under 0.05 (5% significance level). This fact provides strong evidence against the null hypothesis, indicating that the hybrid method performs better (▲) than the IWO, ABC, PSO, GSA, GA, and HS optimization techniques. These results are statistically validated and determine that they have not eventually occurred (i.e., due to the common noise contained in the data).

Respecting the comparison between the hybrid method and the DE algorithm, the hybrid approach presents a better (▲) performance than DE in functions f_1, f_2 and f_5. Nevertheless, in function f_3, both methods present a similar performance. This case can be observed from the column HYBRID versus DE, where the p-value of function f_3 is higher than 0.05 (►). This result exposes that there is no statistical discrepancy regarding accuracy between the hybrid approach and the DE algorithm when they solve function. From the same comparison, it is also visible that the DE method produces better results than the hybrid approach for function f_4. As a summary, the results of the Wilcoxon test confirm that the hybrid algorithm performs better than the other methods in most of the unimodal functions considering 30 dimensions.

In the case of the results for 50 dimensions shown in Table 3.5, all p-values in the columns, Hybrid versus IWO, Hybrid versus ABC, Hybrid versus PSO, Hybrid versus GSA, Hybrid versus GA, and Hybrid versus HS, demonstrate that the hybrid approach performances better than its competitors from a statistical point of view for most of the functions. However, in function f_4, according to the Wilcoxon analysis, the DE method produces the best results in terms of accuracy.

Table 3.6 presents the results of the Wilcoxon analysis over the AB data corresponding to the optimization of unimodal functions in 100 dimensions. From the information of Table 3.6, it is clear that the hybrid method performs better than the other algorithms in all functions. After this study, it can be concluded that the hybrid method operates in high-dimensional problems without substantially reducing its good performance.

3.6.2 Multimodal Test Functions

Differently to unimodal functions, multimodal functions incorporate multiple local optima. For this reason, they are harder to optimize. In this experiment, the results of the hybrid method are compared with IWO, ABC, PSO, GSA, GA, HS, and DE when they solve multimodal functions. Multimodal functions correspond to functions from f_6 to f_{34} in Table 3.20. In this kind of function, the number of local minima increases exponentially as the number of dimensions in the function grows. Therefore, the test exhibits the capacity of each method to detect the global optimum when several local optima are present. In the experiments, each multimodal function is tested in 30, 50, and 100 dimensions. The results, averaged over 30 different executions, are registered in Tables 3.7, 3.8, and 3.9 for 30, 50, and 100 dimensions, respectively. They present the optimization results regarding the Average Best-so-far (**AB**) solution, the Median Best-so-far (**MB**) solution, and the Standard Deviation (**SD**) of the best-so-far solutions. In the tables, the best indicators are highlighted in boldface.

Table 3.7 shows the performance result for multimodal functions from f_6 to f_{34} considering 30 dimensions. For functions $f_6, f_7, f_8, f_9, f_{15}, f_{16}, f_{17}, f_{22}, f_{23}, f_{24}, f_{25}, f_{27}, f_{28}, f_{29}, f_{30}, f_{31}, f_{32}, f_{33}$ and f_{34} the hybrid method exhibits a better performance than IWO, ABC, PSO, GSA, GA, HS, and DE. In the case of functions $f_{12}, f_{14}, f_{18}, f_{21}$ and f_{26} the GSA optimization technique demonstrates to teach the best indicators compared to the other algorithms. On the other hand, in the case of the functions f_{10}, f_{11}, f_{13}, f_{19} and f_{20} the DE method obtains the best indicators compared to IWO, ABC, PSO, GSA, GA, HS, and the hybrid approach. The remainder of the optimization techniques achieves distinct levels of precision. The results obtained in the experiment reflexes a significant difference in the performance of all algorithms. This fact is associated with the different mechanisms used by each method to implement a trade-off between exploration and exploitation.

Table 3.8 exposes the resulting indicators for multimodal functions when they are operated in 50 dimensions. Under this experiment, the hybrid algorithm increases the number of functions in which it obtains a better result than the other methods. Therefore, the hybrid algorithm overcomes the other optimization techniques in functions $f_6, f_7, f_8, f_9, f_{13}, f_{14}, f_{15}, f_{16}, f_{17}, f_{18}, f_{22}, f_{23}, f_{25}, f_{27}, f_{28}, f_{29}, f_{30}, f_{31}, f_{32}, f_{33}$ and f_{34}. The DE method produced the best indicators for the multimodal functions f_{10} and f_{24}. In the case of function f_{21}, the GSA method obtains better performance. The ABC technique presents the best performance results in solving functions f_{11}, f_{12}, f_{20} and f_{26}. Finally, the IWO method yields the best indexes than the other algorithms in function f_{19}. Since the number of local minima increases as the number of dimensions grows, this experiment reflexes the differences among the optimization methods for avoiding being trapped in local optima.

The performance indexes for multimodal functions in 100 dimensions are presented in Table 3.9. After an analysis of the results, it is clear that a similar behavior of the optimization methods is maintained. Under this experiment, the hybrid algorithm presents the best results in functions $f_6, f_7, f_8, f_9, f_{10}, f_{11}, f_{15}, f_{16}, f_{17}, f_{20}, f_{22}, f_{23}, f_{25}, f_{27}, f_{28}, f_{29}, f_{30}, f_{31}, f_{32}, f_{33}$ and f_{34}. On the other hand, the DE algorithms maintain the best indicators for the multimodal functions f_{12}, f_{13} and f_{18}. For functions f_{14}, f_{19} and f_{26}, the GSA method obtains better performance. The ABC technique presents the best performance results in solving the function f_{21}.

The p-values produced by the Wilcoxon test after the analysis of the AB values from Tables 3.7, 3.8, and 3.9 are exhibit in Tables 3.10, 3.11, and 3.12, respectively. Table 3.7 presents the p-values for the case of multimodal functions from f_6 to f_{34} when they are operated in 30 dimensions. The results statistically demonstrate that the hybrid algorithm obtains the best indexes

Table 3.7 Results of the multimodal benchmark functions (AII) with 30 dimensions considering 50,000 function evaluations

Function		Hybrid	IWO	ABC	PSO	GSA	GA	HS	DE
$f_6(\boldsymbol{x})$	AB	**3.6E–15**	1.9E+01	1.2E+00	1.7E+01	1.1E–08	7.7E–02	1.3E+01	5.3E–07
	MD	**3.6E–15**	1.9E+01	1.2E+00	2.0E+01	1.4E–08	7.4E–02	1.3E+01	5.1E–07
	SD	**0.0E+00**	1.8E–01	4.6E–01	7.3E+00	1.9E–09	2.6E–02	7.4E–01	1.1E–07
$f_7(\boldsymbol{x})$	AB	**4.8E–03**	3.4E+03	5.7E+02	2.1E+03	1.3E–01	8.2E–01	1.0E+03	9.6E–01
	MD	**3.2E–03**	3.3E+03	5.9E+02	1.9E+03	9.7E–01	7.5E–01	1.0E+03	7.6E–01
	SD	**4.6E–03**	9.2E+02	2.0E+02	1.8E+03	8.6E–01	3.4E–01	4.0E+02	7.0E–01
$f_8(\boldsymbol{x})$	AB	**1.5E–32**	6.5E+01	5.1E+00	5.9E+00	1.9E–16	1.3E–03	1.5E+01	2.4E–13
	MD	**1.5E–32**	6.2E+01	4.9E+00	4.5E+00	1.5E–02	1.3E–03	1.5E+01	2.2E–13
	SD	**1.1E–47**	1.6E+01	2.1E+00	5.9E+00	8.3E–02	6.0E–04	2.2E+00	1.1E–13
$f_9(\boldsymbol{x})$	AB	**2.4E+01**	6.5E+05	7.1E+02	2.2E+05	2.5E+01	1.1E+02	3.4E+04	3.1E+01
	MD	**2.6E+01**	6.4E+05	6.8E+02	2.2E+05	4.2E+01	1.1E+02	3.4E+04	2.7E+01
	SD	**1.2E–01**	2.0E+05	3.9E+02	1.0E+05	2.5E+01	4.7E+01	9.8E+03	1.2E+01
$f_{10}(\boldsymbol{x})$	AB	5.7E+01	2.6E+40	3.4E+24	7.8E+02	4.1E+01	1.0E+00	9.2E+11	**9.0E–07**
	MD	6.0E–36	5.6E+32	8.4E+21	6.8E+02	1.0E+02	1.0E+00	2.2E+07	**8.9E–07**
	SD	1.2E+02	1.4E+41	1.2E+25	4.0E+02	4.0E+01	2.7E–01	3.3E+12	**2.8E–07**
$f_{11}(\boldsymbol{x})$	AB	2.9E+03	1.0E+04	-5.3E+121	5.0E+03	8.5E+03	2.7E+03	2.5E+03	**2.5E+01**
	MD	2.8E+03	1.0E+04	-3.5E+119	5.0E+03	9.7E+03	2.7E+03	2.6E+03	**3.8E+00**
	SD	4.6E+02	5.7E+02	2.0E+122	6.5E+02	5.4E+02	3.8E+02	4.1E+02	**4.6E+01**
$f_{12}(\boldsymbol{x})$	AB	1.2E+02	2.1E+87	9.9E+77	7.5E+32	**5.4E–07**	8.9E+03	1.1E+76	4.3E+05
	MD	4.3E+01	3.1E+86	5.4E+75	6.1E+17	**5.5E–07**	1.4E+03	3.3E+74	1.1E+24

(continued)

Table 3.7 (continued)

Function		Hybrid	IWO	ABC	PSO	GSA	GA	HS	DE
	SD	2.2E+02	6.0E+87	4.8E+78	4.1E+33	**1.2E–07**	3.5E+04	4.9E+76	5.8E+24
$f_{13}(\boldsymbol{x})$	AB	6.7E–01	9.4E+05	1.0E+02	2.9E+04	6.7E–01	1.0E+01	4.0E+04	**1.2E–14**
	MD	6.7E–01	9.7E+05	7.3E+01	8.3E+01	6.9E–01	9.9E+00	3.7E+04	**1.0E–14**
	SD	5.4E–06	2.4E+05	6.9E+01	6.7E+04	5.8E–02	3.2E+00	1.5E+04	**5.3E–15**
$f_{14}(\boldsymbol{x})$	AB	2.2E–11	6.2E+07	1.7E+03	5.1E+02	**4.5E–46**	4.0E+02	3.7E+02	2.8E+01
	MD	2.0E–11	6.3E+02	1.6E+03	4.7E+02	**9.1E–47**	4.2E+02	3.7E+02	5.2E+01
	SD	7.8E–12	1.8E+08	5.5E+02	1.6E+02	**7.6E–46**	1.1E+02	5.6E+01	1.3E+01
$f_{15}(\boldsymbol{x})$	AB	**-1.2E+03**	-8.2E+02	-7.6E+02	-1.1E+03	**-1.2E+03**	-1.7E+03	-1.1E+03	1.1E+00
	MD	**-1.1E+03**	-8.1E+02	-7.6E+02	**-1.1E+03**	**-1.1E+03**	-1.7E+03	**-1.1E+03**	9.8E–01
	SD	**1.8E–01**	5.3E+01	3.0E+01	3.0E+01	3.2E+01	4.7E+01	2.0E+01	7.0E+01
$f_{16}(\boldsymbol{x})$	AB	**0.0E+00**	6.7E+04	3.3E–02	6.7E+02	**0.0E+00**	1.0E–01	5.1E+03	1.7E–13
	MD	**0.0E+00**	6.8E+04	**0.0E+00**	**0.0E+00**	**0.0E+00**	**0.0E+00**	5.1E+03	1.5E–13
	SD	**0.0E+00**	8.4E+03	1.8E–01	2.5E+03	**0.0E+00**	3.1E–01	1.0E+03	7.6E–14
$f_{17}(\boldsymbol{x})$	AB	**7.1E+00**	3.3E+01	1.2E+01	1.5E+01	7.7E+00	2.0E+01	1.6E+01	2.8E+01
	MD	**8.3E+00**	3.3E+01	1.2E+01	1.2E+01	8.6E+00	2.0E+01	1.6E+01	2.6E+01
	SD	**4.0E–01**	4.1E+00	5.2E–01	5.5E+00	4.8E–01	9.7E–01	1.3E+00	3.4E+00
$f_{18}(\boldsymbol{x})$	AB	1.1E–01	2.6E+01	1.8E+00	1.0E+00	**8.6E–10**	2.3E+00	8.3E+00	6.0E–01
	MD	1.0E–01	2.6E+01	1.8E+00	7.0E–01	**8.0E–10**	2.3E+00	8.3E+00	1.3E+00
	SD	2.5E–02	1.3E+00	1.9E–01	1.8E+00	**2.9E–10**	3.2E–01	7.2E–01	3.6E–01
$f_{19}(\boldsymbol{x})$	AB	4.0E+04	5.6E+08	1.5E+07	1.0E+02	7.1E+01	1.4E+02	1.1E+07	**5.3E–07**

(continued)

Table 3.7 (continued)

Function		Hybrid	IWO	ABC	PSO	GSA	GA	HS	DE
	MD	7.1E+01	5.7E+08	1.4E+07	8.8E+01	7.2E+01	1.4E+02	1.0E+07	**5.1E–07**
	SD	6.2E+04	1.2E+08	3.3E+06	3.6E+01	1.6E+00	5.3E+00	2.6E+06	**1.3E–07**
$f_{20}(\boldsymbol{x})$	AB	1.4E+03	1.0E+09	1.7E+06	1.1E+02	1.1E+02	1.4E+02	1.3E+07	**1.4E–14**
	MD	1.1E+02	9.8E+08	1.6E+06	1.0E+02	1.1E+02	1.4E+02	1.3E+07	**1.1E–14**
	SD	2.9E+03	2.0E+08	5.5E+05	1.0E+01	4.4E+00	7.0E+00	4.9E+06	**7.8E–15**
$f_{21}(\boldsymbol{x})$	AB	3.6E+01	2.0E+02	2.2E+02	1.4E+02	**7.2E–13**	2.6E–01	1.2E+02	6.0E+00
	MD	3.4E+01	2.0E+02	2.2E+02	1.4E+02	**6.0E–13**	1.4E–01	1.2E+02	1.5E+01
	SD	8.3E+00	3.6E+01	1.2E+01	3.3E+01	**3.9E–13**	3.3E–01	1.1E+01	4.1E+00
$f_{22}(\boldsymbol{x})$	AB	**2.0E+00**	3.2E+00	-2.2E+04	1.3E+02	1.3E+02	**2.0E+00**	2.8E+02	1.5E–46
	MD	**2.0E+00**	2.7E+00	-1.5E+03	3.7E+01	8.3E+02	**2.0E+00**	2.5E+02	2.0E–47
	SD	5.5E–03	1.4E+00	5.5E+04	2.3E+02	6.9E+02	1.6E–01	1.7E+02	**3.1E–46**
$f_{23}(\boldsymbol{x})$	AB	**2.0E+00**	3.2E+00	9.5E–01	4.2E+02	8.8E+01	**2.0E+00**	2.7E+02	1.3E+00
	MD	**2.0E+00**	2.8E+00	7.7E–01	2.3E+01	1.2E+03	**2.0E+00**	2.7E+02	9.0E–01
	SD	**3.8E–14**	1.1E+00	4.7E–01	1.5E+03	1.0E+03	1.2E–13	1.2E+02	1.1E+00
$f_{24}(\boldsymbol{x})$	AB	2.7E–07	8.9E–02	4.0E–03	**0.0E+00**	3.9E–29	4.4E–11	3.0E–03	1.9E–13
	MD	9.4E–08	8.9E–02	3.5E–03	**0.0E+00**	5.3E–19	2.2E–11	2.9E–03	1.7E–13
	SD	6.0E–07	3.5E–02	2.4E–03	**0.0E+00**	4.2E–19	6.4E–11	1.1E–03	9.4E–14
$f_{25}(\boldsymbol{x})$	AB	**0.0E+00**	8.5E+05	2.8E–03	7.7E–48	5.9E–100	**0.0E+00**	1.4E–20	2.8E+01
	MD	**0.0E+00**	5.3E+04	2.6E–03	5.2E–67	6.7E–73	**0.0E+00**	6.6E–23	2.7E+01
	SD	**0.0E+00**	1.5E+06	2.1E–03	4.2E–47	3.4E–72	**0.0E+00**	3.9E–20	3.9E+00

(continued)

Table 3.7 (continued)

Function		Hybrid	IWO	ABC	PSO	GSA	GA	HS	DE
$f_{26}(\boldsymbol{x})$	AB	9.5E+01	2.0E+11	5.3E+02	2.5E–01	**8.3E–10**	1.5E+02	4.3E+09	2.7E+04
	MD	8.4E+01	2.0E+11	4.6E+02	3.9E–02	**7.7E–10**	1.2E+02	3.9E+09	2.9E+06
	SD	6.4E+01	3.8E+10	2.0E+02	7.8E–01	**3.5E–10**	8.0E+01	1.6E+09	2.8E+06
$f_{27}(\boldsymbol{x})$	AB	**3.4E–14**	1.1E+05	5.2E+00	1.3E–02	5.7E–07	8.8E–01	3.2E+03	8.2E–07
	MD	**3.4E–14**	1.1E+05	4.7E+00	1.4E–02	6.3E+00	8.5E–01	3.2E+03	7.7E–07
	SD	**2.1E–15**	2.9E+04	2.3E+00	8.9E–03	8.2E+00	2.6E–01	1.2E+03	2.0E–07
$f_{28}(\boldsymbol{x})$	AB	**-3.0E+01**	-2.7E+01	7.5E+02	-2.8E+01	**-3.0E+01**	-5.0E+01	-2.7E+01	1.8E+01
	MD	**-3.0E+01**	-2.8E+01	6.4E+02	-2.8E+01	**-3.0E+01**	-5.0E+01	-2.7E+01	2.7E+00
	SD	**0.0E+00**	1.2E+00	3.6E+02	1.5E+00	1.6E–34	7.0E–04	5.5E–01	4.4E+01
$f_{29}(\boldsymbol{x})$	AB	**2.7E–189**	9.4E–08	-9.6E+00	5.6E–13	1.9E–21	2.3E–14	4.3E–02	6.8E+80
	MD	**6.3E–193**	2.3E–33	-9.6E+00	7.6E–15	1.4E–09	9.5E–15	3.6E–02	3.4E+80
	SD	**0.0E+00**	3.6E–07	4.8E–01	2.6E–12	4.2E–09	3.3E–14	1.7E–02	7.9E+80
$f_{30}(\boldsymbol{x})$	AB	**0.0E+00**	1.9E+01	8.5E–01	2.0E+01	4.5E–16	3.8E–02	1.8E+01	2.4E+00
	MD	**0.0E+00**	1.9E+01	7.0E–01	2.0E+01	4.5E–16	3.5E–02	1.8E+01	9.4E–01
	SD	**0.0E+00**	1.8E–01	4.8E–01	7.8E–04	1.1E–16	1.5E–02	3.0E–01	3.9E+00
$f_{31}(\boldsymbol{x})$	AB	**-1.2E+03**	6.5E+01	4.8E+00	2.0E+02	3.4E+01	1.4E–04	2.5E+02	0.0E+00
	MD	**-1.7E+03**	6.2E+01	4.4E+00	2.0E+02	2.8E+01	1.2E–04	2.5E+02	0.0E+00
	SD	4.5E–02	1.6E+01	1.6E+00	4.8E+01	1.7E+01	1.2E–04	1.9E+01	**0.0E+00**
$f_{32}(\boldsymbol{x})$	AB	**3.4E–01**	6.5E+05	6.1E+02	1.7E+06	4.3E+02	5.3E+01	1.0E+06	1.0E+01
	MD	**3.3E–01**	6.4E+05	6.0E+02	1.6E+06	2.2E+02	7.2E+01	1.0E+06	1.0E+01
	SD	**7.8E–02**	2.0E+05	2.0E+02	4.7E+05	4.5E+02	3.5E+01	1.4E+05	4.8E–01

(continued)

Table 3.7 (continued)

Function		Hybrid	IWO	ABC	PSO	GSA	GA	HS	DE
$f_{33}(\boldsymbol{x})$	AB	**1.6E–32**	1.8E+07	4.1E+00	1.8E+08	1.2E+02	1.4E+01	1.9E+08	4.1E–01
	MD	**1.6E–32**	1.9E+07	3.8E+00	1.5E+08	1.3E+02	9.7E+00	1.9E+08	4.0E–01
	SD	**5.6E–48**	2.8E+06	2.2E+00	6.9E+07	3.5E+01	1.3E+01	1.9E+07	3.8E–02
$f_{34}(\boldsymbol{x})$	AB	**1.3E–32**	4.7E+39	1.6E+25	8.8E+108	9.7E+03	3.1E–01	1.2E+90	7.2E+01
	MD	**1.3E–32**	1.9E+33	7.8E+22	5.4E+70	9.8E+03	2.8E–01	3.1E+81	7.1E+01
	SD	**5.6E–48**	2.6E+40	4.8E+25	4.8E+109	4.8E+02	1.4E–01	6.2E+90	2.7E+00

Bold values represent the main results

Table 3.8 Results of the multimodal benchmark functions (AII) with 50 dimensions considering 80,000 function evaluations

Function		Hybrid	IWO	ABC	PSO	GSA	GA	HS	DE
$f_6(\boldsymbol{x})$	AB	**6.5E–15**	1.9E+01	8.4E+00	1.9E+01	1.4E–08	4.1E–02	1.4E+01	1.6E–06
	MD	**7.1E–15**	1.9E+01	8.3E+00	2.0E+01	1.4E–08	4.0E–02	1.4E+01	1.7E–06
	SD	**1.3E–15**	1.7E–01	1.0E+00	3.2E+00	1.7E–09	1.3E–02	5.1E–01	2.7E–07
$f_7(\boldsymbol{x})$	AB	**6.6E–03**	6.1E+03	8.6E+03	7.3E+03	1.1E+00	3.7E–01	2.6E+03	1.7E+02
	MD	**5.1E–03**	6.2E+03	8.7E+03	7.2E+03	9.2E–01	3.7E–01	2.5E+03	1.6E+02
	SD	**5.6E–03**	1.5E+03	1.7E+03	3.4E+03	7.9E–01	1.0E–01	7.4E+02	8.3E+01
$f_8(\boldsymbol{x})$	AB	**1.5E–32**	1.2E+02	1.5E+02	3.8E+01	3.6E–02	4.0E–04	3.2E+01	9.4E–12
	MD	**1.5E–32**	1.2E+02	1.5E+02	3.7E+01	6.8E–16	3.4E–04	3.1E+01	8.5E–12
	SD	**1.1E–47**	2.5E+01	1.8E+01	1.2E+01	1.2E–01	2.6E–04	5.1E+00	3.9E–12
$f_9(\boldsymbol{x})$	AB	**4.8E+01**	1.3E+06	2.8E+05	5.9E+05	5.8E+01	8.5E+01	7.9E+04	5.1E+01
	MD	**4.3E+01**	1.3E+06	2.8E+05	5.6E+05	4.7E+01	8.0E+01	7.6E+04	4.6E+01
	SD	**1.4E–01**	2.8E+05	6.5E+04	2.1E+05	2.7E+01	5.5E+01	1.8E+04	1.5E+01
$f_{10}(\boldsymbol{x})$	AB	3.9E+01	7.1E+64	2.3E+56	2.2E+03	2.3E+02	5.9E–01	3.7E+21	**6.8E–06**
	MD	6.5E–06	7.6E+62	9.8E+53	2.4E+03	2.3E+02	5.8E–01	1.2E+13	**5.3E–33**
	SD	8.2E+01	2.3E+65	1.1E+57	5.8E+02	4.5E+01	2.0E–01	2.1E+22	**1.7E–06**
$f_{11}(\boldsymbol{x})$	AB	6.8E+03	1.8E+04	**1.0E+00**	1.0E+04	1.7E+04	2.7E+03	4.4E+03	6.0E+03
	MD	6.8E+03	1.8E+04	**1.0E+00**	1.0E+04	1.8E+04	2.7E+03	4.4E+03	6.2E+03
	SD	5.8E+02	7.0E+02	**3.6E–01**	8.1E+02	4.9E+02	5.4E+02	5.5E+02	5.4E+02
$f_{12}(\boldsymbol{x})$	AB	1.0E+03	1.8E+02	**2.7E–01**	2.1E+116	6.0E+42	5.4E+02	2.1E+132	4.5E+00
	MD	5.7E+02	1.8E+02	**2.7E–01**	1.5E+105	1.1E+35	2.8E+02	1.1E+129	3.0E+00
	SD	1.2E+03	1.5E+01	**1.1E–01**	1.1E+117	1.9E+43	5.8E+02	1.1E+133	3.0E+00

(continued)

Table 3.8 (continued)

Function		Hybrid	IWO	ABC	PSO	GSA	GA	HS	DE
$f_{13}(\boldsymbol{x})$	AB	**6.7E–01**	2.2E+04	8.7E+03	3.4E+05	8.0E–01	8.9E+00	1.4E+05	2.7E–01
	MD	**6.7E–01**	2.3E+04	9.1E+03	1.8E+05	**6.7E–01**	8.7E+00	1.4E+05	2.7E–01
	SD	**2.1E–04**	2.4E+03	1.8E+03	3.3E+05	4.5E–01	2.8E+00	3.5E+04	5.3E–04
$f_{14}(\boldsymbol{x})$	AB	**5.5E–10**	7.4E+03	2.8E+05	1.3E+03	1.0E+02	3.8E+02	6.7E+02	5.0E+02
	MD	**5.3E–10**	7.3E+03	2.9E+05	1.3E+03	9.3E+01	3.9E+02	6.7E+02	5.0E+02
	SD	**8.9E–11**	1.6E+03	6.2E+04	2.3E+02	2.3E+01	8.1E+01	7.9E+01	5.1E+01
$f_{15}(\boldsymbol{x})$	AB	**-2.0E+03**	1.4E+02	1.4E+05	-1.7E+03	-1.8E+03	-1.7E+03	-1.7E+03	-1.0E+03
	MD	**-2.0E+03**	1.3E+02	1E+05	-1.7E+03	-1.8E+03	-1.7E+03	-1.7E+03	-1.0E+03
	SD	**3.0E–09**	2.3E+01	4.7E+04	4.1E+01	4.3E+01	3.4E+01	3.0E+01	9.4E–09
$f_{16}(\boldsymbol{x})$	AB	**0.0E+00**	1.6E+06	1.9E+55	1.0E+04	1.0E–01	0.2E+00	1.1E+04	0.1E+00
	MD	**0.0E+00**	1.6E+06	7.0E+52	1.0E+04	0.0E+00	0.4E+00	1.1E+04	0.2E+00
	SD	**0.0E+00**	3.5E+05	3.9E+55	8.5E+03	3.1E–01	1.0E+00	2.1E+03	0.8E+00
$f_{17}(\boldsymbol{x})$	AB	**1.7E+00**	9.5E+07	8.4E+00	4.8E+01	1.7E+01	1.9E+01	3.5E+01	2.0E+01
	MD	**1.7E+00**	9.3E+07	8.3E+00	4.2E+01	1.7E+01	1.9E+01	3.4E+01	2.0E+01
	SD	**6.1E–01**	9.7E+06	1.0E+00	2.0E+01	5.5E–01	1.3E+00	3.1E+00	6.8E–01
$f_{18}(\boldsymbol{x})$	AB	**1.5E–01**	3.2E+64	1.1E+00	8.3E+00	2.6E+00	2.4E+00	1.2E+01	6.5E–01
	MD	**2.0E–01**	6.6E+61	1.0E+00	1.0E+01	2.5E+00	2.4E+00	1.2E+01	6.4E–01
	SD	**5.1E–02**	1.1E+65	4.1E–01	4.7E+00	5.3E–01	2.8E–01	9.7E–01	6.7E–02
$f_{19}(\boldsymbol{x})$	AB	2.5E+05	**1.9E+01**	6.8E+01	4.5E+03	1.2E+02	1.3E+02	2.6E+07	1.9E+04
	MD	1.2E+02	**1.9E+01**	6.6E+01	1.6E+03	1.2E+02	1.3E+02	2.5E+07	2.4E+03
	SD	4.5E+05	**1.9E–01**	1.9E+01	9.1E+03	1.3E+00	3.6E+00	6.6E+06	3.3E+04

(continued)

Table 3.8 (continued)

Function		Hybrid	IWO	ABC	PSO	GSA	GA	HS	DE
$f_{20}(\boldsymbol{x})$	AB	5.7E+03	1.8E+02	**2.7E–01**	7.8E+02	2.0E+02	1.1E+02	3.4E+07	2.4E+03
	MD	1.8E+02	1.8E+02	**2.4E–01**	5.3E+02	2.0E+02	1.1E+02	3.2E+07	7.2E+02
	SD	1.5E+04	2.0E+01	**1.4E–01**	6.8E+02	4.0E+00	5.0E+00	9.9E+06	3.5E+03
$f_{21}(\boldsymbol{x})$	AB	7.7E+01	2.2E+04	8.6E+03	3.4E+02	**4.2E–02**	3.0E+01	2.3E+02	1.7E+02
	MD	7.6E+01	2.2E+04	8.7E+03	3.3E+02	**4.2E–02**	2.9E+01	2.3E+02	1.7E+02
	SD	1.4E+01	2.3E+03	1.7E+03	6.1E+01	**2.2E–02**	5.6E+00	1.9E+01	7.9E+00
$f_{22}(\boldsymbol{x})$	AB	**2.0E+00**	9.5E–08	1.5E+02	9.1E+08	2.0E+07	2.0E+00	2.5E+05	1.0E+00
	MD	**2.0E+00**	9.6E–08	1.5E+02	2.1E+06	4.1E+06	2.0E+00	1.2E+05	1.0E+00
	SD	**0.0E+00**	2.5E–08	1.8E+01	4.7E+09	3.8E+07	1.7E–14	4.0E+05	0.0E+00
$f_{23}(\boldsymbol{x})$	AB	**2.0E+00**	7.1E+03	2.8E+05	1.1E+09	1.5E+07	2.0E+00	2.4E+05	1.0E+00
	MD	**2.0E+00**	6.8E+03	2.8E+05	2.1E+06	3.4E+06	2.0E+00	1.8E+05	1.0E+00
	SD	**0.0E+00**	2.2E+03	6.5E+04	4.8E+09	2.3E+07	8.2E–15	2.0E+05	0.0E+00
$f_{24}(\boldsymbol{x})$	AB	8.8E–05	1.3E+02	1.3E+05	1.0E+00	2.8E–25	2.5E–12	4.4E–03	**0.0E+00**
	MD	6.9E–05	1.3E+02	1.3E+05	1.0E+00	6.4E–32	1.7E–12	4.5E–03	**0.0E+00**
	SD	6.9E–05	2.3E+01	4.6E+04	0.2E+00	1.5E–24	2.3E–12	1.2E–03	**0.1E–06**
$f_{25}(\boldsymbol{x})$	AB	**0.0E+00**	1.6E+06	2.3E+56	4.3E–50	9.9E–102	1.0E+00	3.6E–29	1.6E–306
	MD	**0.0E+00**	1.7E+06	9.8E+53	2.7E–89	2.7E–114	1.0E+00	6.7E–36	0.0E+00
	SD	**0.0E+00**	2.8E+05	1.1E+57	2.4E–49	3.9E–101	0.3E+00	2.0E–28	0.0E+00
$f_{26}(\boldsymbol{x})$	AB	2.0E+03	9.4E+07	**8.7E+00**	1.6E+02	4.1E+06	4.4E+01	1.1E+10	2.0E+03
	MD	1.9E+03	9.3E+07	**8.6E+00**	8.4E+01	3.1E+06	3.7E+01	1.1E+10	2.0E+03
	SD	7.0E+02	7.1E+06	**1.1E+00**	3.6E+02	3.5E+06	3.0E+01	3.0E+09	2.7E+02

(continued)

Table 3.8 (continued)

Function		Hybrid	IWO	ABC	PSO	GSA	GA	HS	DE
$f_{27}(\boldsymbol{x})$	AB	**1.1E–13**	5.0E+64	1.0E+00	1.8E+00	1.4E+01	4.2E–01	7.4E+03	9.2E+00
	MD	**1.1E–13**	1.5E+62	1.0E+00	1.1E+00	1.2E+01	3.9E–01	6.9E+03	9.2E+00
	SD	**8.9E–15**	2.3E+65	3.2E–01	2.8E+00	9.9E+00	1.3E–01	1.9E+03	2.4E+00
$f_{28}(\boldsymbol{x})$	AB	**-5.0E+01**	1.8E+04	5.7E+01	-4.4E+01	-3.0E+01	-5.0E+01	-4.5E+01	-4.3E+01
	MD	**-5.0E+01**	1.8E+04	5.6E+01	-4.4E+01	-4.0E+01	-5.0E+01	-4.5E+01	-4.2E+01
	SD	**0.0E+00**	8.3E+02	2.0E+01	2.8E+00	0.1E+00	3.2E–04	8.0E–01	6.5E–01
$f_{29}(\boldsymbol{x})$	AB	**1.7E–179**	1.9E+01	2.4E–01	1.8E–08	3.2E–08	2.3E–15	1.2E–01	9.0E–26
	MD	**1.0E–180**	1.9E+01	2.4E–01	3.5E–09	2.5E–09	4.5E–16	1.2E–01	7.5E–26
	SD	**0.0E+00**	1.4E–01	1.3E–01	3.9E–08	9.0E–08	8.0E–15	3.1E–02	8.7E–26
$f_{30}(\boldsymbol{x})$	AB	**0.0E+00**	1.7E+02	8.4E+03	8.1E+01	5.7E+00	8.3E–02	9.8E+01	1.8E–10
	MD	**0.0E+00**	1.8E+02	8.2E+03	9.0E+01	4.8E+00	7.9E–02	9.9E+01	1.5E–10
	SD	**0.0E+00**	1.9E+01	1.7E+03	6.9E+01	2.1E+00	3.8E–02	1.4E+01	1.2E–10
$f_{31}(\boldsymbol{x})$	AB	**-2.0E+04**	9.3E+07	9.9E–01	-1.1E+04	-3.4E+03	-1.8E+04	-1.7E+04	-1.5E+04
	MD	**-2.0E+04**	9.3E+07	9.7E–01	-1.0E+04	-3.3E+03	-1.8E+04	-1.7E+04	-1.5E+04
	SD	**4.8E–02**	1.0E+07	3.5E–01	1.4E+03	6.2E+02	5.3E+02	5.6E+02	5.0E+02
$f_{32}(\boldsymbol{x})$	AB	**4.3E–02**	2.9E+65	6.6E+01	3.0E+02	2.9E+01	8.1E+01	2.3E+02	1.8E+02
	MD	**4.1E–02**	4.0E+62	6.1E+01	3.1E+02	3.0E+01	8.4E+01	2.3E+02	1.8E+02
	SD	**1.8E–02**	1.1E+66	3.1E+01	4.4E+01	5.6E+00	1.4E+01	1.9E+01	7.7E+00
$f_{33}(\boldsymbol{x})$	AB	**7.3E–15**	1.8E+04	2.8E–01	2.0E+01	1.6E–08	4.2E–02	1.4E+01	1.8E–06
	MD	**8.0E–15**	1.8E+04	2.6E–01	2.0E+01	1.5E–08	4.1E–02	1.4E+01	1.7E–06
	SD	**1.4E–15**	6.6E+02	1.3E–01	1.5E–03	3.5E–09	1.3E–02	5.5E–01	3.5E–07

(continued)

Table 3.8 (continued)

Function		Hybrid	IWO	ABC	PSO	GSA	GA	HS	DE
$f_{34}(\boldsymbol{x})$	AB	**1.3E–32**	2.2E+04	3.0E+05	1.4E+07	2.7E–01	3.7E–03	3.2E+07	2.6E–10
	MD	**1.3E–32**	2.2E+04	2.9E+05	5.0E+00	1.1E–02	1.7E–03	3.3E+07	2.3E–10
	SD	**5.6E–48**	2.2E+03	7.7E+04	7.5E+07	7.2E–01	4.6E–03	9.3E+06	1.4E–10

Bold values represent the main results

Table 3.9 Results of the multimodal benchmark functions (AII) with 100 dimensions considering 160,000 function evaluations

Function		Hybrid	IWO	ABC	PSO	GSA	GA	HS	DE
$f_6(\boldsymbol{x})$	AB	**7.1E–15**	1.9E+01	2.1E+01	2.0E+01	2.6E–08	4.5E–02	1.5E+01	6.1E–05
	MD	**7.1E–15**	1.9E+01	2.1E+01	2.0E+01	2.2E–08	4.5E–02	1.5E+01	6.1E–05
	SD	**0.0E+00**	1.0E–01	1.1E–01	2.3E–01	1.2E–08	1.0E–02	2.3E–01	6.3E–06
$f_7(\boldsymbol{x})$	AB	**1.9E–02**	9.1E+03	7.3E+04	3.4E+04	4.1E+00	8.7E–01	9.5E+03	5.4E+03
	MD	**1.3E–02**	9.2E+03	7.3E+04	3.4E+04	3.4E+00	8.6E–01	9.4E+03	5.2E+03
	SD	**2.6E–02**	2.1E+03	7.9E+03	8.0E+03	2.6E+00	1.9E–01	2.0E+03	9.5E+02
$f_8(\boldsymbol{x})$	AB	**1.5E–32**	2.0E+02	5.7E+02	1.7E+02	2.5E–01	8.6E–04	9.0E+01	2.0E–07
	MD	**1.5E–32**	1.9E+02	5.7E+02	1.7E+02	1.8E–01	8.2E–04	8.8E+01	2.1E–07
	SD	**1.1E–47**	2.3E+01	4.6E+01	4.2E+01	2.0E–01	2.9E–04	1.1E+01	7.2E–08
$f_9(\boldsymbol{x})$	AB	**9.6E+01**	2.4E+06	7.8E+04	2.3E+06	1.4E+02	1.3E+02	3.1E+05	9.8E+01
	MD	**9.6E+01**	2.4E+06	7.9E+04	2.2E+06	1.0E+02	1.0E+02	3.1E+05	9.9E+01
	SD	**1.3E–01**	3.9E+05	6.8E+03	4.4E+05	5.3E+01	6.0E+01	5.0E+04	2.2E+00
$f_{10}(\boldsymbol{x})$	AB	**2.4E–03**	1.1E+141	7.3E+04	1.0E+75	4.8E+02	1.1E+00	2.5E+43	2.3E+02
	MD	**2.4E–03**	4.3E+137	7.3E+04	4.9E+03	5.0E+02	1.1E+00	1.5E+31	3.2E+02
	SD	**6.0E–04**	3.3E+141	7.9E+03	5.5E+75	8.6E+01	2.4E–01	1.4E+44	2.2E+02
$f_{11}(\boldsymbol{x})$	AB	**1.8E+02**	3.8E+04	8.2E+02	2.5E+04	3.7E+04	5.5E+03	9.9E+03	2.1E+04
	MD	**1.8E+02**	3.8E+04	8.3E+02	2.5E+04	3.7E+04	5.6E+03	9.9E+03	2.1E+04
	SD	**7.6E+00**	9.6E+02	5.1E+01	1.9E+03	8.7E+02	6.1E+02	7.0E+02	4.3E+02
$f_{12}(\boldsymbol{x})$	AB	2.5E+03	2.6E+02	5.7E+02	1.3E+02	2.7E+107	2.1E+03	4.9E+271	**6.0E–05**
	MD	2.1E+03	2.6E+02	5.8E+02	1.1E+02	1.5E+95	1.8E+03	2.0E+269	**6.0E–05**
	SD	2.0E+03	2.5E+01	5.1E+01	5.0E+01	1.5E+108	1.5E+03	6.6E+04	**6.3E–06**

(continued)

Table 3.9 (continued)

Function		Hybrid	IWO	ABC	PSO	GSA	GA	HS	DE
$f_{13}(\boldsymbol{x})$	AB	6.7E–01	8.0E+04	8.1E+04	2.5E+04	4.9E+00	3.3E+01	1.2E+06	**5.5E–10**
	MD	6.7E–01	8.0E+04	8.2E+04	2.6E+04	3.4E+00	3.2E+01	1.2E+06	**5.4E–10**
	SD	2.2E–03	5.9E+03	5.3E+03	7.5E+03	4.8E+00	9.5E+00	1.4E+05	**1.2E–10**
$f_{14}(\boldsymbol{x})$	AB	1.5E–08	9.6E+03	7.0E+04	3.3E+04	**5.7E–26**	1.0E+03	1.5E+03	2.3E+02
	MD	1.5E–08	9.3E+03	7.1E+04	3.3E+04	**5.2E–27**	9.9E+02	1.5E+03	2.3E+02
	SD	3.2E–09	1.9E+03	7.9E+03	8.3E+03	**2.1E–25**	1.4E+02	1.2E+02	3.6E+01
$f_{15}(\boldsymbol{x})$	AB	**-3.8E+03**	2.0E+02	2.1E+01	1.8E+02	-3.6E+03	-3.4E+03	-3.4E+03	5.2E+03
	MD	**-3.8E+03**	2.0E+02	2.1E+01	1.9E+02	-3.6E+03	-3.4E+03	-3.4E+03	5.1E+03
	SD	5.4E–01	2.2E+01	**1.1E–01**	4.3E+01	5.9E+01	7.6E+01	5.0E+01	1.1E+03
$f_{16}(\boldsymbol{x})$	AB	**0.0E+00**	2.5E+06	5.7E+02	2.0E+06	7.4E+00	0.3E+00	3.1E+04	2.0E–07
	MD	**0.0E+00**	2.5E+06	5.8E+02	2.1E+06	2.0E+00	0.4E+00	3.1E+04	1.9E–07
	SD	**0.0E+00**	3.7E+05	5.9E+01	4.1E+05	1.1E+01	0.1E+00	3.4E+03	8.0E–08
$f_{17}(\boldsymbol{x})$	AB	**3.9E+01**	8.3E+08	8.2E+04	2.1E+08	4.9E+01	5.1E+01	1.4E+02	9.8E+01
	MD	**3.9E+01**	8.4E+08	8.0E+04	2.2E+08	5.9E+01	5.1E+01	1.4E+02	9.6E+01
	SD	**7.9E–01**	6.2E+07	7.2E+03	5.6E+07	9.6E–01	3.0E+00	1.1E+01	1.2E+01
$f_{18}(\boldsymbol{x})$	AB	2.1E–01	3.2E+141	1.0E+00	5.9E+94	5.2E+00	4.4E+00	1.9E+01	**2.6E–04**
	MD	2.0E–01	1.7E+137	1.0E+00	3.5E+48	5.2E+00	4.4E+00	1.9E+01	**2.5E–04**
	SD	3.1E–02	1.6E+142	3.4E–01	3.2E+95	4.7E–01	6.4E–01	9.8E–01	**6.9E–05**
$f_{19}(\boldsymbol{x})$	AB	3.2E+06	3.7E+04	7.2E+04	2.5E+04	**2.6E–03**	2.6E+02	1.1E+08	2.7E+02
	MD	2.4E+02	3.8E+04	7.3E+04	2.5E+04	**2.6E–03**	2.6E+02	1.1E+08	2.5E+02
	SD	4.0E+06	1.2E+03	1.0E+04	1.8E+03	**4.9E–04**	5.2E+00	1.4E+07	2.8E+01

(continued)

Table 3.9 (continued)

Function		Hybrid	IWO	ABC	PSO	GSA	GA	HS	DE
$f_{20}(\boldsymbol{x})$	AB	**1.6E+01**	1.9E+01	2.1E+01	2.0E+01	4.1E+02	2.2E+02	1.6E+08	2.1E+04
	MD	**1.5E+01**	1.9E+01	2.1E+01	2.0E+01	4.1E+02	2.2E+02	1.6E+08	2.1E+04
	SD	**2.4E–04**	8.8E–02	1.1E–01	1.0E–03	2.3E+01	8.8E+00	2.3E+07	4.9E+02
$f_{21}(\boldsymbol{x})$	AB	2.1E+02	2.6E+02	**5.1E–10**	2.0E+01	6.8E+01	9.1E–02	5.7E+02	5.7E+02
	MD	2.1E+02	2.6E+02	**5.2E–10**	2.0E+01	6.9E+01	8.8E–02	5.7E+02	5.8E+02
	SD	2.6E+01	2.2E+01	**1.2E–10**	9.5E–04	9.4E+00	3.8E–02	2.7E+01	4.8E+01
$f_{22}(\boldsymbol{x})$	AB	**2.0E+00**	8.0E+04	8.3E+04	1.1E+02	6.1E+22	3.0E+00	4.6E+16	6.8E–08
	MD	**2.0E+00**	8.0E+04	8.3E+04	1.0E+02	1.3E+21	3.4E+00	1.1E+16	6.2E–08
	SD	**2.5E–16**	5.6E+03	5.1E+03	5.0E+01	2.9E+23	5.8E–14	9.6E+16	2.1E–08
$f_{23}(\boldsymbol{x})$	AB	**2.0E+00**	2.7E–08	1.2E+00	2.7E+04	2.5E+23	3.0E+00	1.0E+17	7.4E–26
	MD	**2.0E+00**	2.6E–08	1.2E+00	2.7E+04	2.6E+22	3.4E+00	2.0E+16	1.2E–26
	SD	**8.9E–15**	5.7E–09	3.2E–01	7.5E+03	5.2E+23	5.8E–14	1.5E+17	1.6E–25
$f_{24}(\boldsymbol{x})$	AB	3.0E–03	9.5E+03	7.3E+04	1.5E–01	**1.1E–29**	3.7E–12	9.4E–03	5.1E+03
	MD	3.2E–03	9.5E+03	7.4E+04	1.4E–03	**0.0E+00**	2.2E–12	9.4E–03	4.9E+03
	SD	1.8E–03	2.2E+03	8.3E+03	3.4E–01	**5.8E–29**	3.6E–12	1.7E–03	1.0E+03
$f_{25}(\boldsymbol{x})$	AB	**0.0E+00**	1.9E+02	2.1E+01	3.2E+04	1.4E–153	1.0E+00	1.5E–56	2.0E–07
	MD	**0.0E+00**	1.9E+02	2.1E+01	3.1E+04	9.7E–175	1.5E+00	1.1E–61	1.7E–07
	SD	**0.0E+00**	1.6E+01	8.3E–02	1.0E+04	7.5E–153	0.2E+00	6.8E–56	1.0E–07
$f_{26}(\boldsymbol{x})$	AB	5.7E+04	2.2E+06	5.8E+02	1.8E+02	**9.7E+01**	2.2E+02	4.0E+10	5.0E+06
	MD	5.6E+04	2.3E+06	5.8E+02	1.8E+02	**9.6E+01**	1.8E+02	4.0E+10	3.9E+06
	SD	7.3E+03	4.2E+05	3.3E+01	4.0E+01	**5.1E+00**	1.7E+02	7.0E+09	4.1E+06

(continued)

Table 3.9 (continued)

Function		Hybrid	IWO	ABC	PSO	GSA	GA	HS	DE
$f_{27}(\boldsymbol{x})$	AB	**7.5E–13**	8.3E+08	8.2E+04	2.0E+06	4.8E+01	9.5E–01	3.6E+04	2.6E–04
	MD	**7.5E–13**	8.3E+08	8.3E+04	1.9E+06	4.2E+01	9.1E–01	3.7E+04	2.3E–04
	SD	**2.2E–14**	5.6E+07	6.2E+03	4.7E+05	1.9E+01	1.8E–01	6.2E+03	7.9E–05
$f_{28}(\boldsymbol{x})$	AB	**-1.0E+02**	9.9E+139	1.0E+00	1.9E+08	-0.8E+02	-0.9E+02	-8.5E+01	2.4E–03
	MD	**-1.0E+02**	3.3E+138	1.1E+00	1.9E+08	-0.3E+02	-0.6E+02	-8.5E+01	2.3E–03
	SD	**9.4E–09**	2.9E+140	2.6E–01	5.4E+07	6.5E–08	3.2E–04	1.1E+00	3.5E–04
$f_{29}(\boldsymbol{x})$	AB	**1.3E–110**	3.8E+04	7.2E+04	5.1E+95	3.6E–07	1.8E–15	5.6E–01	2.1E+04
	MD	**4.2E–111**	3.8E+04	7.0E+04	1.7E+45	2.8E–07	7.1E–16	5.6E–01	2.1E+04
	SD	**3.0E–110**	1.0E+03	7.7E+03	2.8E+96	3.1E–07	2.9E–15	1.1E–01	5.2E+02
$f_{30}(\boldsymbol{x})$	AB	**0.0E+00**	3.1E+140	2.1E+01	2.5E+04	8.3E+00	1.0E–01	2.8E+02	5.9E–05
	MD	**0.0E+00**	4.3E+137	2.1E+01	2.5E+04	8.4E+00	8.9E–02	2.9E+02	5.8E–05
	SD	**0.0E+00**	9.0E+140	7.0E–02	1.8E+03	2.6E+00	6.0E–02	2.2E+01	6.6E–06
$f_{31}(\boldsymbol{x})$	AB	**-1.6E+03**	2.2E+02	5.5E+02	1.7E+02	-5.3E+03	-3.6E+04	-3.2E+04	9.7E+01
	MD	**-1.5E+03**	2.1E+02	5.6E+02	1.7E+02	-5.4E+03	-3.6E+04	-3.1E+04	9.6E+01
	SD	**8.8E–02**	2.6E+01	3.9E+01	5.5E+01	7.7E+02	5.9E+02	8.0E+02	4.7E+00
$f_{32}(\boldsymbol{x})$	AB	**2.5E–04**	2.8E+06	7.9E+04	2.1E+06	6.6E+01	9.9E–02	5.8E+02	2.3E+02
	MD	**2.5E–04**	2.9E+06	8.0E+04	2.0E+06	6.6E+01	9.8E–02	5.9E+02	2.2E+02
	SD	**7.5E–05**	4.2E+05	6.3E+03	3.8E+05	1.0E+01	3.7E–02	2.3E+01	2.8E+01
$f_{33}(\boldsymbol{x})$	AB	**6.5E–33**	3.7E+04	5.7E+02	2.5E+04	5.1E–01	2.9E–04	5.1E+07	5.9E–05
	MD	**6.5E–33**	3.8E+04	5.8E+02	2.5E+04	4.6E–01	1.1E–04	5.2E+07	5.8E–05
	SD	**4.1E–34**	9.9E+02	4.3E+01	1.9E+03	2.9E–01	9.1E–04	9.9E+06	7.1E–06

(continued)

Table 3.9 (continued)

Function		Hybrid	IWO	ABC	PSO	GSA	GA	HS	DE
$f_{34}(\boldsymbol{x})$	AB	**5.8E–32**	1.9E+01	8.1E+04	2.0E+01	8.7E+00	5.5E–03	1.6E+08	5.5E–10
	MD	**5.8E–32**	1.9E+01	8.1E+04	2.0E+01	6.6E+00	3.6E–03	1.6E+08	5.4E–10
	SD	**1.7E–32**	1.2E–01	6.3E+03	1.1E–03	7.4E+00	5.2E–03	2.5E+07	1.2E–10

Bold values represent the main results

Table 3.10 *p*-values produced by the Wilcoxon test comparing Hybrid versus IWO, Hybrid versus ABC, Hybrid versus PSO, Hybrid versus GSA, Hybrid versus GA, Hybrid versus HS, and Hybrid versus DE over the AB values from Table 3.7 ($n = 30$)

Function	Hybrid versus IWO	Hybrid versus ABC	Hybrid versus PSO	Hybrid versus GSA	Hybrid versus GA	Hybrid versus HS	Hybrid versus DE
$f_6(\mathbf{x})$	1.2E–12▲	1.2E–12▲	1.2E–12▲	1.2E–12▲	1.2E–12▲	1.2E–12▲	1.2E–12▲
$f_7(\mathbf{x})$	1.1E–06▲	3.0E–11▲	4.5E–09▲	3.0E–11▲	3.0E–11▲	1.2E–12▲	3.0E–11▲
$f_8(\mathbf{x})$	1.2E–12▲	1.2E–12▲	1.2E–12▲	1.2E–12▲	1.2E–12▲	1.2E–12▲	1.2E–12▲
$f_9(\mathbf{x})$	3.0E–11▲	3.0E–11▲	3.0E–11▲	3.0E–11▲	3.0E–11▲	3.0E–11▲	3.0E–11▲
$f_{10}(\mathbf{x})$	2.0E–03▲	1.1E–10▲	6.7E–11▲	1.6E–03▲	2.3E–04▲	3.0E–11▲	2.0E–03▼
$f_{11}(\mathbf{x})$	5.6E–10▲	3.0E–11▲	3.0E–11▲	3.0E–11▲	3.0E–11▲	8.5E–09▲	3.0E–11▼
$f_{12}(\mathbf{x})$	3.0E–11▲	3.0E–11▲	3.0E–11▲	3.0E–11▼	3.0E–11▲	3.0E–11▲	7.7E–04▲
$f_{13}(\mathbf{x})$	6.6E–01▲	3.0E–11▲	3.0E–11▲	3.0E–11▲	3.0E–11▲	3.0E–11▲	3.0E–11▼
$f_{14}(\mathbf{x})$	3.0E–11▲	3.0E–11▲	3.0E–11▲	3.0E–11▼	3.0E–11▲	3.0E–11▲	3.0E–11▲
$f_{15}(\mathbf{x})$	3.0E–11▲	3.0E–11▲	3.0E–11▲	3.0E–11▲	3.0E–11▲	3.0E–11▲	3.0E–11▲
$f_{16}(\mathbf{x})$	1.2E–12▲	1.2E–12▲	1.2E–12▲	1.1E–12▲	1.2E–12▲	1.2E–12▲	1.2E–12▲
$f_{17}(\mathbf{x})$	3.0E–11▲	1.7E–12▲	2.2E–09▲	1.2E–12▲	3.2E–12▲	3.0E–11▲	3.0E–11▲
$f_{18}(\mathbf{x})$	1.9E–11▲	1.9E–11▲	1.9E–11▲	1.9E–11▼	1.9E–11▲	1.9E–11▲	1.9E–11▲
$f_{19}(\mathbf{x})$	3.0E–11▲	3.0E–11▲	3.0E–11▲	3.0E–11▲	3.0E–11▲	3.0E–11▲	3.0E–11▼
$f_{20}(\mathbf{x})$	3.0E–11▲	3.0E–11▲	2.7E–06▲	2.9E–11▲	2.0E–03▲	3.0E–11▲	3.0E–11▼
$f_{21}(\mathbf{x})$	7.1E–01▲	3.0E–11▲	3.0E–11▲	3.0E–11▼	3.0E–11▲	3.0E–11▲	3.0E–11▲
$f_{22}(\mathbf{x})$	3.0E–11▲	3.0E–11▲	3.0E–11▲	3.0E–11▲	3.0E–11▲	3.0E–11▲	3.0E–11▲
$f_{23}(\mathbf{x})$	1.2E–12▲	1.6E–11▲	6.4E–09▲	3.0E–11▲	5.3E–09▲	3.0E–11▲	3.0E–11▲
$f_{24}(\mathbf{x})$	1.2E–12▲	3.0E–11▲	2.7E–11▲	3.0E–11▲	2.0E–11▲	3.0E–11▲	3.0E–11▲
$f_{25}(\mathbf{x})$	3.0E–11▲	3.0E–11▲	1.2E–12▲	5.6E–10▲	3.0E–11▲	3.0E–11▲	3.0E–11▲
$f_{26}(\mathbf{x})$	3.0E–11▲	3.0E–11▲	3.0E–11▲	3.0E–11▼	1.2E–12▲	3.0E–11▲	2.3E–06▲
$f_{27}(\mathbf{x})$	2.9E–11▲	2.9E–11▲	2.9E–11▲	2.9E–11▲	2.9E–11▲	2.9E–11▲	2.9E–11▲
$f_{28}(\mathbf{x})$	3.0E–11▲	3.0E–11▲	3.0E–11▲	3.0E–11▲	3.0E–11▲	3.0E–11▲	3.0E–11▲
$f_{29}(\mathbf{x})$	3.0E–11▲	3.0E–11▲	3.0E–11▲	1.2E–12▲	3.0E–11▲	3.0E–11▲	3.0E–11▲
$f_{30}(\mathbf{x})$	1.2E–12▲	1.2E–12▲	1.2E–12▲	1.2E–12▲	1.2E–12▲	1.2E–12▲	1.2E–12▲
$f_{31}(\mathbf{x})$	3.0E–11▲	3.0E–11▲	2.8E–11▲	3.0E–11▲	3.0E–11▲	3.0E–11▲	3.0E–11▲
$f_{32}(\mathbf{x})$	1.8E–10▲	3.0E–11▲	3.0E–11▲	2.3E–02▲	3.0E–11▲	3.0E–11▲	1.2E–12▲
$f_{33}(\mathbf{x})$	1.2E–12▲	1.2E–12▲	1.2E–12▲	1.2E–12▲	1.2E–12▲	1.2E–12▲	1.2E–12▲
$f_{34}(\mathbf{x})$	1.2E–12▲	1.2E–12▲	1.2E–12▲	1.2E–12▲	1.2E–12▲	1.2E–12▲	1.2E–12▲

Table 3.11 *p*-values produced by the Wilcoxon test comparing Hybrid versus IWO, Hybrid versus ABC, Hybrid versus PSO, Hybrid versus GSA, Hybrid versus GA, Hybrid versus HS, and Hybrid versus DE over the AB values from Table 3.8 ($n = 50$)

Function	Hybrid versus IWO	Hybrid versus ABC	Hybrid versus PSO	Hybrid versus GSA	Hybrid versus GA	Hybrid versus HS	Hybrid versus DE
$f_6(\boldsymbol{x})$	5.1E–12▲	5.1E–12▲	5.1E–12▲	5.1E–12▲	5.1E–12▲	5.1E–12▲	5.1E–12▲
$f_7(\boldsymbol{x})$	3.0E–11▲	3.0E–11▲	2.8E–11▲	3.0E–11▲	4.5E–11▲	3.0E–11▲	3.0E–11▲
$f_8(\boldsymbol{x})$	1.2E–12▲	1.2E–12▲	1.2E–12▲	1.2E–12▲	1.2E–12▲	1.2E–12▲	1.2E–12▲
$f_9(\boldsymbol{x})$	3.0E–11▲	3.0E–11▲	7.6E–07▲	3.0E–11▲	3.0E–11▲	3.0E–11▲	3.0E–11▲
$f_{10}(\boldsymbol{x})$	3.0E–11▲	3.0E–11▲	3.0E–11▲	6.8E–05▲	4.1E–05▲	3.0E–11▲	6.8E–05▼
$f_{11}(\boldsymbol{x})$	3.0E–11▲	3.0E–11▼	3.0E–11▲	3.0E–11▲	3.0E–11▲	3.0E–11▲	3.0E–11▲
$f_{12}(\boldsymbol{x})$	3.0E–11▲	3.0E–11▼	3.0E–11▲	3.0E–11▲	1.9E–07▲	4.2E–10▲	3.0E–11▲
$f_{13}(x)$	3.0E–11▲	3.0E–11▲	3.0E–11▲	3.0E–11▲	3.0E–11▲	3.0E–11▲	3.0E–11▲
$f_{14}(\boldsymbol{x})$	3.0E–11▲	3.0E–11▲	3.0E–11▲	3.0E–11▲	3.0E–11▲	3.0E–11▲	3.0E–11▲
$f_{15}(\boldsymbol{x})$	3.0E–11▲	3.0E–11▲	3.0E–11▲	3.0E–11▲	3.0E–11▲	3.0E–11▲	3.0E–11▲
$f_{16}(\boldsymbol{x})$	1.2E–12▲	1.2E–12▲	1.2E–12▲	1.2E–12▲	1.2E–12▲	1.2E–12▲	1.2E–12▲
$f_{17}(\boldsymbol{x})$	3.0E–11▲	3.0E–11▲	1.9E–03▲	3.2E–12▲	1.2E–12▲	3.0E–11▲	1.2E–12▲
$f_{18}(\boldsymbol{x})$	2.9E–11▲	2.9E–11▲	2.9E–11▲	2.9E–11▲	2.9E–11▲	2.9E–11▲	2.9E–11▲
$f_{19}(\boldsymbol{x})$	3.0E–11▼	3.0E–11▲	3.0E–11▲	3.0E–11▲	2.9E–11▲	3.0E–11▲	3.0E–11▲
$f_{20}(\boldsymbol{x})$	3.0E–11▲	3.0E–11▼	3.2E–05▲	3.0E–11▲	3.0E–11▲	3.0E–11▲	3.5E–02▲
$f_{21}(\boldsymbol{x})$	3.0E–11▲	3.0E–11▲	3.0E–11▲	3.0E–11▼	8.1E–10▲	3.0E–11▲	3.0E–11▲
$f_{22}(\boldsymbol{x})$	3.0E–11▲	3.0E–11▲	3.0E–11▲	3.0E–11▲	3.0E–11▲	3.0E–11▲	3.0E–11▲
$f_{23}(\boldsymbol{x})$	3.0E–11▲	3.0E–11▲	2.7E–11▲	3.0E–11▲	2.2E–11▲	3.0E–11▲	1.2E–12▲
$f_{24}(\boldsymbol{x})$	3.0E–11▲	3.0E–11▲	2.9E–11▲	3.0E–11▲	9.0E–12▲	3.0E–11▲	1.2E–12▼
$f_{25}(\boldsymbol{x})$	2.7E–11▲	2.7E–11▲	1.3E–07▲	2.0E–07▲	2.7E–11▲	2.7E–11▲	0.0721▶
$f_{26}(\boldsymbol{x})$	3.0E–11▲	3.0E–11▼	3.0E–11▲	3.0E–11▲	1.2E–12▲	3.0E–11▲	3.0E–11▲
$f_{27}(\boldsymbol{x})$	3.0E–11▲	3.0E–11▲	3.0E–11▲	3.0E–11▲	3.0E–11▲	3.0E–11▲	3.0E–11▲
$f_{28}(\boldsymbol{x})$	3.0E–11▲	3.0E–11▲	3.0E–11▲	3.0E–11▲	3.0E–11▲	3.0E–11▲	3.0E–11▲
$f_{29}(\boldsymbol{x})$	3.0E–11▲	3.0E–11▲	3.0E–11▲	1.2E–12▲	3.0E–11▲	3.0E–11▲	0.0607▶
$f_{30}(\boldsymbol{x})$	1.2E–12▲	1.2E–12▲	1.2E–12▲	1.2E–12▲	1.2E–12▲	1.2E–12▲	1.2E–12▲
$f_{31}(\boldsymbol{x})$	3.0E–11▲	3.0E–11▲	3.0E–11▲	3.0E–11▲	3.0E–11▲	3.0E–11▲	3.0E–11▲
$f_{32}(\boldsymbol{x})$	3.0E–11▲	3.0E–11▲	3.0E–11▲	3.0E–11▲	3.0E–11▲	3.0E–11▲	3.0E–11▲
$f_{33}(\boldsymbol{x})$	1.2E–12▲	1.2E–12▲	1.2E–12▲	1.2E–12▲	1.2E–12▲	1.2E–12▲	1.2E–12▲
$f_{34}(\boldsymbol{x})$	1.2E–12▲	1.2E–12▲	1.2E–12▲	1.2E–12▲	1.2E–12▲	1.2E–12▲	1.2E–12▲

Table 3.12 *p*-values produced by the Wilcoxon test comparing Hybrid versus IWO, Hybrid versus ABC, Hybrid versus PSO, Hybrid versus GSA, Hybrid versus GA, Hybrid versus HS, and Hybrid versus DE over the AB values from Table 3.9 ($n = 100$)

Function	Hybrid versus IWO	Hybrid versus ABC	Hybrid versus PSO	Hybrid versus GSA	Hybrid versus GA	Hybrid versus HS	Hybrid versus DE
$f_6(\boldsymbol{x})$	1.2E–12▲	1.2E–12▲	1.2E–12▲	1.2E–12▲	1.2E–12▲	1.2E–12▲	1.2E–12▲
$f_7(\boldsymbol{x})$	3.0E–11▲	3.0E–11▲	2.2E–11▲	3.0E–11▲	5.2E–07▲	3.0E–11▲	3.0E–11▲
$f_8(\boldsymbol{x})$	1.2E–12▲	1.2E–12▲	1.2E–12▲	1.2E–12▲	1.2E–12▲	1.2E–12▲	1.2E–12▲
$f_9(\boldsymbol{x})$	3.0E–11▲	3.0E–11▲	3.0E–11▲	3.0E–11▲	3.0E–11▲	6.8E–05▲	3.0E–11▲
$f_{10}(\boldsymbol{x})$	3.0E–11▲	3.0E–11▲	3.0E–11▲	3.8E–01▲	3.8E–01▲	3.0E–11▲	3.8E–01▲
$f_{11}(\boldsymbol{x})$	3.0E–11▲	3.0E–11▲	2.7E–02▲	3.0E–11▲	3.0E–11▲	3.0E–11▲	3.0E–11▲
$f_{12}(\boldsymbol{x})$	3.0E–11▲	4.0E–04▲	3.0E–11▲	3.0E–11▲	9.8E–08▲	3.0E–11▲	3.0E–11▼
$f_{13}(\boldsymbol{x})$	3.0E–11▲	3.0E–11▲	3.0E–11▲	3.0E–11▲	3.0E–11▲	3.0E–11▲	3.0E–11▼
$f_{14}(\boldsymbol{x})$	3.0E–11▲	3.0E–11▲	3.0E–11▲	3.0E–11▼	3.0E–11▲	3.0E–11▲	3.0E–11▲
$f_{15}(\boldsymbol{x})$	3.0E–11▲	3.0E–11▲	3.0E–11▲	3.0E–11▲	3.0E–11▲	3.0E–11▲	3.0E–11▲
$f_{16}(\boldsymbol{x})$	1.2E–12▲	1.2E–12▲	1.2E–12▲	1.2E–12▲	1.2E–12▲	1.2E–12▲	1.2E–12▲
$f_{17}(\boldsymbol{x})$	3.0E–11▲	3.0E–11▲	3.0E–11▲	5.1E–10▲	1.2E–12▲	3.0E–11▲	3.0E–11▲
$f_{18}(\boldsymbol{x})$	2.1E–11▲	2.1E–11▲	2.1E–11▲	2.1E–11▲	2.1E–11▲	2.1E–11▲	2.1E–11▼
$f_{19}(\boldsymbol{x})$	3.0E–11▲	3.0E–11▲	4.2E–09▲	3.0E–11▼	3.0E–11▲	3.0E–11▲	3.0E–11▲
$f_{20}(\boldsymbol{x})$	2.7E–02▲	2.7E–02▲	2.7E–02▲	3.0E–11▲	3.0E–11▲	3.0E–11▲	3.0E–11▲
$f_{21}(\boldsymbol{x})$	3.0E–11▲	3.0E–11▼	2.3E–11▲	3.0E–11▲	5.4E–09▲	3.0E–11▲	3.0E–11▲
$f_{22}(\boldsymbol{x})$	1.1E–07▲	1.1E–07▲	7.7E–03▲	9.2E–05▲	3.0E–11▲	1.1E–07▲	3.0E–11▲
$f_{23}(\boldsymbol{x})$	3.0E–11▲	3.0E–11▲	3.0E–11▲	3.0E–11▲	1.9E–11▲	3.0E–11▲	3.0E–11▲
$f_{24}(\boldsymbol{x})$	1.1E–06▲	3.0E–11▲	3.0E–11▲	3.0E–11▼	2.5E–11▲	3.0E–11▲	3.0E–11▲
$f_{25}(\boldsymbol{x})$	3.0E–11▲	3.0E–11▲	3.0E–11▲	2.5E–03▲	3.0E–11▲	3.0E–11▲	3.0E–11▲
$f_{26}(\boldsymbol{x})$	3.0E–11▲	3.0E–11▲	1.2E–10▲	3.0E–11▼	1.2E–12▲	3.0E–11▲	3.0E–11▲
$f_{27}(\boldsymbol{x})$	3.0E–11▲	3.0E–11▲	3.0E–11▲	3.0E–11▲	3.0E–11▲	3.0E–11▲	3.0E–11▲
$f_{28}(\boldsymbol{x})$	3.0E–11▲	3.0E–11▲	3.0E–11▲	3.0E–11▲	3.0E–11▲	3.0E–11▲	3.0E–11▲
$f_{29}(\boldsymbol{x})$	3.0E–11▲	3.0E–11▲	3.0E–11▲	2.7E–11▲	3.0E–11▲	3.0E–11▲	3.0E–11▲
$f_{30}(\boldsymbol{x})$	1.2E–12▲	1.2E–12▲	1.2E–12▲	1.2E–12▲	1.2E–12▲	1.2E–12▲	1.2E–12▲
$f_{31}(\boldsymbol{x})$	3.0E–11▲	3.0E–11▲	3.0E–11▲	3.0E–11▲	3.0E–11▲	3.0E–11▲	3.0E–11▲
$f_{32}(\boldsymbol{x})$	2.4E–01▲	3.0E–11▲	6.4E–05▲	3.0E–11▲	3.0E–11▲	3.0E–11▲	3.0E–11▲
$f_{33}(\boldsymbol{x})$	3.0E–11▲	3.0E–11▲	3.0E–11▲	3.0E–11▲	3.0E–11▲	3.0E–11▲	3.0E–11▲
$f_{34}(\boldsymbol{x})$	3.0E–11▲	3.0E–11▲	3.0E–11▲	3.0E–11▲	3.0E–11▲	3.0E–11▲	3.0E–11▲

in comparison to IWO, ABC, PSO, GSA, GA, HS, and DE in functions $f_6, f_7, f_8, f_9, f_{15}, f_{16}, f_{17}, f_{22}, f_{23}, f_{24}, f_{25}, f_{27}, f_{28}, f_{29}, f_{30}, f_{31}, f_{32}, f_{33}$ and f_{34}. In the case of the comparison between GSA and the hybrid method, the GSA method maintains a better (▲) performance in functions $f_{12}, f_{14}, f_{18}, f_{21}$ and f_{26}. On the other hand, in functions f_{10}, f_{11}, f_{13}, f_{19} and f_{20} the hybrid method produce worse results (▼) than the DE algorithm.

In Table 3.11, the Wilcoxon results of the AB data from Table 3.8 are presented. According to these results, the hybrid algorithm obtains the best indicators in functions $f_6, f_7, f_8, f_9, f_{13}, f_{14}, f_{15}, f_{16}, f_{17}, f_{18}, f_{22}, f_{23}, f_{27}, f_{28}, f_{30}, f_{31}, f_{32}, f_{33}$ and f_{34}. However, there is no statistical evidence that the hybrid method gets better solutions than the DE algorithm for functions f_{25} and f_{29}. In the case of functions f_{10} and f_{24}, the DE optimization technique produces the best performance results. On the other hand, the ABC performs better than the other methods in functions f_{11}, f_{12}, f_{20} and f_{26}. Finally, in function f_{21}, GSA demonstrates to be the best method.

Table 3.12 presents the results of the Wilcoxon analysis over the AB data corresponding to the optimization of multimodal functions in 100 dimensions. From the information of Table 3.12, it is clear that the hybrid method obtain better solutions than the other algorithms in functions $f_6, f_7, f_8, f_9, f_{10}, f_{11}, f_{15}, f_{16}, f_{17}, f_{20}, f_{22}, f_{23}, f_{25}, f_{27}, f_{28}, f_{29}, f_{30}, f_{31}, f_{32}, f_{33}$ and f_{34}. On the other hand, the Wilcoxon test demonstrates that the DE algorithms maintain the best indicators for the multimodal functions f_{12}, f_{13} and f_{18}. For functions f_{14}, f_{19} and f_{26}, the GSA method presents statistical evidence of its better performance. The ABC technique presents the best performance results in solving the function f_{21}. In general, it can be concluded that the hybrid method presents an excellent capacity to solve functions that contains multiple local optima. This fact can be attributed to its interesting capacities for balancing the exploration and exploitation of the search space through the integration of different approaches.

3.6.3 Composite Test Functions

In this section, composite functions are used to analyze the effectiveness of the search strategy implemented in the hybrid method. Composite functions, exhibited in Table 3.21, are models that are generated as a combination of different multimodal functions. Under such conditions, its behavior is notably more complex than single multimodal functions. Specific details of their implementation can be consulted in [30].

In the analysis, the results of the hybrid method are contrasted with those obtained by IWO, ABC, PSO, GSA, GA, HS, and DE, considering the composite functions described in Appendix A from f_{35} to f_{38}. The results have been collected from the operation of the composite functions in 30, 50, and 100 dimensions. The produced indexes obtained from 30 distinct executions are exhibited in Tables 3.13, 3.14, and 3.15, corresponding to the cases of 30, 50, and 100 dimensions, respectively.

Table 3.13 Results of the composite benchmark functions (AIII) with 30 dimensions considering 50,000 function evaluations

Function		Hybrid	IWO	ABC	PSO	GSA	GA	HS	DE
$f_{35}(\boldsymbol{x})$	AB	**1.7E–41**	9.5E+01	8.9E–05	1.3E+02	3.9E–05	8.8E–05	2.0E+02	6.7E–01
	MD	**1.2E–41**	9.7E+01	8.0E–05	1.3E+02	4.0E–13	7.4E–05	2.0E+02	6.7E–01
	SD	**1.9E–41**	1.4E+01	5.5E–05	5.0E+01	2.2E–04	6.0E–05	1.8E+01	4.9E–03
$f_{36}(\boldsymbol{x})$	AB	**2.9E+01**	7.4E+03	3.7E–03	2.8E+04	2.0E–08	4.9E–03	3.3E+04	3.1E–11
	MD	**2.9E+01**	7.6E+03	3.0E–03	2.7E+04	5.2E–09	3.5E–03	3.4E+04	2.5E–11
	SD	**1.3E–11**	1.0E+03	2.1E–03	9.9E+03	2.9E–08	3.9E–03	3.1E+03	4.6E–11
$f_{37}(\boldsymbol{x})$	AB	**3.2E+01**	1.3E–07	2.6E–03	4.6E–01	8.4E–01	1.5E–08	2.0E–02	1.7E+02
	MD	**3.2E+01**	1.3E–07	2.1E–03	3.3E–01	6.2E–01	3.3E–10	1.6E–02	1.6E+02
	SD	**1.5E–09**	4.5E–08	2.1E–03	4.6E–01	6.5E–01	4.3E–08	9.3E–03	2.1E+01
$f_{38}(\boldsymbol{x})$	AB	**2.9E+01**	3.4E+03	5.3E+02	2.9E+04	8.6E–16	1.9E–01	2.9E+04	1.2E+03
	MD	**2.9E+01**	3.3E+03	5.3E+02	2.9E+04	3.7E–16	1.7E–01	2.9E+04	1.2E+03
	SD	**1.6E–15**	9.2E+02	2.1E+02	8.1E+03	2.4E–15	6.6E–02	4.5E+03	1.1E–02

Bold values represent the main results

Table 3.14 Results of the composite benchmark functions (AIII) with 50 dimensions considering 80,000 function evaluations

Function		Hybrid	IWO	ABC	PSO	GSA	GA	HS	DE
$f_{35}(\boldsymbol{x})$	AB	**1.1E–35**	1.1E–07	2.9E+05	8.0E+04	3.2E–08	1.8E–01	8.0E+03	7.4E–07
	MD	**1.2E–35**	1.1E–07	2.9E+05	7.5E+04	1.2E–08	1.6E–01	7.8E+03	7.2E–07
	SD	**4.1E–36**	3.4E–08	7.5E+04	3.8E+04	7.5E–08	8.7E–02	1.4E+03	1.8E–07
$f_{36}(\boldsymbol{x})$	AB	**4.9E+01**	7.2E+03	**4.9E+01**	5.2E+02	1.3E+02	1.2E+05	7.2E+02	**4.9E+01**
	MD	**4.9E+01**	6.8E+03	**4.9E+01**	5.1E+02	1.3E+02	1.1E+05	7.3E+02	**4.9E+01**
	SD	**2.9E–10**	2.1E+03	3.5E+04	2.6E+02	2.0E+01	2.6E–01	4.1E+01	4.3E–09
$f_{37}(\boldsymbol{x})$	AB	**5.4E+01**	1.3E+02	5.4E+55	7.4E+02	4.3E+02	1.1E+03	3.3E+07	1.1E+03
	MD	**5.4E+01**	1.2E+02	6.8E+53	7.6E+02	4.3E+02	1.1E+03	3.3E+07	1.1E+03
	SD	**9.1E–09**	2.1E+01	2.3E+56	1.4E+02	8.4E+01	1.5E+02	8.9E+06	1.0E+02
$f_{38}(\boldsymbol{x})$	AB	**4.9E+01**	1.4E+06	8.7E+00	2.5E+03	**4.9E+01**	1.5E+02	6.3E+02	**4.9E+01**
	MD	**4.9E+01**	1.3E+06	8.8E+00	2.6E+03	**4.9E+01**	1.5E+02	6.4E+02	**4.9E+01**
	SD	**1.0E–14**	3.2E+05	8.9E–01	1.0E+03	3.3E+01	3.2E–01	5.2E+01	1.4E–06

Bold values represent the main results

In Table 3.13, the resulting indexes for composite functions in 30 dimensions are presented. According to Table 3.13, the hybrid method maintains superior performance in comparison to the other methods. Table 3.14 shows the performance results in the case of 50 dimensions. From the indicators presented in Table 3.14, it is clear that the hybrid algorithm gets the best indexes in all functions. However, its results are like those produced by other algorithms in some composite models. In the case

Table 3.15 Results of the composite benchmark functions (AIII) with 100 dimensions considering 160,000 function evaluations

Function		Hybrid	IWO	ABC	PSO	GSA	GA	HS	DE
$f_{35}(\boldsymbol{x})$	AB	**1.0E–23**	3.1E+02	7.9E+04	1.3E+02	3.3E–04	3.8E–01	2.3E+04	7.7E–08
	MD	**9.5E–24**	3.1E+02	8.0E+04	1.3E+02	1.7E–05	3.6E–01	2.3E+04	7.7E–08
	SD	**2.8E–24**	2.9E+01	7.7E+03	5.5E+01	7.1E–04	1.6E–01	3.8E+03	1.9E–08
$f_{36}(\boldsymbol{x})$	AB	**9.9E+01**	8.7E+04	1.1E+00	3.0E+04	2.7E+02	1.2E+02	1.7E+03	2.2E–26
	MD	**9.9E+01**	8.8E+04	1.1E+00	3.1E+04	2.7E+02	2.1E+02	1.7E+03	8.6E–27
	SD	**5.8E–07**	5.2E+03	2.8E–01	7.7E+03	2.6E+01	1.7E–01	1.0E+02	3.9E–06
$f_{37}(\boldsymbol{x})$	AB	**1.1E+02**	3.6E+02	7.2E+04	2.4E+02	1.4E+03	3.7E+03	1.6E+08	5.1E+03
	MD	**1.1E+02**	3.4E+01	7.4E+04	3.4E+01	1.4E+03	3.7E+03	1.6E+08	5.1E+03
	SD	**6.0E–06**	9.3E–05	9.1E+03	6.4E–02	2.3E+02	4.1E+02	3.1E+07	8.5E+02
$f_{38}(\boldsymbol{x})$	AB	**9.9E+01**	1.3E+04	**9.9E+01**	3.4E+04	3.0E+02	2.1E+01	1.5E+03	1.8E+02
	MD	**9.9E+01**	1.2E+04	**9.9E+01**	3.4E+04	3.0E+02	2.1E+01	1.5E+03	1.4E+02
	SD	**3.2E–14**	3.1E+03	1.4E–01	8.1E+03	4.4E+01	2.3E–01	1.1E+02	8.0E–03

Bold values represent the main results

of the function f_{36}, the ABC, DE, and the hybrid algorithm obtains the same results. On the other hand, under the function f_{38}, the methods GSA, DE, and the hybrid maintain the same performance. Finally, Table 3.15 shows the results for the case of composite functions in 100 dimensions. The information in the table demonstrates that the hybrid method obtains better results than the other optimization techniques, except for the case of the function f_{38} where its performance is similar to the produced by GSA. This remarkable behavior is attributed to the interesting capacities of the algorithm to balance exploration and exploitation through the integration of different evolutionary computing paradigms.

In order to statistically validate the results of Tables 3.13, 3.14, and 3.15, the Wilcoxon test is applied. It only analyzes the differences among the methods corresponding to the resulting AB data. Tables 3.16, 3.17, and 3.18 presents the Wilcoxon results for 30, 50, and 100 dimensions, respectively.

Table 3.16 p-values produced by the Wilcoxon test comparing Hybrid versus IWO, Hybrid versus ABC, Hybrid versus PSO, Hybrid versus GSA, Hybrid versus GA, Hybrid versus HS, and Hybrid versus DE over the AB values from Table 3.13 ($n = 30$)

Function	Hybrid versus IWO	Hybrid versus ABC	Hybrid versus PSO	Hybrid versus GSA	Hybrid versus GA	Hybrid versus HS	Hybrid versus DE
$f_{35}(\boldsymbol{x})$	3.0E–11▲	3.0E–11▲	3.0E–11▲	3.0E–11▲	3.0E–11▲	3.0E–11▲	3.0E–11▲
$f_{36}(\boldsymbol{x})$	3.0E–11▲	3.0E–11▲	5.6E–10▲	3.0E–11▲	3.0E–11▲	3.0E–11▲	3.0E–11▲
$f_{37}(\boldsymbol{x})$	3.0E–11▲	3.0E–11▲	3.0E–11▲	3.0E–11▲	3.0E–11▲	3.0E–11▲	3.0E–11▲
$f_{38}(\boldsymbol{x})$	1.2E–12▲	1.2E–12▲	1.2E–12▲	1.2E–12▲	1.2E–12▲	1.2E–12▲	1.2E–12▲

Table 3.17 p-values produced by the Wilcoxon test comparing Hybrid versus IWO, Hybrid versus ABC, Hybrid versus PSO, Hybrid versus GSA, Hybrid versus GA, Hybrid versus HS, and Hybrid versus DE over the AB values from Table 3.14 ($n = 50$)

Function	Hybrid versus IWO	Hybrid versus ABC	Hybrid versus PSO	Hybrid versus GSA	Hybrid versus GA	Hybrid versus HS	Hybrid versus DE
$f_{35}(\boldsymbol{x})$	3.0E–11▲	3.0E–11▲	3.0E–11▲	3.0E–11▲	3.0E–11▲	3.0E–11▲	3.0E–11▲
$f_{36}(\boldsymbol{x})$	3.0E–11▲	3.0E–11►	3.0E–11▲	3.0E–11▲	3.0E–11▲	3.0E–11▲	3.0E–11►
$f_{37}(\boldsymbol{x})$	3.0E–11▲	3.0E–11▲	3.0E–11▲	3.0E–11▲	3.0E–11▲	3.0E–11▲	3.0E–11▲
$f_{38}(\boldsymbol{x})$	1.2E–12▲	1.2E–12▲	1.2E–12▲	1.2E–12►	1.2E–12▲	1.2E–12▲	1.2E–12►

Table 3.18 p-values produced by the Wilcoxon test comparing Hybrid versus IWO, Hybrid versus ABC, Hybrid versus PSO, Hybrid versus GSA, Hybrid versus GA, Hybrid versus HS, and Hybrid versus DE over the AB values from Table 3.15 ($n = 100$)

Function	Hybrid versus IWO	Hybrid versus ABC	Hybrid versus PSO	Hybrid versus GSA	Hybrid versus GA	Hybrid versus HS	Hybrid versus DE
$f_{35}(\boldsymbol{x})$	3.0E–11▲	3.0E–11▲	2.4E–11▲	3.0E–11▲	3.0E–11▲	3.0E–11▲	3.0E–11▲
$f_{36}(\boldsymbol{x})$	3.0E–11▲	3.0E–11▲	4.0E–04▲	3.0E–11▲	3.0E–11▲	3.0E–11▲	3.0E–11▲
$f_{37}(\boldsymbol{x})$	3.0E–11▲	3.0E–11▲	3.0E–11▲	3.0E–11▲	3.0E–11▲	3.0E–11▲	3.0E–11▲
$f_{38}(\boldsymbol{x})$	1.0E–11▲	1.0E–11►	1.0E–11▲	1.0E–11▲	1.0E–11▲	1.0E–11▲	1.0E–11▲

Table 3.19 Unimodal test functions

Name	Function	D	Dim	Minimum
Sphere	$f_1(\mathbf{x}) = \sum_{i=1}^{n} x_i^2$	$[-5, 5]^n$	$n = 30$ $n = 50$ $n = 100$	$f(\mathbf{x}^*) = 0; \mathbf{x}^* = (0, \ldots, 0)$
Sum squares	$f_2(\mathbf{x}) = \sum_{i=1}^{n} i x_i^2$	$[-10, 10]^n$	$n = 30$ $n = 50$ $n = 100$	$f(\mathbf{x}^*) = 0; \mathbf{x}^* = (0, \ldots, 0)$
Sum of different powers	$f_3(\mathbf{x}) = \sum_{i=1}^{n} \lvert x_i \rvert^{i+1}$	$[-1, 1]^n$	$n = 30$ $n = 50$ $n = 100$	$f(\mathbf{x}^*) = 0; \mathbf{x}^* = (0, \ldots, 0)$
Schwefel 2	$f_4(\mathbf{x}) = \sum_{i=1}^{n} \left(\sum_{j=1}^{i} x_i\right)^2$	$[-100, 100]^n$	$n = 30$ $n = 50$ $n = 100$	$f(\mathbf{x}^*) = 0; \mathbf{x}^* = (0, \ldots, 0)$
Rotated hyper-ellipsoid	$f_5(\mathbf{x}) = \sum_{i=1}^{n} \sum_{j=1}^{i} x_j^2$	$[-65.5, 65.5]^n$	$n = 30$ $n = 50$ $n = 100$	$f(\mathbf{x}^*) = 0; \mathbf{x}^* = (0, \ldots, 0)$

Table 3.20 Multimodal test functions

Name	Function	D	Dims	Minimum
Ackley	$f_6(\mathbf{x}) = -20e^{-0.2\sqrt{\frac{1}{n}\sum_{i=1}^{n} x_i^2}} - e^{\frac{1}{n}\sum_{i=1}^{n}\cos(2\pi x_i)} + 20 + e$	$[-30, 30]^n$	$n = 30$ $n = 50$ $n = 100$	$f(\mathbf{x}^*) = 0; \mathbf{x}^* = (0, \ldots, 0)$
Powell	$f_{754}(\mathbf{x}) = \sum_{i=1}^{n/4}\left[(x_{4i-3} + 10x_{4i-2})^2 + 5(x_{4i-1} - x_{4i})^2 + (x_{4i-2} - 2x_{4i-1})^4 + 10(x_{4i-3} - x_{4i})^4\right]$	$[-4, 5]^n$	$n = 30$ $n = 50$ $n = 100$	$f(\mathbf{x}^*) = 0; \mathbf{x}^* = (0, \ldots, 0)$
Levy	$f_8(\mathbf{x}) = \sin^2(\pi y_1) + \sum_{i=1}^{n-1}(y_i - 1)^2\left[1 + 10\sin^2(\pi y_i + 1)\right] + (y_n - 1)^2\left[1 + \sin^2(2\pi y_n)\right];$ $y_i = 1 + \left(\frac{x_i + 1}{4}\right)$	$[-10, 10]^n$	$n = 30$ $n = 50$ $n = 100$	$f(\mathbf{x}^*) = 0; \mathbf{x}^* = (1, \ldots, 1)$
Rosenbrock	$f_9(\mathbf{x}) = \sum_{i=1}^{n-1}\left[100(x_{i+1} - x_i^2)^2 + (x_i - 1)^2\right]$	$[-5, 10]^n$	$n = 30$ $n = 50$ $n = 100$	$f(\mathbf{x}^*) = 0; \mathbf{x}^* = (1, \ldots, 1)$

(continued)

Table 3.20 (continued)

Name	Function	D	Dims	Minimum
Schwefel 22	$f_{10}(\mathbf{x}) = \sum_{i=1}^{n} \lvert x_i \rvert + \prod_{i=1}^{n} \lvert x_i \rvert$	$[-100, 100]^n$	$n = 30$ $n = 50$ $n = 100$	$f(\mathbf{x}^*) = 0; \mathbf{x}^* = (0, \ldots, 0)$
Schwefel 26A	$f_{11}(\mathbf{x}) = 418.9829n - \sum_{i=1}^{n} x_i \sin(\sqrt{\lvert x_i \rvert})$	$[-500, 500]^n$	$n = 30$ $n = 50$ $n = 100$	$f(\mathbf{x}^*) = 0;$ $\mathbf{x}^* = (420.96, \ldots, 420.96)$
Perm 2	$f_{12}(\mathbf{x}) = \sum_{i=1}^{n} \left[\sum_{j=1}^{n} (j + \beta) \left(x_j{}^i - \frac{1}{j} \right) \right]^2$	$[-n, n]^n$	$n = 30$ $n = 50$ $n = 100$	$f(\mathbf{x}^*) = 0;$ $\mathbf{x}^* = \left(1, \frac{1}{2}, \ldots, \frac{1}{n}\right)$
Dixon Price	$f_{13}(\mathbf{x}) = (x_i - 1)^2 + \sum_{i=2}^{n} i(2x_i^2 - x_{i-1})^2$	$[-10, 10]^n$	$n = 30$ $n = 50$ $n = 100$	$f(\mathbf{x}^*) = 0; \mathbf{x}^* = 2^{-\frac{2^i - 2}{2^i}}$ for $i = 1, \ldots, n$

(continued)

Table 3.20 (continued)

Name	Function	D	Dims	Minimum
Zakharov	$f_{14}(\mathbf{x}) = \sum_{i=1}^{n} x_i^2 + \left(\sum_{i=1}^{n} 0.5ix_i\right)^2 + \left(\sum_{n=1}^{n} 0.5ix_i\right)^4$	$[-5, 10]^n$	$n =$ 30 $n =$ 50 $n =$ 100	$f(\mathbf{x}^*) = 0$; $\mathbf{x}^* = (0, \ldots, 0)$
Styblinski Tang	$f_{15}(\mathbf{x}) = \frac{1}{2} \sum_{i=1}^{n} \left(x_i^4 - 16x_i^2 + 5x_i\right)$	$[-5, 5]^n$	$n =$ 30 $n =$ 50 $n =$ 100	$f(\mathbf{x}^*) = -39.16599n$; $\mathbf{x}^* = (-2.903, \ldots, 2.903)$
Step	$f_{16}(\mathbf{x}) = \sum_{i=1}^{n} (\lfloor x_i \rfloor + 0.5)^2$	$[-100, 100]^n$	$n =$ 30 $n =$ 50 $n =$ 100	$f(\mathbf{x}^*) = 0$; $\mathbf{x}^* = (0.5, \ldots, 0.5)$
Quartic	$f_{17}(\mathbf{x}) = \sum_{i=1}^{n} \left\{(ix_i)^4 + rand[0, 1)\right\}$	$[-1.28, 1.28]^n$	$n =$ 30 $n =$ 50 $n =$ 100	$f(\mathbf{x}^*) = 0$; $\mathbf{x}^* = (0, \ldots, 0)$

(continued)

Table 3.20 (continued)

Name	Function	D	Dims	Minimum
Salomon	$f_{18}(\mathbf{x}) = 1 - \cos\left(2\pi\sqrt{\sum_{i=1}^{n} x_i^2}\right) + 0.1\sqrt{\sum_{i=1}^{n} x_i^2}$	$[-100, 100]^n$	$n =$ 30 $n =$ 50 $n =$ 100	$f(\mathbf{x}^*) = 0; \mathbf{x}^* = (0, \ldots, 0)$
Penalty 1A	$f_{19}(\mathbf{x}) = \frac{\pi}{30}\left\{\begin{array}{c} 10\sin^2(\pi y_1) \\ + \sum_{i=1}^{n-1} (y_i - 1)^2\left[1 + 10\sin^2(\pi y_i + 1)\right] \\ +(y_n - 1)^2 \end{array}\right\} + \sum_{i=1}^{n} u(x_i, 10, 100, 4);$ $y_i = 1 + \frac{x_i + 1}{4};$ $u(x_i, a, k, m) = \begin{cases} k(x_i - a)^m, x_i > a \\ 0, -a \le x_i \le a \\ k(-x_i - a)^m, x_i < -a \end{cases}$	$[-50, 50]^n$	$n =$ 30 $n =$ 50 $n =$ 100	$f(\mathbf{x}^*) = 0;$ $\mathbf{x}^* = (-1, \ldots, -1)$

(continued)

Table 3.20 (continued)

Name	Function	D	Dims	Minimum
Penalty 2A	$f_{20}(\mathbf{x}) = 0.1\left\{\begin{array}{c}\sin^2(3\pi x_1) \\ +\sum_{i=1}^{n-1}(x_i-1)^2\left[1+\sin^2(3\pi x_{i+1})\right] \\ +(x_n-1)^2\left[1+\sin^2(2\pi x_n)\right]\end{array}\right\} + \sum_{i=1}^{n} u(x_i, 5, 100, 4)$; $u(x_i, a, k, m) = \begin{cases} k(x_i-a)^m, x_i > a \\ 0, -a \le x_i \le a \\ k(-x_i-a)^m, x_i < -a \end{cases}$	$[-50, 50]^n$	$n =$ 30 $n =$ 50 $n =$ 100	$f(\mathbf{x}^*) = 0$; $\mathbf{x}^* = (1, \ldots, 1)$
Rastrigin A	$f_{21}(\mathbf{x}) = 10n + \sum_{i=1}^{n}\left[{x_i}^2 - 10\cos(2\pi x_i)\right]$	$[-5.12, 5.12]^n$	$n =$ 30 $n =$ 50 $n =$ 100	$f(\mathbf{x}^*) = 0$; $\mathbf{x}^* = (0, \ldots, 0)$
Mishra 1	$f_{22}(\mathbf{x}) = (1 + x_n)^{x_n}$; $x_n = n - \sum_{i=1}^{n-1} x_i$	$[0, 1]^n$	$n =$ 30 $n =$ 50 $n =$ 100	$f(\mathbf{x}^*) = 2$; $\mathbf{x}^* = (1, \ldots, 1)$

(continued)

Table 3.20 (continued)

Name	Function	D	Dims	Minimum
Mishra 2	$f_{23}(\mathbf{x}) = (1 + x_n)^{x_n}; x_n = n - \sum_{i=1}^{n-1} \frac{(x_i + x_{i+1})}{2}$	$[0, 1]^n$	$n = 30$ $n = 50$ $n = 100$	$f(\mathbf{x}^*) = 2; \mathbf{x}^* = (1, \ldots, 1)$
Mishra 11	$f_{24}(\mathbf{x}) = \left[\frac{1}{n}\sum_{i=1}^{n} \lvert x_i\rvert - \left(\prod_{i=1}^{n} \lvert x_i\rvert\right)^{\frac{1}{n}}\right]^2$	$[-10, 10]^n$	$n = 30$ $n = 50$ $n = 100$	$f(\mathbf{x}^*) = 0; \mathbf{x}^* = (0, \ldots, 0)$
Multimodal	$f_{25}(\mathbf{x}) = \sum_{i=1}^{n} \lvert x_i\rvert \prod_{i=1}^{n} \lvert x_i\rvert$	$[-10, 10]^n$	$n = 30$ $n = 50$ $n = 100$	$f(\mathbf{x}^*) = 0; \mathbf{x}^* = (0, \ldots, 0)$
Qing	$f_{26}(\mathbf{x}) = \sum_{i=1}^{n} \left(x_i^2 - i\right)^2$	$[-500, 500]^n$	$n = 30$ $n = 50$ $n = 100$	$f(\mathbf{x}^*) = 0;$ $\mathbf{x}^* = \left(\pm\sqrt{i}, \ldots, \pm\sqrt{i}\right)$

(continued)

Table 3.20 (continued)

Name	Function	D	Dims	Minimum
Quintic	$f_{27}(\mathbf{x}) = \sum_{i=1}^{n}\left(\left\|x_i^5 - x_i^4 + 4x_i^3 + 2x_i^2 - 10x_i - 4\right\|\right)$	$[-10, 10]^n$	$n = 30$ $n = 50$ $n = 100$	$f(\mathbf{x}^*) = 0$; $\mathbf{x}^* = (-1, \ldots, -1)$
Vincent	$f_{28}(\mathbf{x}) = -\sum_{i=1}^{n} \sin(10\log x_i)$	$[0.25, 10]^n$	$n = 30$ $n = 50$ $n = 100$	$f(\mathbf{x}^*) = -n$; $\mathbf{x}^* = (7.7062, \ldots, 7.7062)$
Infinity	$f_{29}(\mathbf{x}) = \sum_{i=1}^{n} x_i^6\left[\sin\left(\frac{1}{x_i}\right) + 2\right]$	$[-1, 1]^n$	$n = 30$ $n = 50$ $n = 100$	$f(\mathbf{x}^*) = 0$; $\mathbf{x}^* = (0, \ldots, 0)$
Griewank	$f_{30}(\mathbf{x}) = \frac{1}{4000}\sum_{i=1}^{n} {x_i}^2 - \prod_{i=1}^{n}\cos\left(\frac{x_i}{\sqrt{i}}\right) + 1$	$[-600, 600]^n$	$n = 30$ $n = 50$ $n = 100$	$f(\mathbf{x}^*) = 0$; $\mathbf{x}^* = (0, \ldots, 0)$

(continued)

Table 3.20 (continued)

Name	Function	D	Dims	Minimum
Schwefel 26B	$f_{31}(\mathbf{x}) = \sum_{i=1}^{n} -x_i \sin\left(\sqrt{\lvert x_i \rvert}\right)$	$[-500, 500]^n$	$n = 30$ $n = 50$ $n = 100$	$f(\mathbf{x}^*) = -418.9829n$; $\mathbf{x}^* = (-300, \ldots, 300)$
Rastrigin B	$f_{32}(\mathbf{x}) = \sum_{i=1}^{n}\left[{x_i}^2 - 10\cos(2\pi x_i) + 10\right]$	$[-5.12, 5.12]^n$	$n = 30$ $n = 50$ $n = 100$	$f(\mathbf{x}^*) = 0$; $\mathbf{x}^* = (-2, \ldots, -2)$
Penalty 1B	$f_{33}(\mathbf{x}) = \frac{\pi}{n}\left\{10\sin(\pi y_1) + \sum_{i=1}^{n-1}(y_i - 1)^2\left[1 + 10\sin^2(\pi y_i + 1)\right] + (y_n - 1)^2\right\} + \sum_{i=1}^{n} u(x_i, 10, 100, 4)$ $y_i = 1 + \frac{x_i + 1}{4}$ $u(x_i, a, k, m) = \begin{cases} k(x_i - a)^m, x_i > a \\ 0, -a < x_i < a \\ k(-x_i - a)^m, x_i < -a \end{cases}$	$[-50, 50]^n$	$n = 30$ $n = 50$ $n = 100$	$f(\mathbf{x}^*) = 0$; $\mathbf{x}^* = (-30, \ldots, -30)$

(continued)

Table 3.20 (continued)

Name	Function	D	Dims	Minimum
Penalty 2B	$f_{34}(\mathbf{x}) = 0.1\{\sin^2(3\pi x_i) + \sum_{i=1}^{n}(x_i - 1)^2[1 + \sin^2(3\pi x_i + 1)] + (x_n - 1)^2[1 + \sin^2(2\pi x_n)]\} + \sum_{i=1\vert}^{n} u(x_i, 5, 100, 4)$	$[-50, 50]^n$	$n = 30$ $n = 50$ $n = 100$	$f(\mathbf{x}^*) = 0$; $\mathbf{x}^* = (-100, \ldots, 100)$

Table 3.21 Composite test functions considered in the experimental study

Name	Function	D	Dims	Minimum
Fx35	$f_{35}(\mathbf{x}) = f_1(\mathbf{x}) + f_{10}(\mathbf{x}) + f_{21}(\mathbf{x})$	$[-100, 100]^n$	$n = 30$ $n = 50$ $n = 100$	$f(\mathbf{x}^*) = 0$; $\mathbf{x}^* = (0, \ldots, 0)$
Fx36	$f_{36}(\mathbf{x}) = f_9(\mathbf{x}) + f_{21}(\mathbf{x}) + f_{30}(\mathbf{x})$	$[-100, 100]^n$	$n = 30$ $n = 50$ $n = 100$	$f(\mathbf{x}^*) = n - 1$; $\mathbf{x}^* = (0, \ldots, 0)$
Fx37	$f_{37}(\mathbf{x}) = f_4(\mathbf{x}) + f_6(\mathbf{x}) + f_9(\mathbf{x}) + f_{20}(\mathbf{x})$	$[-100, 100]^n$	$n = 30$ $n = 50$ $n = 100$	$f(\mathbf{x}^*) = (1.1n) - 1$; $\mathbf{x}^* = (0, \ldots, 0)$
Fx38	$f_{38}(\mathbf{x}) = f_6(\mathbf{x}) + f_9(\mathbf{x}) + f_{10}(\mathbf{x}) + f_{21}(\mathbf{x}) + f_{30}(\mathbf{x})$	$[-100, 100]^n$	$n = 30$ $n = 50$ $n = 100$	$f(\mathbf{x}^*) = n - 1$; $\mathbf{x}^* = (0, \ldots, 0)$

Table 3.16 presents the Wilcoxon results for composite functions in the case of 30 dimensions. According to the information in the table, the hybrid method maintains the best performance in all functions. Additionally, Table 3.17 shows the *p*-values of composite functions in 50 dimensions. They demonstrate that the hybrid method obtains the best values in all functions. However, its results are similar to those produced by other algorithms in some composite models. In the case of the function f_{36}, the ABC, DE, and the hybrid algorithm obtains the same results. On the other hand, under the function f_{38}, the methods GSA, DE, and the hybrid maintain the same performance. Finally, Table 3.18 exhibits the resulting p-values after applying the Wilcoxon test under the operation of composite functions in 100 dimensions. The results statistically demonstrate that the hybrid method is better than the other algorithms, except for the case of the function f_{38} where its performance is similar to the produced by GSA.

3.6.4 Benchmark Functions

Tables 3.19, 3.20, and 3.21 describe the benchmark functions used in the experiments, where $f(\mathbf{x}^*)$ is the optimum value of the function, $\mathbf{x}^*$ the optimum position and $\mathbf{D}$ the search space (a subset of $\mathbb{R}^n$).

3.6.5 Convergence Evaluation

The analysis of the final solution cannot entirely define the effectiveness of a search strategy in an optimization algorithm. Therefore, in this part, the convergence of the analyzed methods is examined. The objective of this test is to assess the velocity with which each optimization algorithm attains its best solution.

In the analysis, the evolution of the optimization process for each algorithm on a particular function is evaluated. The test considers the behavior of each method when it solves the function in 100 dimensions. Six distinct functions have been selected for the study: two unimodal functions (f_1, f_2), two multimodal functions (f_{29}, f_{30}), and two composite functions (f_{35}, f_{37}).

In the experiment, due to its complexity, the optimization of functions in 100 dimensions has been selected. Figure 3.4 shows the evolution curves of all algorithms when they optimize the unimodal functions f_1 and f_1, considering 100 dimensions. In Fig. 3.5, the convergence results for multimodal functions f_{29} and f_{30} are presented. Finally, in Fig. 3.6, the convergence graphs for composite functions f_{35} and f_{37} are exhibit. From the tables, the hybrid method converges faster than IWO, ABC, PSO, GSA, GA, HS, and DE. On the other hand, the other methods maintain evident difficulties to appropriately converge. Most of them, in several executions, never reach an acceptable solution during the evolution process, as can be seen in Figs. 3.4, 3.5, and 3.6.

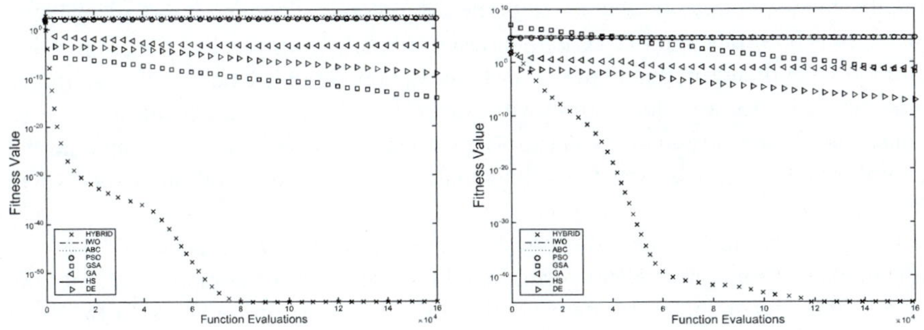

(a) (b)

Fig. 3.4 Convergence test results for functions. **a** f_1 and **b** f_2 in 100 dimensions

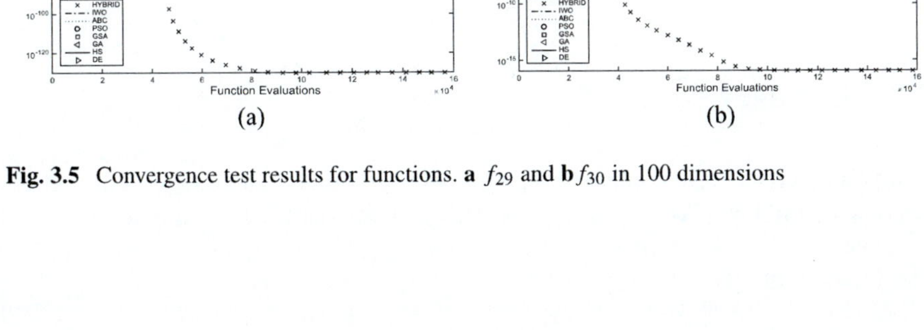

(a) (b)

Fig. 3.5 Convergence test results for functions. **a** f_{29} and **b** f_{30} in 100 dimensions

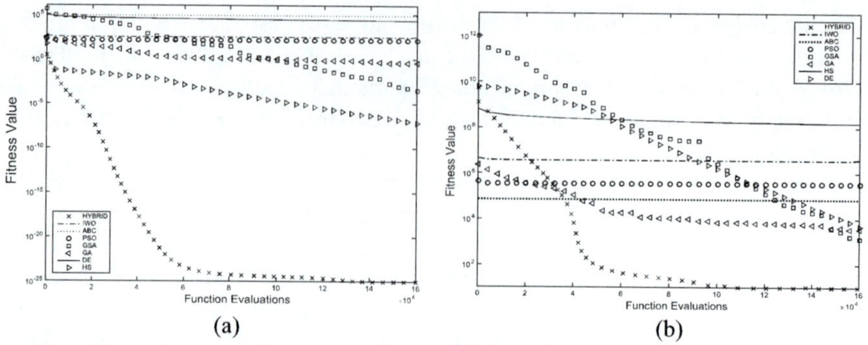

Fig. 3.6 Convergence test results for functions. **a** f_{35} and **b** f_{37} in 100 dimensions

3.6.6 Computational Complexity

Finally, in this section, the evaluation of the computational complexity of all analyzed algorithms is conducted. EC techniques are, essentially, complicated procedures that contain several deterministic rules and stochastic operations. Under such conditions, it is impossible to carry out a complexity analysis from a deterministic perspective. In consequence, the computational effort (CE) is utilized to estimate the computational complexity. CE expresses the CPU time consumed by an optimization method when it is executed.

To calculate the CE, the test suggested in [30] has been considered. In the experiment, the computational effort is computed through the evaluation of different time measurements conducted in the execution of a known function (f_{24}). In the analysis, all optimization algorithms have been operated in MATLAB, considering an HP computer with a Core i7 processor, running Windows 10 operating system with 8 Gb of memory.

The resulting CE values for IWO, ABC, PSO, GSA, GA, HS, DE, and the Hybrid method are 52.88, 75.21, 37.89, 61.21, 50.97, 72.10, 67.24, and 51.07 respectively. A smaller CE value means that the algorithm maintains a low complexity. This fact indicates that the algorithm reaches a faster execution velocity under the same conditions. The resulting CE values reveal that although the hybrid method is slightly more complicated than PSO and GA, their CE values are relatively similar. Furthermore, the hybrid method is significantly less computationally complex than HS, IWO, ABC, DE, and GSA.

3.7 Summary

In this chapter, a hybrid method for solving optimization problems is presented. The approach combines (A) the explorative characteristics of the IWO method, (B) the probabilistic models of EDA, and (C) the dispersion capacities of a mixed Gaussian-Cauchy distribution to produce its own search strategy. With these mechanisms, the presented method conducts an optimization strategy over search areas that deserve a special interest according to a probabilistic model and the fitness value of the existent solutions.

In the presented method, each individual of the population generates new elements around its own location, dispersed according to the mixed distribution. The number of new elements depends on the relative fitness value of the individual regarding the complete population. After this process, a group of promising solutions is selected from the set compound by the (1) new elements and the (2) original individuals. Based on the chosen solutions, a probabilistic model is built from which a certain number of members (3) is sampled. Then, all the individuals of the sets (1), (2), and (3) are joined in a single group and ranked in terms of their fitness values. Finally, the best elements of the group are selected to replace the original population. This process is repeated until a termination criterion has been reached.

The performance of the hybrid method has been analyzed, facing seven popular EC methods utilizing a set of 38 benchmark functions. The results have been statistically validated within a non-parametric framework to eradicate the random effect. Experimental results demonstrate that the hybrid algorithm can produce better and consistent solutions over its competitors.

References

1. Alba E, Dorronsoro B (2005) The exploration/exploitation tradeoff in dynamic cellular genetic algorithms. IEEE Trans Evol Comput 9(3):126–142
2. Bäck T (1996) Evolutionary algorithms in theory and practice. Oxford Univ. Press, New York
3. Basak A, Maity D, Das S (2013) A differential invasive weed optimization algorithm for improved global numerical optimization. Appl Math Comput 219:6645–6668
4. Beigvand D, Abdi H, La Scala M (2017) Hybrid gravitational search algorithm-particle swarm optimization with time-varying acceleration coefficients for large scale CHPED problem. Energy 126(2017):841–853
5. Blum C, Puchinger J, Raidl G, Roli A (2011) Hybrid metaheuristics in combinatorial optimization: a survey. Applied Soft Computing 11:4135–4151
6. Blum C, Blesa MJ, Roli A, Sampels M (2008) Hybrid metaheuristics—An emerging approach to optimization, vol 114 of studies in computational intelligence. Springer
7. Chellapilla K (1998) Combining mutation operators in evolutionary programming. IEEE Trans Evol Comput 2(3):91–96
8. Chen C-H, Chen YP (2007) Real-coded ECGA for economic dispatch. In: Genetic and Evolutionary Computation Conference, GECCO-2007, pp 1920–1927
9. Cuevas E, Cienfuegos M, Zaldívar D, Pérez-Cisneros M (2013) A swarm optimization algorithm inspired in the behavior of the social-spider. Expert Syst Appl 40(16):6374–6384
10. Cuevas E, Echavarría A, Ramírez-Ortegón M (2014) An optimization algorithm inspired by the States of Matter that improves the balance between exploration and exploitation. Appl Intell 40(2):256–272
11. Cuevas E, Gálvez J, Hinojosa S, Avalos O, Zaldívar D, Pérezcisneros MA (2014) (2014) Comparison of evolutionary computation techniques for IIR model identification. J Appl Math 2014:827206

12. Cuevas E, González M, Zaldivar D, Pérez-Cisneros M, García G (2012) An algorithm for global optimization inspired by collective animal behaviour. Discr Dyn Nat Soc art no 638275
13. Díaz P, Pérez-Cisneros M, Cuevas E, Hinojosa S, Zaldivar D (2018) An improved crow search algorithm applied to energy problems. Energies 11(3):571
14. Ducheyne E, DeBaets B, De Wulf R (2004). Probabilistic models for linkage learning in forest management. In: Jin Y (ed) Knowledge incorporation in evolutionary computation, Springer, pp 177–194
15. Ehrgott M, Gandibleux X (2008) Hybrid Metaheuristics for multi-objective combinatorial optimization, vol 114 of Blum et al. [14], pp 221–259 (Chapter 8)
16. Garcia S, Molina D, Lozano M, Herrera F (2008) A study on the use of non-parametric tests for analyzing the evolutionary algorithms' behavior: a case study on the CEC'2005 Special session on real parameter optimization. J Heurist. https://doi.org/10.1007/s10732-008-9080-4
17. Garg H (2016) A hybrid PSO-GA algorithm for constrained optimization problems. Appl Math Comput 274:292–305
18. Geem ZW, Kim JH, Loganathan GV (2001) A new heuristic optimization algorithm: harmony search. Simulations 7:60–68
19. Goldberg DE, Korb B, Deb K (1989) Messy genetic algorithms: motivation, analysis, and first results. Complex Syst. 3(5):493–530
20. Grosan C, Abraham A (2007) Hybrid evolutionary algorithms: methodologies, architectures, and reviews. Stud Comput Intell (SCI) 75:1–17
21. Han M, Liu Ch, Xing J (2014) An evolutionary membrane algorithm for global numerical optimization problems. Inf Sci 276:219–241
22. Holland JH (1975) Adaptation in natural and artificial systems. University of Michigan Press, Michigan
23. Holland JH (1992) Adaptation in natural and artificial systems: an introductory analysis with applications to biology control and artificial intelligence. MIT Press, Cambridge, MA, USA. ISBN 0262082136
24. Yu JJQ, Li VOK (2015) A social spider algorithm for global optimization. Appl Soft Comput 30:614–627
25. Ji Y, Zhang K-C, Qu S-J (2007) A deterministic global optimization algorithm. Appl Math Comput 185:382–387
26. Karaboga D (2005) An idea based on honeybee swarm for numerical optimization. Computer Engineering Department and Engineering Faculty, Erciyes University, Talas
27. Kennedy J, Eberhart RC (1995) Particle swarm optimization. In: Proceedings of IEEE international conference on neural networks, vol 4, pp 1942–1948
28. Li D, Zhao H, Weng XW, Han T (2016) A novel nature-inspired algorithm for optimization: Virus colony search. Adv Eng Softw 92:65–88
29. Li Z, Wang W, Yan Y, Li Z (2015) PS–ABC: A hybrid algorithm based on particle swarm and artificial bee colony for high-dimensional optimization problems. Expert Syst Appl 42(22):8881–8895
30. Liang JJ, Qu B-Y, Suganthan PN (2015) Problem definitions and evaluation criteria for the CEC 2015 special session and competition on single objective real parameter numerical optimization. Technical report 201311. Computational Intelligence Laboratory, Zhengzhou University, Zhengzhou China, and Nanyang Technological University, Singapore
31. Lipinski P (2007) ECGA versus BOA in discovering stock market trading experts. In: Genetic and evolutionary computation conference, GECCO-2007, pp 531–538
32. Mallahzadeh AR, Es'haghi S, Alipour A (2009) Design of an E-shaped MIMO antenna using the IWO algorithm for wireless application at 5.8 GHz. Progr Electromag Res PIER 90:187–203
33. Mehrabian AR, Lucas C (2006) A novel numerical optimization algorithm inspired from weed colonization. Ecol Inf 1:355–366
34. Mehrabian AR, Yousefi-Koma A (2007) Optimal positioning of piezoelectric actuators on a smart fin using bio-inspired algorithms. Aerosp Sci Technol 11:174–182
35. Meng Z, Jeng-Shyang P (2016) Monkey king evolution: a new memetic evolutionary algorithm and its application in vehicle fuel consumption optimization. Knowl-Based Syst 97:144–157

36. Mühlenbein H, Paaß GH (1996) From recombination of genes to the estimation of distributions I. Binary parameters. In: Eiben A, Bäck T, Shoenauer M, Schwefel H (eds) Parallel problem solving from nature. Springer, Verlag, Berlin, pp 178–187
37. Mühlenbein H, Schlierkamp-Voosen D (1993) Predictive models for the breeder genetic algorithm I Continuous Parameter Optimization. Evol. Comput. 1(1):25–49
38. Ou-Yang C, Utamima A (2013) Hybrid estimation of distribution algorithm for solving single row facility layout problem. Comput Ind Eng 66:95–103
39. Paenke I, Jin Y, Branke J (2009) Balancing population- and individual-level adaptation in changing environments. Adapt Behav 17(2):153–174
40. Pardalos PM, Romeijn HE, Tuy H (2000) Recent developments and trends in global optimization. J Comput Appl Math 124:209–228
41. Rashedi E, Nezamabadi-pour H, Saryazdi S (2009) GSA: a gravitational search algorithm. Inf Sci 179:2232–2248
42. Rudolph (1997) Local convergence rates of simple evolutionary algorithms with Cauchy mutations. IEEE Trans Evol Comput 1:249–258
43. Santana R, Larrañaga P, Lozano JA (2008) Protein folding in simplified models with estimation of distribution algorithms . IEEE Evol Comput 12:418–438
44. Storn R, Price K (1995) Differential evolution—A simple and efficient adaptive scheme for global optimization over continuous spaces. Technical report TR-95–012. ICSI, Berkeley, CA
45. Tan KC, Chiam SC, Mamun AA, Goh CK (2009) Balancing exploration and exploitation with an adaptive variation for evolutionary multi-objective optimization. Eur J Oper Res 197:701–713
46. Trivedi A, Srinivasan D, Biswas S, Reindl T (2016) A genetic algorithm—differential evolution based hybrid framework: a case study on unit commitment scheduling problem. Inf Sci 354:275–300
47. Wang Y, Li B (2009) A self-adaptive mixed distribution based uni-variate estimation of distribution algorithm for large scale global optimization, nature-inspired algorithms for optimisation. Chiong R (ed) Studies in computational intelligence, vol 193, pp 171–198
48. Wilcoxon F (1945) Individual comparisons by ranking methods. Biometrics 1:80–83
49. Wolpert DH, Macready WG (1997) No free lunch theorems for optimization. IEEE Trans Evol Comput 1(1):67–82
50. Yang XS (2009) Firefly algorithms for multimodal optimization, in Stochastic algorithms: foundations and applications. In: SAGA 2009 lecture notes in computer sciences, vol 5792, pp 169–178
51. Yang X-S (2010) Engineering optimization: an introduction with metaheuristic application. Wiley, USA
52. Yao X, Liu Y (1996) Fast evolutionary programming. In: Fogel LJ, Angeline PJ, Back T (eds) Proceedings of fifth annual conf. evolutionary programming (EP'96). MIT Press, Cambridge, MA, pp 451–460
53. Yao X, Liu Y, Liu G (1999) Evolutionary programming made faster. IEEE Trans Evol Comput 3(2):82–102
54. Yu T-L, Santarelli S, Goldberg DE (2006) Military antenna design using a simple genetic algorithm and hBOA. In: Pelikan M, Sastry K, Cantú-Paz E (eds) Scalable optimization via probabilistic modeling: from algorithms to applications. Springer, pp 275–289
55. Zhang J, Wu Y, Guo Y, Wang Bo, Wang H, Liu H (2016) A hybrid harmony search algorithm with differential evolution for day-ahead scheduling problem of a microgrid with consideration of power flow constraints. Appl Energy 183:791–804
56. Zhang X, Wang Y, Cui G, Niu Y, Xu J (2009) Application of a novel IWO to the design of encoding sequences for DNA computing. Comput Math Appl 57:2001–2008

Chapter 4
Corner Detection Algorithm Based on Cellular Neural Networks (CNN) and Differential Evolution (DE)

Corner detection represents one of the most important steps to identify features in images. Due to their powerful local processing capabilities, Cellular Nonlinear/Neural Networks (CNN) are commonly utilized in image processing applications such as image edge detection, image encoding and image hole filling. CNN perform well for locating corner features in binary images. However, their use in grayscale images has not been considered due to their design difficulties. In this chapter, a corner detector based on CNN for grayscale images is presented. In the approach, the original processing scheme of the CNN is modified to include a nonlinear operation for increasing the contrast of the local information in the image. With this adaptation, the final CNN parameters that allow the appropriate detection of corner points are estimated through the Differential evolution algorithm by using standard training images. Different test images have been used to evaluate the performance of the presented corner detector. Its results are also compared with popular corner methods from the literature. Computational simulations demonstrate that the presented CNN approach presents competitive results in comparison with other algorithms in terms of accuracy and robustness.

4.1 Introduction

In computer vision, visual features play a significant role. They refer to structures which define specific characteristics of the image [1]. According to their characterization level, visual features can be classified into two classes [2]: interest points and descriptors. An interest point represents the position of a local maximum with regard to any image function such as "cornerness", "curvature", etc. On the other hand, a descriptor is a vector of values that characterize an image region around an interest point. Such values involve information which could be as simple as the neighbor

© The Author(s), under exclusive license to Springer Nature Switzerland AG 2021

E. Cuevas et al., *Recent Metaheuristic Computation Schemes in Engineering*,
Studies in Computational Intelligence 948,
https://doi.org/10.1007/978-3-030-66007-9_4

pixel values, or it could be more complex, such as the histogram of local gradient orientations.

In the literature, corners are considered the most common interesting points in digital images [3]. Since corners are robust and invariant in several context, they are useful in several image applications such as pattern recognition [4], image stitching [5], 3D reconstruction [6], to name a few. A corner can be defined as the point at which two different edge directions occur in the local neighborhood. Several corner detectors have been proposed in the literature. These detectors can be broadly divided into three schemes [7]: gradient-based methods, template-based methods, and contour-based methods. Gradient-based approaches detect corners at locations where a significant grayscale variation is found in all directions. The most representative algorithms of this scheme include the method suggested by Harris and Stephens [8] who introduced a rotation invariant corner detector, along with the algorithm presented by Mikolajczyk and Schmid [9] to identify corners features invariant to some image transformations. In gradient-based techniques, the computation of the first and the second-order derivatives is used as the main process. As a result, they are computationally expensive and very susceptible to noise. Template-based methods find corners by comparing the intensity of neighboring pixels with a predefined template. One of the most typical approaches from this scheme is the SUSAN algorithm [10]. The SUSAN method applies a mask, called 'USAN', for comparing the intensity corresponding to the center of the mask, with the greyscale value of its neighbors within the template. Therefore, corner points correspond to those points that represent the local minima in the USAN map. Despite the relatively low computational complexity, template-based schemes are prone to noisy information and present an inadequate performance in texturized images. Contour-based schemes are based on the curvature presented in the segments of the image edge map. These methods smooth the edge-segments with Gaussian functions at different scales. Subsequently, they determine the curvature on each point of the smoothed curves. The positions with the maximal values are considered potential corner points. Some examples of these methods involve the absolute curvature scale space (ACSS) [11], the fast corner detector based on the chord-to point distance accumulation (F-CPDA) [12], the corner detection and classification using anisotropic directional derivative representations (DCAR) [13] and the corner method via angle difference of principal directions of anisotropic Gaussian derivatives (ANDD) [14]. In general, contour-based schemes present a low probability of detecting false corners. However, contour-based detectors experiment three principal problems. First, the estimation of the curvature is highly susceptible to the local variation and noise on the edge segment. Second, for the determination of the curvature value, it is required the computation of higher order derivatives in edge point-locations that produce errors and unstable results. Third, contour-based schemes require appropriate Gaussian smoothing-scale selection. This is considered a difficult task since inappropriate Gaussian scales deliver a poor corner detection performance.

Different to interesting points, descriptors (or blob detectors) characterize invariant image regions in an image [2]. These approaches maintain an abundant amount of works in the literature. The most popular descriptors include the Scale

Invariant Feature Transform (SIFT) [15] and the Speed Up Robust Feature (SURF) [16]. Recently, several techniques and concepts of machine learning have been incorporated to descriptors to improve their performance producing interesting approaches such as Learned Invariant Feature Transform (LIFT) [17], The FAST method [18] and others [19]. In this work, the presented algorithm is conceived as an interest point detector. For this reason, its results and conclusions are related and compared with similar methods (corner detectors) and not with descriptors.

On the other hand, Cellular neural network (CNN) [20, 21] represents a parallel computational scheme which aggregates a set of identical processing units. In this scheme, each unit corresponds to a nonlinear dynamic system that is locally connected with its neighbor units. CNN maintains two essential characteristics: real-time processing ability and local association [22]. As a result, CNN can operate at high speed (in real time) by using several digital architectures, covering a wide range of applications characterized by their spatial dynamics such as control systems [23], data fusion [24] and prediction [25]. Due to their powerful local processing capabilities, Cellular Neural Networks (CNN) are also commonly utilized in image processing contexts with applications in edge detection, image encoding, image hole filling, and so on [26–29]. CNN have been already proposed to detect corner features in binary images [30]; however, their use in grayscale images has not been considered due to their design difficulties.

In CNN, processing units are characterized by multiple input–single output elements whose dynamic behavior is determined by 19 parameters distributed in three structural elements called feedback template **A**, control template **B** and bias I. Since CNN converge to only one of its attractors defined by its templates (local rules), they can operate with noise data or incomplete information, which is typical in real world applications [31–33]. The estimation of the cloning templates **A** and **B** along with the bias I represent the main problem in the design of a CCN model [34, 35]. There exist three different techniques for calculating such elements: the Intuitive, the direct and the learning-based method [36–38]. In the intuitive approach, the parameters are calculated by exploring the effect of different local rules. The main disadvantage of the intuitive technique is the execution of a significant number of experiments to find the correct dynamical behavior. In the direct method, the parameters are determined by using the matrices that model the exact required behavior, its description employs unknown functions; therefore such method is seldom used. The most widely adopted way of estimating the CNN design parameters is the learning-based method. In this technique, a learning algorithm is applied to adjust the design parameters until the desired CNN behavior is approximated.

Currently, there exists a wide interest in the use of evolutionary algorithms as learning methods [22, 39–41]. In general terms, they have proved to produce better results than those based on classical learning methods regarding robustness and accuracy [42, 43]. Under this approach, the learning process is transformed into an optimization problem. Where, individuals, in the context of evolutionary algorithms, correspond to possible solutions in the learning scenario. Each candidate solution is evaluated by an objective function to assess its quality with regard to the learning task. Conducted by the values of this function, the group of candidate solutions is evolved by using the evolutionary algorithm until the best possible solution of the learning problem has been reached. Among all evolutionary algorithms, the differential evolution (DE) method, introduced by Price and Storn [44], represents one of the most efficient techniques for solving numerical optimization problems. The main characteristics of DE are its simplicity, implementation easiness, fast convergence and robustness. Such properties have motivated its use in a wide range of applications in the literature [45–49].

In this chapter, a corner detector based on CNN for grayscale images is presented. In the approach, the original processing scheme of the CNN is modified to include a nonlinear operation for increasing the contrast of the local information in the image. With this adaptation, the final CNN parameters that allow the appropriate detection of corner points are estimated through the Differential evolution algorithm by using standard training images. Different test images have been used to evaluate the performance of the presented corner detector. Its results are also compared with popular corner methods from the literature. Computational simulations demonstrate that the presented CNN approach presents competitive results in comparison with other algorithms in terms of accuracy and robustness.

The chapter is organized as follows: Sect. 4.2 explains the architecture of the CNN while Sect. 4.3 presents the main characteristics of the DE method. Section 4.4 exposes the learning scenario for the CNN parameter estimation and the adaptation of the nonlinear processing conducted by the cloning template. Section 4.5 presents the experimental results including a comparison and analysis. Finally, in Sect. 4.6, the conclusion is established.

4.2 Cellular Nonlinear/Neural Network (CNN)

A CNN involves an arrangement of $M \times N$ locally-connected processing units. In a CNN, each unit represents a nonlinear dynamical system which is modeled by the state equation given in Eq. (4.1).

$$\dot{x}_{ij} = -x_{ij} + \sum_{k,l \in S_{ij}(r)} a_{kl} y_{kl} + \sum_{k,l \in S_{ij}(r)} b_{kl} u_{kl} + z_{ij} \tag{4.1}$$

where

$$\begin{gathered} y_{ij} = f(x_{ij}) = \frac{1}{2}\left(\left|x_{ij}+1\right| - \left|x_{ij}-1\right|\right) \\ i = 1, 2, \ldots, M; \quad j = 1, 2, \ldots, N; 1 \le k \le M; 1 \le l \le N; \end{gathered} \tag{4.2}$$

In Eqs. (4.1) and (4.2), the input, output, state and threshold, are represented by $u_{ij}, y_{ij}, x_{ij}, z_{ij}$, respectively. Each unit x_{ij} also maintains a feed-forward connection $b_{kl}u_{ij}$ that corresponds to the input effect. On the other hand, there exists a feedback

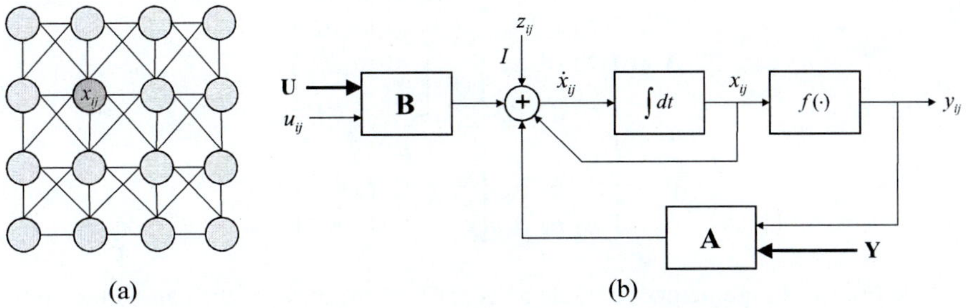

Fig. 4.1 Schematic representation of a CNN (**a**) and its processing model (**b**)

link $a_{kl}y_{ij}$ that represents the output impact produced by the neighbor units. Figure 4.1 shows a schematic representation of a CNN (a) and its processing model (b).

In CNN, every processing unit x_{ij} (from row i and column j) is connected locally to each of its neighbor units considering an area of influence $S_{ij}(r)$ of distance r, according to the following model:

$$S_{ij}(r) = \{x_{kl} : \max(|k - i|, |l - j|) \leq r\},$$
$$1 \leq k \leq M; 1 \leq l \leq N; \tag{4.3}$$

In order to illustrate this concept, assuming as an example, a 3×3 area of influence, it maintains a distance $r = 1$, where x_{ij} is connected to its first neighbor elements x_{kl} which correspond to $\{k, l\} =(i + 1, j + 1), (i + 1, j), (i + 1, j - 1), (i, j + 1), (i, j - 1), (i - 1, j + 1), (i - 1, j), (i - 1, j - 1)$.

The dynamical behavior of a CNN with a 3×3 area of influence is determined by setting 19 different real numbers such as one for threshold $z_{ij}(I)$, nine for the cloning feed-forward template $b_{kl}(\mathbf{B})$, and nine for the cloning feed-back template $a_{kl}(\mathbf{A})$.

Considering the Eqs. (4.1) and (4.2), the total stability in all trajectories for a standard CNN can be determined according to the following assumptions:

$$\left.\begin{array}{l}\mathbf{A}(i, j; k, l) = \mathbf{A}(k, l; i, j)\\ \mathbf{B}(i, j; k, l) = \mathbf{B}(k, l; i, j)\end{array}\right\}, \left|x_{ij}(0)\right| \leq 1, \left|u_{ij}(0)\right| \leq 1 \tag{4.4}$$

Therefore, a complete stable CNN [39, 40, 50] is designed by selecting the cloning templates as follows:

$$\begin{array}{l}a_1 = a_9\ a_2 = a_8\ a_3 = a_7\ a_4 = a_6\\ b_1 = b_9\ b_2 = b_8\ b_3 = b_7\ b_4 = b_6\end{array} \tag{4.5}$$

Under such conditions, the cloning templates are designed considering the following configuration:

$$\mathbf{A} = \begin{bmatrix} a_1 & a_2 & a_3 \\ a_4 & a_5 & a_6 \\ a_7 & a_8 & a_9 \end{bmatrix} = \begin{bmatrix} a_1 & a_2 & a_3 \\ a_4 & a_5 & a_4 \\ a_3 & a_2 & a_1 \end{bmatrix}$$

$$\mathbf{B} = \begin{bmatrix} b_1 & b_2 & b_3 \\ b_4 & b_5 & b_6 \\ b_7 & b_8 & a_9 \end{bmatrix} = \begin{bmatrix} b_1 & b_2 & b_3 \\ b_4 & b_5 & b_4 \\ b_3 & b_2 & b_1 \end{bmatrix} \tag{4.6}$$

Therefore, 11 parameters integrate a vector **T** that represents the parameters to be estimated in order to characterize the CNN model.

$$\mathbf{T} = [a_1, a_2, a_3, a_4, a_5, b_1, b_2, b_3, b_4, b_5, z] \tag{4.7}$$

4.3 Differential Evolution Method

Differential evolution [44] is a population-based and simple optimization method for finding the global solution of multi-modal functions. DE is fast, easy to use, adaptable for discrete optimization, and quite powerful in nonlinear constraint optimization considering complex penalty functions.

Similar to Genetic Algorithms (GA), it uses the mutation and crossover operations, along with a selection procedure. The most important difference between DE and GA is that GA considers the process of crossover as the best way to exchange information among the candidates for producing better solutions. On the other hand, the DE method uses the operation of mutation as its central operator. DE also utilizes a non-uniform crossover process. This operator incites the use of produced candidate solutions from one parent more frequently than from others. This non-uniform crossover process allows efficiently managing information about of successful solution combinations. Therefore, with this operator, the search strategy concentrates on the most promising sections of the search space. DE introduces an interesting mutation operation which is not only effective, but also simple. This mutation uses the differences of selected random pairs of solutions to build new search positions.

In this chapter, the DE variant known as DE/best/l/exp [51, 52] has been used. The standard DE method starts by defining a population of N_p d-dimensional vectors whose values are uniformly distributed with random elements between the pre-specified lower initial parameter bound p_e^{low} ($e \in 1, \ldots, d$) and the upper initial parameter bound p_e^{high}.

$$\begin{aligned} p_e^h(k) &= p_e^{low} + \text{rand}(0,1) \cdot (p_e^{high} - p_e^{low}) \\ e &= 1, 2, \ldots, d; \quad h = 1, 2, \ldots, N_p; k = 0 \end{aligned} \tag{4.8}$$

The index k represents the execution number, while e and h correspond to the decision variable and the vector number in the population, respectively. Therefore, $p_e^h(k)$ symbolizes the e-th variable of the h-th population vector at iteration k.

To produce a new candidate solution (trial solution), DE mutates the best solution from the current population by combining a scaled difference between two randomly selected vectors of the population.

$$t_e^h = p_e^{best}(k) + F \cdot (p_e^{r_1}(k) - p_e^{r_2}(k));$$
$$r_1, r_2 \in \{1, 2, \ldots, N_p\}; \tag{4.9}$$

where t_e^h represents the mutant element. Indices r_1 and r_2 corresponds to the vectors which are randomly selected. However, both indices must be different ($r_1 \neq r_2$). The mutation scale parameter F is a positive number within the interval (0, 1.0).

To enhance the diversity of the population, the crossover process is applied. This is conducted between the mutant element t_e^h and the original elements of the population $p_e^h(k)$. As a result, a new element $q_e^h(k)$ is produced. $q_e^h(k)$ is calculated by considering the following operation:

$$q_e^h(k) = \begin{cases} t_e^h & \text{if rand(0,1)} \leq CR \text{ or } e = e_{rand} \\ p_e^h & \text{otherwise} \end{cases}, \; e_{rand} \in \{1, 2, \ldots, d\} \tag{4.10}$$

where $e_{rand} \in 1, 2, \ldots, d$. The parameter *CR* represents the crossover parameter which is a positive number within the interval ($0.0 \leq CR \leq 1.0$). This factor rules the portion of elements that the mutant vector is providing to the final candidate solution. The trial element $q_e^h(k)$ always maintains mutant elements according to the randomly selected number e_{rand}, confirming that the new solutions diverge by at least one decision variable from the original solution.

As a final step, a greedy selection process is considered to obtain the best quality solutions. Therefore, if the value of the objective function obtained by the $q_e^h(k)$ is better than or equal to the quality of the element $p_e^h(k)$, then $q_e^h(k)$ substitutes $p_e^h(k)$ in the next iterations. Otherwise, $p_e^h(k)$ is maintained. This process can be modeled as follows:

$$p_e^h(k+1) = \begin{cases} q_e^h(k) & \text{if } f(\mathbf{q}^h) \leq f(\mathbf{p}^h) \\ p_e^h(k) & \text{otherwise} \end{cases} \tag{4.11}$$

where $f(\cdot)$ represents the objective function, while $\mathbf{q}^h$ and $\mathbf{p}^h$ correspond to the complete crossover and original vectors, respectively. The whole process is repeated until a certain termination criterion is reached.

4.4 Learning Scenario for the CNN

Due to their powerful local processing capabilities, Cellular Neural Networks (CNN) are commonly utilized in image processing applications. CNN perform well for locating corner features in binary images [53]. However, they abruptly fail when typical grayscale images are considered. In this chapter, a corner detector based on CNN for grayscale images is introduced. To design a CNN system with these characteristics two modifications are necessary: (4.1) Adaptation of the cloning template processing and (4.2) estimation of the CNN parameters trough learning.

4.4.1 Adaptation of the Cloning Template Processing

The CNN input u_{ij} represents in our application domain the pixel grayscale value of an $M \times N$ intensity image $\mathbf{R}(i,j)$. To operate under the CNN scheme, such values must be normalized to have the range $-1 \leq u_{ij} \leq 1$, where -1 represents "white" and $+1$ corresponds to "black".

Assume a 3×3 cloning template $\mathbf{B}$ and the input sub-image $\mathbf{U}$ with the following configuration:

$$\mathbf{B} = \begin{bmatrix} b_{-1,-1} & b_{-1,0} & b_{-1,1} \\ b_{0,-1} & b_{0,0} & b_{0,1} \\ b_{1,-1} & b_{1,0} & b_{1,1} \end{bmatrix}, \mathbf{U} = \begin{bmatrix} u_{i-1,j-1} & u_{i-1,j} & u_{i-1,j+1} \\ u_{i,j-1} & u_{ij} & u_{i,j+1} \\ u_{i+1,j-1} & u_{i+1,j} & u_{i+1,j+1} \end{bmatrix} \quad (4.12)$$

In the standard CNN scheme, the influence of the input information in the CNN model (Eq. 4.1) is computed as follows:

$$\mathbf{B} * \mathbf{U} = \sum_{a=-1}^{1} \sum_{b=-1}^{1} b_{a,b} u_{i+a,j+a}, \quad (4.13)$$

To increase the contrast of the local information, in this approach, the input values of $\mathbf{U}$ are nonlinear processed to generate new input values $\widehat{\mathbf{U}}$. To produce the new input values $\widehat{\mathbf{U}}$, a new matrix with contrast information $\mathbf{C}$ is calculated. The elements of $\mathbf{C}$ represent the gray-level differences between the center pixel and its neighbor components. Under this operation, $\mathbf{C}$ can be defined as follows:

$$\mathbf{C} = \begin{bmatrix} c_{1,1} & c_{1,2} & c_{1,3} \\ c_{2,1} & c_{2,2} & c_{2,3} \\ c_{3,1} & c_{3,2} & c_{3,3} \end{bmatrix} = \begin{bmatrix} u_{i-1,j-1} - u_{ij} & u_{i-1,j} - u_{ij} & u_{i-1,j+1} - u_{ij} \\ u_{i,j-1} - u_{ij} & 0 & u_{i,j+1} - u_{ij} \\ u_{i+1,j-1} - u_{ij} & u_{i+1,j} - u_{ij} & u_{i+1,j+1} - u_{ij} \end{bmatrix}, \quad (4.14)$$

If the input region of $\mathbf{C}$ belongs to a homogeneous area, then $\mathbf{C}$ includes elements near to zero. To obtain robust results in noisy environments, small differences in $\mathbf{C}$

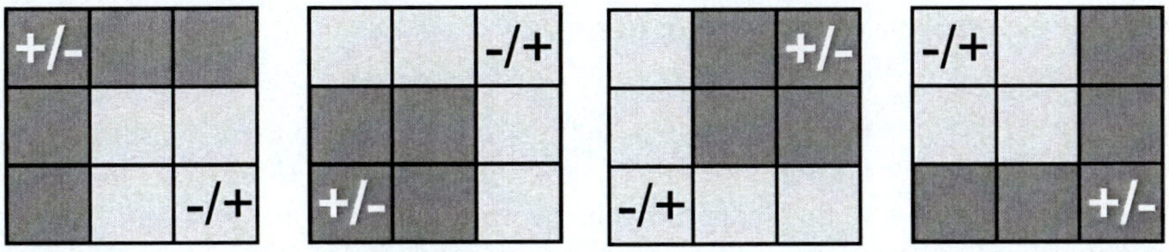

Fig. 4.2 Some possible corner configurations produced by calculating $\widehat{\mathbf{U}}$

are eliminated. This process is modeled as follows:

$$c_{o,s} = \begin{cases} 1 & \text{if } |c_{o,s}| \leq th \\ c_{o,s} & \text{otherwise} \end{cases}, o, s \in \{1, 2, 3\} \tag{4.15}$$

where *th* represents a threshold value that controls the robustness of the approach. The value of one corresponds to a positive number which categorizes the pixel value. Once conducted this operation, the final input information $\widehat{\mathbf{U}}$ is calculated. The main idea is to divide the information of **C**, in case of corner feature, into two parts, one with positive and another with negative differences values. Figure 4.2 illustrates this process considering some possible corner configurations. Therefore,$\widehat{\mathbf{U}}$ is produced by applying the nonlinear function sign(·) to each element $c_{o,s}$ of C considering the following process:

$$\widehat{u}_{o,s} = \text{sign}(c_{o,s}), \tag{4.16}$$

The presented CNN detector should have the capacity to identify corner characteristics in dark or light image regions. Therefore, as a final step, the values of $\widehat{\mathbf{U}}$ must be recalculated in accordance to its central element $c_{2,2}$. This process can be formulated as follows:

$$\widehat{\mathbf{U}} = \begin{cases} -\widehat{\mathbf{U}} \text{ if } c_{2,2} = -1 \\ \widehat{\mathbf{U}} \quad \text{otherwise} \end{cases} \tag{4.17}$$

Under such conditions, the final nonlinear dynamical behavior of each CNN processing unit can be modeled as follows:

$$\dot{x}_{ij} = -x_{ij} + \sum_{k,l \in S_{ij}(r)} a_{kl} y_{kl} + \sum_{k,l \in S_{ij}(r)} b_{kl} \widehat{u}_{kl} + z_{ij} \tag{4.18}$$

where $\widehat{u}_{kl}$ represents the values from $\widehat{\mathbf{U}}$. The same procedure can be generalized for templates of bigger dimensions.

4.4.2 Learning Scenario for the CNN

In this part, the learning scenario in which the DE method is used to estimate the CNN parameters for corner detection is described. Figure 4.3 illustrates the CNN parameter learning setting. Under this scheme, the cloning templates **A** and **B** along with the bias *I* are iteratively adapted so that the final CNN output reaches the output results of an ideal corner detector which, in fact, can effectively identify the corner features in the grayscale image. This ideal corner detector corresponds to the resulting ground-true image produced by a human operator.

Under this scheme, the detection task is translated into an optimization problem where the optimal solution corresponds to the best possible parameters that allow a correct corner detection with the modified CNN system. In this approach, each candidate solution h is coded with a set of 11 elements ($\mathbf{T}^h = [a_1^h, a_2^h, a_3^h, a_4^h, a_5^h, b_1^h, b_2^h, b_3^h, b_4^h, b_5^h, z^h]$) that represents the parameters of the CNN structure (which guaranty its stability). To evaluate the error generated by a candidate solution $\mathbf{T}^h$ an objective function J is defined. This cost function J expresses the similitude between the detection results produced for the candidate solution $\mathbf{T}^h$ and the ground truth. Assuming that the image $ICNN_{\mathbf{T}^h}$ represents the corner detection results provided by the candidate solution $\mathbf{T}^h$ and *IGTR* the ground-truth image, the objective function es modeled as follows:

$$J(\mathbf{T}^h) = \sum_{a=1}^{M} \sum_{b=1}^{N} K_{a,b}, \; K_{a,b} = \begin{cases} 1 \text{ if } ICNN_{\mathbf{T}^h}(a,b) \neq IGTR(a,b) \\ 0 \qquad\qquad \text{otherwise} \end{cases} \tag{4.19}$$

Therefore, the learning process begins by initializing a set of N_p candidate solutions. Each candidate solution is evaluated by the objective function J to assess its quality with regard to the detection task. Conducted by the values of J, the group of

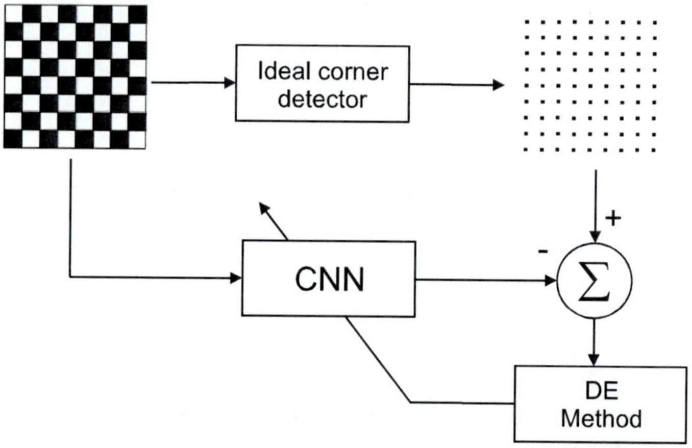

Fig. 4.3 CNN parameter learning setting

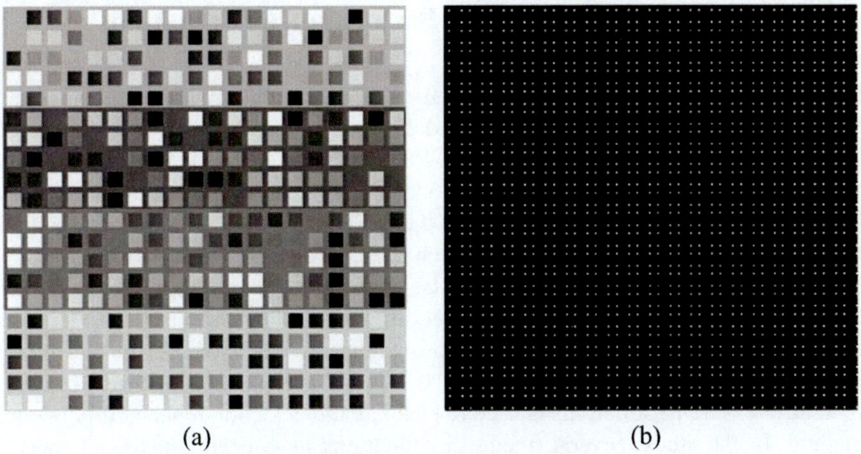

Fig. 4.4 Training images for the estimation of the CNN parameters

N_p candidate solutions is evolved by using the operators of the DE algorithm until the parameters of the CNN that minimize J have been found.

Figure 4.4 illustrates the images for training used in the learning process of the CNN parameters. Figure 4.4a presents the original training image, which involves a 200 × 200 image artificially generated. Each square in the image considers an intensity value selected randomly between 0 and 255 with uniform distribution. On the other hand, Fig. 4.4b is the ground-truth produced by a human expert.

Under these conditions, the learning process is executed considering the following configuration: $N_p = 250$, $F = 0.8$, $CR = 0.7$, $k = 1000$, $th = 0.08$, $Timestep = 0.1$, $iteration = 30$. Once the learning procedure of CNN is completed, the optimized parameters maintain the following values:

$$\mathbf{A} = \begin{bmatrix} -3.5549 & 7.5011 & 3.8973 \\ -1.7861 & 31.8141 & -1.7861 \\ 3.8973 & 7.5011 & -3.5549 \end{bmatrix},$$
$$\mathbf{B} = \begin{bmatrix} -43.8471 & -76.6244 & -21.6907 \\ -75.9302 & 10.9746 & -75.9302 \\ -21.6907 & -76.6244 & -43.8471 \end{bmatrix}$$
$$I = 27.9821 \tag{4.20}$$

Once found such parameters, they are maintained during all experimental section.

4.5 Experimental Results and Performance Evaluation

In this subsection, the performance evaluation of the presented CNN-based approach is presented. In the analysis, our corner detector is compared against four state-of-the-art detectors such as Harris [8], F-CPDA [12], ACSS [11], SUSAN [10] and ANDD [14]. In the comparisons, the code for each algorithm has been obtained from their original sources. In the case of the F-CPDA and ACSS detectors, for a fair comparison, the D-P method [54] has been used to improve the quality of contours.

The results are divided into three parts. In the first experiment (5.1), the corner algorithms are compared considering a set of images with ground truth. Under these conditions, the detection and localization results of each method can be directly assessed in relation to the ideal case. In the second experiment (5.2), a set of images with different transformations is used to evaluate the detection repeatability of each method. In the study, several image modifications have been considered such as affine transformations, JPEG compression, and noise degradation. Finally, in the third experiment (5.3), the computational cost of each method is evaluated. In the evaluation, appropriate performance indexes have been considered in order to be compatible with other results reported in the literature [14, 55, 56].

4.5.1 *Detection and Localization Using Images with Ground Truth*

Under this experiment, the results of each method is evaluated in terms of the number of missed corners (c_m), the number of false corner (c_f) and the localization error of matched coroner pairs (E_L). For the evaluation, it is necessary to have the ground truth elements of test images. Ground truth is a reference solution that shows the number of true corners in the image and their localization.

In the comparison analysis, two different sets C_{DET} and C_{REF} are defined. The set $C_{DET} = \{\hat{\mathbf{c}}_p, p = 1, \ldots, N_{DET}\}$ represents the list of N_{DET} corner locations detected by a particular method. On other hand, the set $C_{REF} = \{\mathbf{c}_q, q = 1, \ldots, N_{REF}\}$ is the group of N_{REF} corner points contained in the ground truth. Assume that $dis(\hat{\mathbf{c}}_p, \mathbf{c}_q)$ corresponds to the distance between the p-th corner $\hat{\mathbf{c}}_p$ in the detected set C_{DET} and the q-th corner position $\mathbf{c}_q$ from the reference set C_{REF}.

For each corner point $\mathbf{c}_q$ from C_{REF}, if the $dis(\hat{\mathbf{c}}_p, \mathbf{c}_q)$ between represents the minimum value for $\forall p \in C_{DET}$ and if $dis(\hat{\mathbf{c}}_p, \mathbf{c}_q) \leq D_{\max}$, the corner $\mathbf{c}_q$ is classified as a *correct* detected by $\hat{\mathbf{c}}_p$, otherwise $\mathbf{c}_q$ is classified as a *missed* corner (c_m). Here, $D_{\max}$ defines the maximum admissible distance for considering the correct detection of a corner point from the ground truth. In the comparisons, this value has been fixed to 4 pixels ($D_{\max} = 4$). In case of a correct detection, the points $\mathbf{c}_q$ and $\hat{\mathbf{c}}_p$ represent a *matched corner pair*. Similarly, all corner points $\hat{\mathbf{c}}_p$ from C_{DET} that have not found a matching correspondence in C_{REF} will be considered as a *false* detected corner points (c_f).

The average localization error is defined as the mean distance of all the matched corner pairs. Assuming that $\{(\hat{\mathbf{c}}_1, \mathbf{c}_1), (\hat{\mathbf{c}}_2, \mathbf{c}_2), \ldots (\hat{\mathbf{c}}_{N_m}, \mathbf{c}_{N_m})\}$ represents the list of matched pairs, the average localization error is formulated as follows:

$$E_L = \sqrt{\frac{1}{N_m} \sum\nolimits_{k=1}^{N_m} dis(\hat{\mathbf{c}}_k, \mathbf{c}_k)} \tag{4.21}$$

For this experiment, a representative set of 10 different images collected from the literature have been used. Figure 4.5 shows the complete set of images. In the figure, each image is labeled as E1.X that refers to the image X from the experiment one. In the study, similar to [11], the ground truth solutions for all images have been manually produced from their original versions. In the generation of reference solutions for a ground truth is usually complicated to determine whether or not a candidate point should be labeled as a corner. For this reason, in this study, only the really obvious corners are considered as reference solutions in the ground truth.

The visual results of the six corner detection methods are depicted in Figs. 4.6, 4.7, 4.8 and 4.9. For the sake of space, only the results for images E1.1, E1.3, E1.4 and E1.8 are displayed. The numerical results in terms of the number of missed corners (c_m), the number of false corner (c_f) and the localization error of matched coroner pairs (E_L) are registered in Table 4.1.

According to the results, the Harris detector produces the worst performance in comparison to the other methods. It detects several false corners as a consequence of the use of gradient operations. Since contour-based corner detectors (F-CPDA, ACSS and ANDD) base their strategy on contour extraction, they maintain a great reduction of false corners. The F-CPDA method produces the least number of false corners, but misses a significant amount of actual corners. Conform Table 4.1, the ANDD detector maintain a better performance than ACSS and F-CPDA methods. It present a better balance between false and missing corners. This behavior is produced by the way in which the ANDD softens and evaluates each contour segment to decide if an edge pixel is a corner point or not. Under the ANDD process, the possibility of not detecting a true maxima is strongly reduced. On the other hand, the SUSAN detector presents a low number of false corners with regard to the harris method, but higher compared with contour-based corner detectors (F-CPDA, ACSS and ANDD). Since the SUSAN algorithm considers as a procedure the comparison of pixel intensities in particular positions (circular templates), the number of false corners increase in texturized regions. In general terms, ANDD approach and the CNN-based detector miss the smallest number of corners. Compared with the CNN-based, the ANDD detector presents slightly more false corners. The reason is that the CNN scheme considers the detection of a corner point only if the CNN reaches one of the attractors defined by its templates (local rules). Under these circumstances, the CNN can operate with noise data or incomplete information, which is typical in a corner detection process.

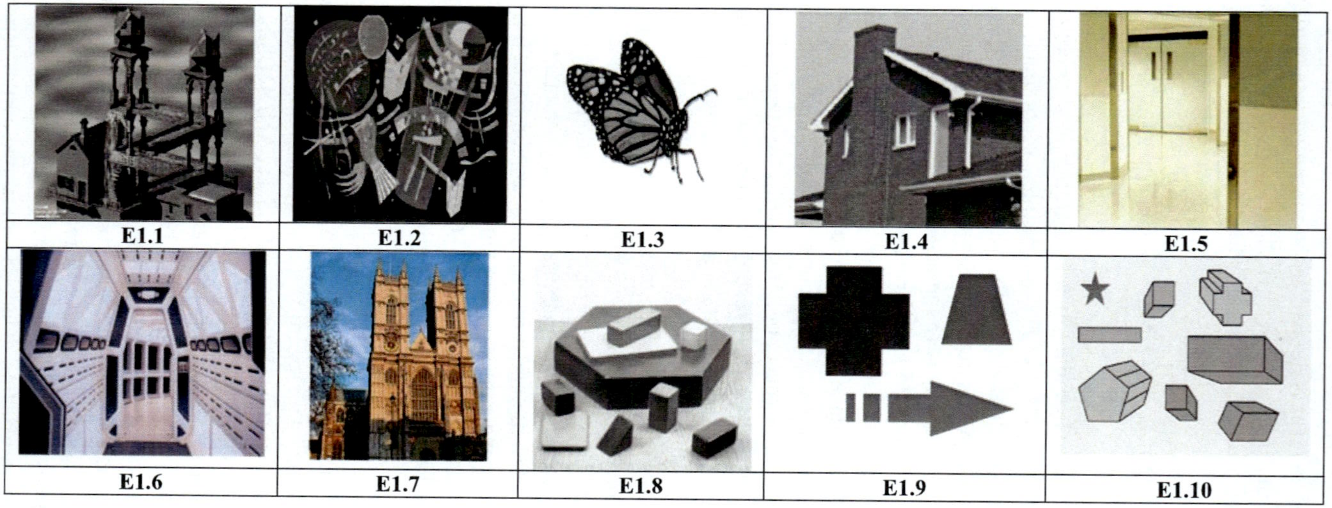

Fig. 4.5 Set of images used to evaluate the performance of each algorithm in terms of the number of missed corners (c_m), the number of false corner (c_f) and the localization error of matched coroner pairs (E_L)

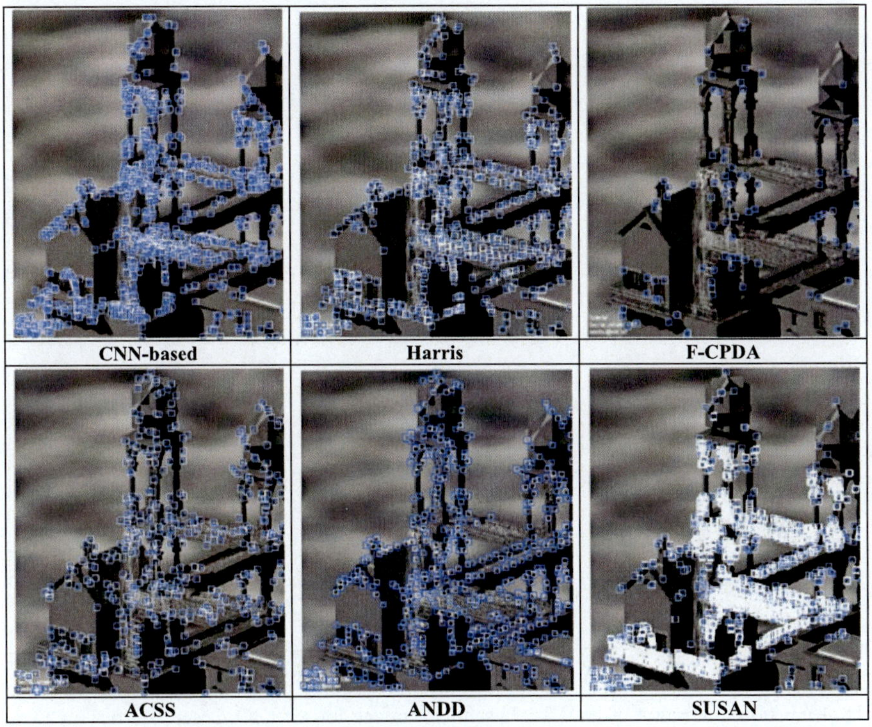

Fig. 4.6 Corner detection of the E1.1 benchmark image for the CNN-based, Harris, F-CPDA, ACSS, ANDD and SUSAN approaches

Considering that reporting a false corner or missing a true corner represent the same cost in performance. The total amount of missed true corner and false corners can be employed as a performance index to evaluate the quality of a corner detector. These performance values can be seen in column COST from Table 4.1. According to these values the CNN-based approach obtain the best cost in comparison with the other methods. Another important metric to assess the detection performance of a corner detector is the localization accuracy. For all images, the Harris and SUSAN detectors obtain the worst accuracy. Contrarily, the F-CPDA detector reaches the best accuracy. This is as a consequence that the algorithm only considers strong and clear corners in its detection. The CNN-based detector and the ANDD algorithm are slightly worse than the F-CPDA detector but better than Harris, ACSS and SUSAN detectors. In conclusion, the presented detector present the best detection performance in terms of the number of missed corners (c_m), the number of false corner (c_f) while it maintains competitive results regarding the localization accuracy.

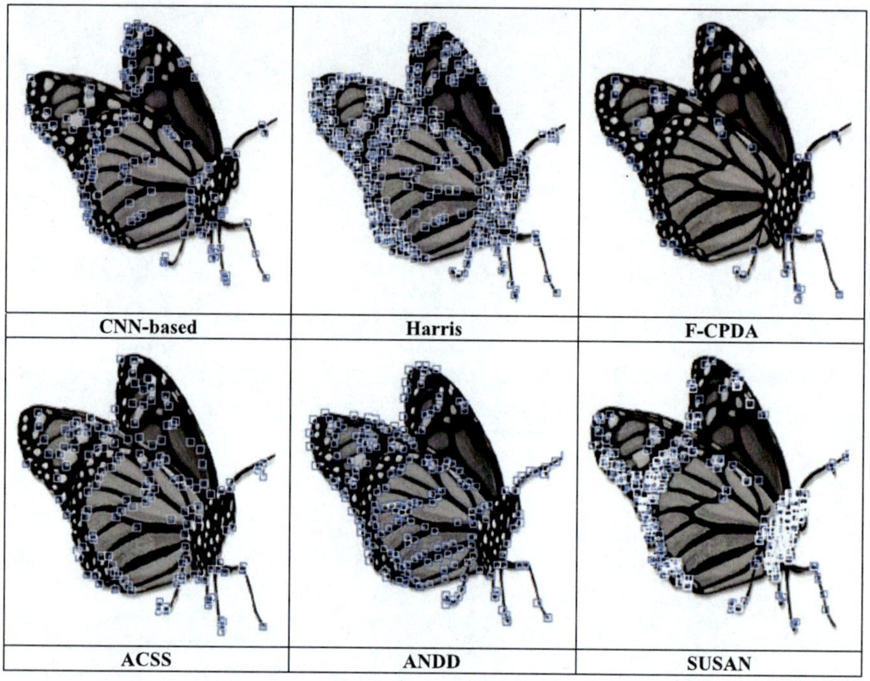

Fig. 4.7 Corner detection of the E1.3 benchmark image for the CNN-based, Harris, F-CPDA, ACSS, ANDD and SUSAN approaches

4.5.2 Repeatability Evaluation Under Image Transformations

One of the most popular applications of corner detection algorithms is the image registration and feature matching. In order to extend the comparison analysis, in this subsection, the robustness of each approach under different image transformations is tested. The repeatability is the performance index employed to evaluate the robustness or stability of each corner detection approach. Repeatability refers to the capacity of a corner algorithm to detect repeated corner point positions in spite of an image transformation. A higher repeatability implies better performance. For the estimation of repeatability, first, the corner detection method under test is applied to the original an transformed images. Then, the number of detected corners N_O in the original image and the number of detected corners N_T in the transformed image are obtained. Finally, the number of matched corner pairs N_M between both images is computed. To decide if two different corners are repeated, the maximum admissible distance has also been fixed to 4 pixels ($D_{\max} = 4$). From these values, the repeatability can be defined as:

$$R = \frac{N_M}{2}\left(\frac{1}{N_O} + \frac{1}{N_T}\right) \tag{4.22}$$

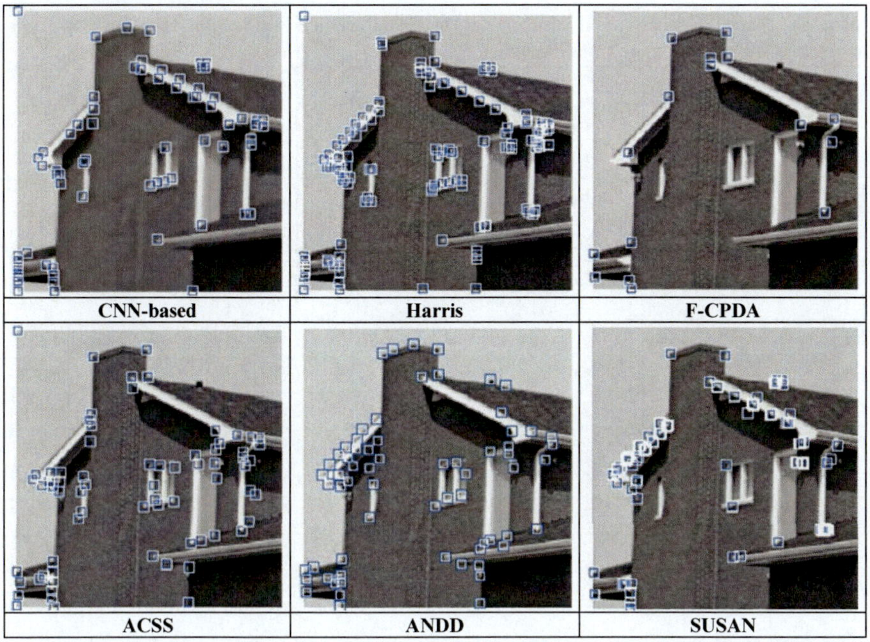

Fig. 4.8 Corner detection of the E1.4 benchmark image for the CNN-based, Harris, F-CPDA, ACSS, ANDD and SUSAN approaches

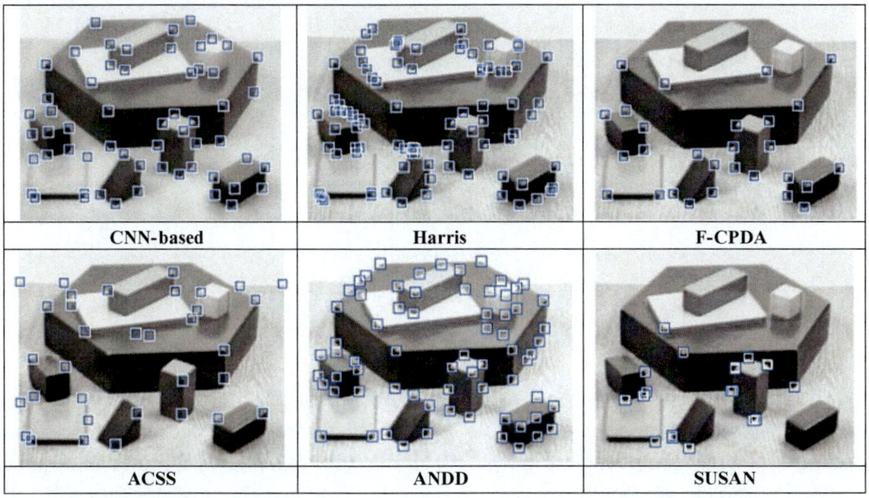

Fig. 4.9 Corner detection of the E1.8 benchmark image for the CNN-based, Harris, F-CPDA, ACSS, ANDD and SUSAN approaches

Table 4.1 Missed Corners (c_m), false Corners (c_f) and average localization error for each corner detector for every test image from the Fig. 4.5

Image name	CNN-based			Harris			F-CPDA		
	c_m	c_f	E_L	c_m	c_f	E_L	c_m	c_f	E_L
E1.1	65	91	1.57	94	179	1.82	172	20	1.47
E1.2	46	66	1.56	74	136	1.86	156	20	1.42
E1.3	26	37	1.69	48	77	1.88	69	9	1.62
E1.4	4	3	1.13	10	11	1.78	19	0	1.06
E1.5	2	2	1.31	5	6	1.78	11	1	1.20
E1.6	58	91	1.43	97	182	1.87	172	28	1.34
E1.7	48	68	1.34	69	133	1.84	147	25	1.21
E1.8	4	3	1.38	11	12	1.85	23	2	1.33
E1.9	2	2	1.37	5	6	1.78	11	1	1.22
E1.10	4	3	1.32	14	15	1.69	27	2	1.13
$\sum$	259	366		427	757		807	108	
Cost	625			1,184			915		

Image name	ACSS			ANDD			SUSAN		
	c_m	c_f	E_L	c_m	c_f	E_L	c_m	c_f	E_L
E1.1	87	121	1.64	73	107	1.60	89	135	1.78
E1.2	64	89	1.68	56	83	1.61	62	113	1.72
E1.3	36	46	1.77	33	43	1.71	39	62	1.79
E1.4	6	5	1.31	4	4	1.17	7	9	1.24
E1.5	3	3	1.37	2	2	1.32	4	5	1.41
E1.6	83	126	1.48	67	110	1.45	78	137	1.50
E1.7	61	97	1.42	46	81	1.32	61	109	1.51
E1.8	6	5	1.55	5	4	1.39	7	9	1.60
E1.9	3	2	1.42	2	2	1.36	4	5	1.51
E1.10	8	7	1.41	4	5	1.34	10	12	1.48
$\sum$	357	501		292	441		361	596	
Cost	858			733			957		

A value of R closer to one means a better robustness of a detector.

For this experiment, a representative set of 50 different images collected from the literature [57] have been used. Figure 4.10 shows a representative number of images from the complete set. Such images are considered to test the repeatability of each corner detector under different image transformations. In our study, several image transformations are considered such as affine transformations, JPEG compressions and noise degradations.

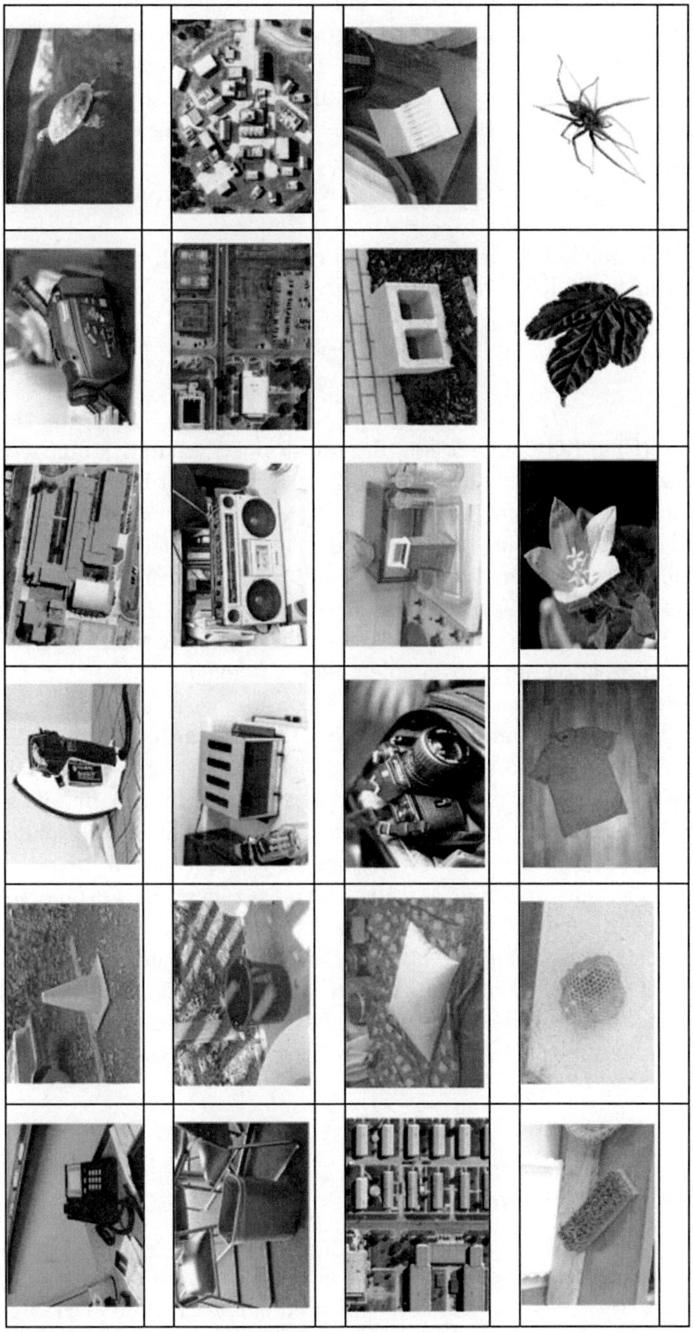

Fig. 4.10 A representative number of the set of images used to evaluate the performance of each algorithm in terms of the average repeatability of the six detectors under affine transforms, JPEG compression, and noise degradation

Four types of affine transformations are examined in the analysis. They include (A) rotation, (B) non-uniform scaling, (C) uniform scaling and (D) shear transform. Therefore, the transformed images are produced according to the following configuration:

(A) Rotation, 20 different angles in the interval [-90°, 90°] in steps of 9°.

(B) Non-uniform scaling transformation $x' = xs_x$, $y' = ys_y$, where $s_x = 1$, s_y within [0.5,1.8], with a resolution of 0.1.

(C) Uniform scaling transformation $s_x = s_y$ within [0.5, 2], with a resolution of 0.1.

(D) Shear transformation, with a shear factor c within $[-1,1]$, resolution of 0.1, according to the following linear transformation:

$$\begin{bmatrix} x' \\ y' \end{bmatrix} = \begin{bmatrix} 1 & c \\ 0 & 1 \end{bmatrix} \begin{bmatrix} x \\ y \end{bmatrix} \quad (4.23)$$

In case of the (E) JPEG compression, the transformed versions are generated considering a quality factor within the interval [5, 100] in steps of 5. Noise degradation (F) is produced in transformed images by using white Gaussian noise with zero mean a standard deviation within [1, 15] in steps of 1.

The results of the six corner detection methods are shown in Fig. 4.11. They present the average repeatability of each corner detector under a particular image transformation. The average repeatability refers to the mean repeatability obtained for a specific transformation over the complete set of 50 images.

Figure 4.11 presents the average repeatability R considering 10(a) Rotation, 10(b) Non-Uniform scaling, 11(c) Uniform scaling, 11(d) Shear transforms, 11(e) JPEG compression and 11(f) Additive white Gaussian noise. According to the Fig. 4.11, the Harris and ANDD detector maintain a good performance in uniform and non-uniform scaling but obtains a bad behavior under the other image transformations. The CNN-based detector obtains the best average repeatability under the rotation, uniform scaling, shear transformation, as well as the noise robustness. On other hand, contour-based corner detectors (F-CPDA, ACSS and ANDD) present a competitive behavior with different performance levels. They present in particular a good performance against JPEG compression. Though the repeatability of the presented CNN-based detector is not best, it is quite close to the best one. In general, the presented detector reaches the best repeatability under the affine transforms and the lossy JPEG compression.

4.5.3 Computational Time Evaluation

Cellular neural networks (CNNs) are a class of parallel information-processing schemes. Consequently, CNN can work at high speed in real time on VLSI circuits [58]. Under such conditions, array computations can be implemented on special

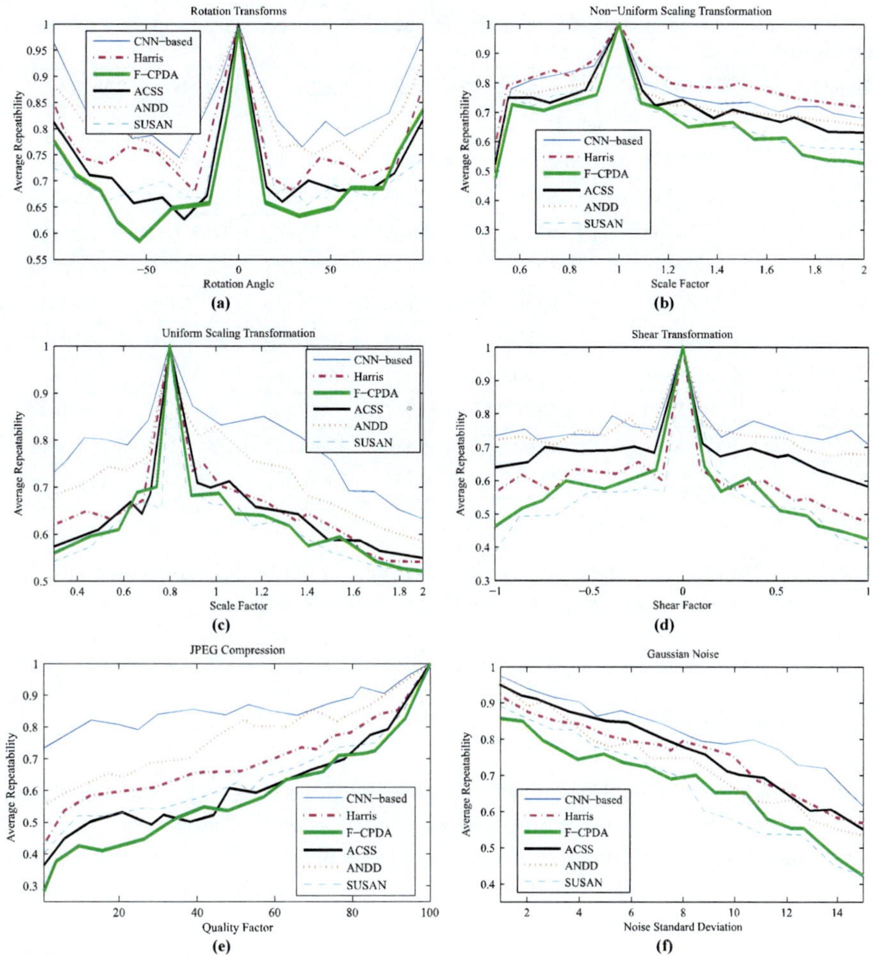

Fig. 4.11 Average repeatability R of the six detectors under the different transformations, **a** rotation, **b** non-uniform scaling, **c** uniform scaling, **d** shear transforms, **e** JPEG compression and **f** additive white Gaussian noise

architectures with speed at least thousand times faster than sequential processors [59]. Nevertheless, the presented CNN-based method has been implemented on a PC in order to analyze the detection performance. Therefore, its computational cost can be compared with Harris, F-CPDA, ACSS, ANDD and SUSAN approaches which are devised for running only in sequential processors.

To evaluate its computational complexity, each detection algorithm is executed 100 times over every image from Fig. 4.5 while its averaged execution time t_{CE} is registered in seconds. Therefore, the processing speed of the presented algorithm and the other comparison methods are exhibited in Table 4.2. Results demonstrate that the CNN-based approach obtains the lower execution times in 7 from 10

Table 4.2 Mean execution time after 100 runs for the CNN-based presented approach and the six comparison detectors

Image name	CNN-based	Harris	F-CPDA	ACSS	ANDD	SUSAN
	t_{CE}	t_{CE}	t_{CE}	t_{CE}	t_{CE}	t_{CE}
E1.1	3.66438	10.26012	4.41987	3.75690	4.91648	8.16452
E1.2	4.76310	12.65492	3.42365	4.95671	5.16913	7.94637
E1.3	5.10364	14.67301	3.40169	4.96341	6.27611	7.94173
E1.4	3.97141	10.97136	4.38947	3.84663	5.94327	9.09752
E1.5	3.31392	7.64308	3.37162	4.74623	4.76193	5.23641
E1.6	10.41913	73.09738	11.68743	15.06924	14.43920	62.19705
E1.7	5.74308	11.27903	7.47631	6.36421	7.13962	8.31642
E1.8	3.31970	8.47312	3.57931	3.67201	4.19337	6.92491
E1.9	15.14391	65.35402	16.79310	19.34122	22.97316	54.93011
E1.10	8.75310	64.51933	10.42153	14.99611	14.60741	63.90125

images. In conclusion, the presented detector present the best performance in terms of computational cost in comparison with the other schemes.

4.6 Conclusions

Due to their powerful local processing capabilities, Cellular Neural Networks (CNN) are commonly utilized in image processing applications such as image edge detection, image encoding, image hole filling, and so on. CNN perform well for locating corner features in binary images. However, their use in gray scale images has not been considered due to their design difficulties.

In this chapter, a corner detector based on CNN for grayscale images is presented. In the approach, the original processing scheme of the CNN is modified to include a nonlinear operation for increasing the contrast of the local information in the image. With this adaptation, the final CNN parameters that allow the appropriate detection of corner points are estimated through the Differential evolution algorithm by using standard training images.

A representative set of images collected from different sources have been employed to analyze the performance of the CNN-based corner detection method. Its results are also compared with popular corner methods from the literature such as Harris, F-CPDA, ACSS, ANDD and SUSAN.

Computational simulations demonstrate that the presented CNN approach presents competitive results than other algorithms in terms of detection and robustness. The reason of this remarkable behavior is that the CNN scheme considers the detection of a corner point only if the CNN reaches one of the attractors defined by its templates (local rules). Under these circumstances, the CNN can operate with noise data or incomplete information, which is typical in a corner detection process.

References

1. Rublee E, Rabaud V, Konolige K, Bradski G (2011) ORB: An efficient alternative to SIFT or SURF. In: 2011 IEEE international conference on computer vision (ICCV), 2011 Nov 6. IEEE, pp 2564–2571
2. Yali L, Shengjin W, Qi T, Xiaoqing D (2015) A survey of recent advances in visual feature detection. Neurocomputing 149:736–751
3. Cui J, Xie J, Liu T, Guo X, Chen Z (2014) Corners detection on finger vein images using the improved Harris algorithm. Optik 125:4668–4671
4. Kitti T, Jaruwan T, Chaiyapon T (2012) An object recognition and identification system using the Harris Corner detection method
5. Zhou Z, Yan M, Chen S (2015) Image registration and stitching algorithm of rice low-altitude remote sensing based on Harris corner self-adaptive detection . Nongye Gongcheng Xuebao/Trans Chin Soc Agric Eng 31(14):186–193
6. Bagchi P, Bhattacharjee D, Nasipuri M (2015) A robust analysis detection and recognition of facial features in 2.5D images. Multimedia Tools Appl 75(18):1–38s
7. Rosten E, Drummond T (2006) Machine learning for high-speed corner detection. In: European conference on computer vision, 2006 May 7. Springer Berlin, Heidelberg, pp 430–443
8. Harris C, Stephens M (1988) A combined corner and edge detector. In: Alvey vision conference, vol. 15. Citeseer, 50
9. Mikolajczyk K, Schmid C (2004) Scale and affine invariant interest point detectors. Int J Comput Vis 60(1):6386–6396
10. Smith SM, Brady JM (1997) Susan a newapproach to lowlevel image processing. Int J Comput Vis 23(1):4578–4589
11. He XC, Yung NHC (2008) Corner detector based on global and local curvature properties. Opt Eng 47(5):057008
12. Awrangjeb M, Lu G (2008) Robust image corner detection based on the chord-to-point distance accumulation technique. IEEE Trans Multimedia 10(6):1059–1072
13. Shui P-L, Zhang W-C (2013) Corner detection and classification using anisotropic directional derivative representations. IEEE Trans Image Proces 22(8):3204–3219
14. Zhang W, Shui P-L (2015) Contour-based corner detection via angle difference of principal directions of anisotropic Gaussian directional derivatives. Pattern Recogn 48:2785–2797
15. Lowe D (2004) Distinctive image features from scale-invariant key points. Int J Comput Vis 60(2):91–110
16. Bay H, Ess A, Tuytelaars T, Gool LV (2008) Surf: speeded up robust features. Comput Vis Image Underst 110(3):346–359
17. Yi KM, Trulls E, Lepetit V, Fua P (2016) Lift: learned invariant feature transform. In: European conference on computer vision. Springer
18. Rosten E, Drummond T (2006) Machine learning for high-speed corner detection. In: Computer vision ECCV. Springer, pp 430–443
19. Schönberger JL, Hans H, Torsten S, Marc P (2017) Comparative evaluation of hand-crafted and learned local features. Computer vision and pattern recognition (CVPR)
20. Chua LO, Yang L (1988) Cellular neural networks: theory. IEEE Trans Circ Syst 35:1257–1272
21. Chua LO (1997) CNN: a version of complexity. Int J Bifurc Chaos 7(10):2219–2425
22. Li H, Liao X, Li C, Huang H, Li C (2011) Edge detection of noisy images based on cellular neural networks. Commun Nonlinear Sci Numer Simulat 16:3746–3759
23. Liu Z, Schurz H, Ansari N, Wang Q (2012) Theoretic design of differential minimax controllers for stochastic cellular neural networks. Neural Netw 26:110–117

24. Chatziagorakis P, Georgios Ch, Sirakoulis, Lygouras JN (2012) Design automation of cellular neural networks for data fusion applications. Microprocess Microsyst 36(1):33–44
25. Starkov SO, Lavrenkov YN (2017) Prediction of the moderator temperature field in a heavy water reactor based on a cellular neural network. Nucl Energy Technol 3(2):133–140
26. Hu X, Feng G, Duan S, Liu L (2017) A memristive multilayer cellular neural network with applications to image processing. IEEE Trans Neural Netw Learn Syst 28(8):1889–1901
27. Ando T, Uwate Y, Nishio Y (2017) Image processing by cellular neural networks with switching two templates. In: 2017 IEEE Asia pacific conference on postgraduate research in microelectronics and electronics (PrimeAsia), pp 41–44
28. Morales-Romero JJ, Gómez-Castañeda F, Moreno-Cadenas JA, Reyes-Barranca MA, Flores-Nava MA (2017) Time-multiplexing cellular neural network in FPGA for image processing. In: 2017 14th international conference on electrical engineering, computing science and automatic Control (CCE), pp 1–5
29. Adhikari SP, Kim H, Yang C, Chua LO (2018) Building cellular neural network templates with a hardware friendly learning algorithm. Neurocomputing 312:276–284
30. Chua L, Roska T (2004) Cellular neural networks and visual computing: foundation and applications. Cambridge University Press
31. Harrer H, Nossek JA, Roska T, Chua LO (199) A current-mode DTCNN universal chip. In: Proceedings of IEEE international symposium on circuits and systems, pp 135–138
32. Cruz JM, Chua LO, Roska T (1994) A fast, complex and efficient test implementation of the CNN universal machine. In: Proceedings of 3th IEEE international workshop on cellular neural networks and their application. Rome, pp 61–66
33. Roska T, Chua LO (1993) The CNN universal machine: an analogic array computer. IEEE Trans Circ Syst II 40(3):163–173
34. Crounse KR, Chua LO (1995). Methods for image processing and pattern formation in cellular neural networks: a tutorial. IEEE Trans Circ Syst I Fundam Theory Appl 42(10):583–601
35. Matsumoto T, Chua LO, Furukawa R (1990) CNN cloning template: Hole filler. IEEE Trans Circ Syst I Fundam Theory Appl 37(5):635–638
36. Wang L, Zhang J, Shao H (2014) Existence and global stability of a periodic solution for a cellular neural network. Commun Nonlinear Sci Numer Simul 19(9):2983–2992
37. Zheng C-D, Jing X-T, Wang Z-S, Feng J (2010) Further results for robust stability of cellular neural networks with linear fractional uncertainty. Commun Nonlinear Sci Numer Simul 15(10):3046–3057
38. Chedjou JC, Kyamakya K (2015) A universal concept based on cellular neural networks for ultrafast and flexible solving of differential equations. IEEE Trans Neural Netw Learn Syst 26(4):749–762
39. Baştürk A, Günay E (2009) Efficient edge detection in digital images using a cellular neural network optimized by differential evolution algorithm. Expert Syst Appl 36:2645–2650
40. Li J, Peng Z (2015) Multi-source image fusion algorithm based on cellular neural networks with genetic algorithm. Opt Int J Light Electron Opt 126(24):5230–5236
41. Cuevas E, Gálvez J, Hinojosa S, Zaldívar D, Pérez-Cisneros M (2014) A comparison of evolutionary computation techniques for IIR model identification. J Appl Maths 827206
42. Giaquinto A, Fornarelli G (2009) PSO-Based cloning template design for CNN associative memories. IEEE Trans Neural Netw 20(11):1837–1841
43. Díaz P, Pérez-Cisneros M, Cuevas E, Hinojosa S, Zaldivar D (2018) An improved crow search algorithm applied to energy problems. Energies 11(3):571
44. Storn R, Price K (1995) Differential evolution—a simple and efficient adaptive scheme for global optimization over continuous spaces. Technical Rep. No. TR-95-012, International Computer Science Institute, Berkley
45. Bureerat S, Pholdee N (2018) Inverse problem based differential evolution for efficient structural health monitoring of trusses. Appl Soft Comput 66:462–472
46. Yang Y-H, Xian-Bin Xu, He S-B, Wang J-B, Wen Y-H (2018) Cluster-based niching differential evolution algorithm for optimizing the stable structures of metallic clusters. Comput Mater Sci 149:416–423

47. Suganthi ST, Devaraj D, Ramar K, Hosimin Thilagar S (2018) An improved differential evolution algorithm for congestion management in the presence of wind turbine generators. Renew Sustain Energy Rev 81, Part 1:635–642
48. Guedes JJ, Favoretto Castoldi M, Goedtel A, Marcos Agulhari C, Sipoli Sanches D (2018) Parameters estimation of three-phase induction motors using differential evolution. Electr Power Syst Res 154:204–212
49. Maggi S (2017) Estimating water retention characteristic parameters using differential evolution. Comput Geotech 86:163–172
50. Chua LO, Roska T, Kozek T, Zarandy A (1993) The CNN paradigm, cellular neural networks. Baffins Lane, Chichester, West Sussex PO191UD. Wiley, England
51. Price KV (1996) Differential evolution: a fast and simple numerical optimizer. In: IEEE biennial conference of the North American fuzzy information processing society, NAFIPS. Berkeley, CA, pp 524–527
52. Price KV, Storn RM, Lampinen J (2005) Differential evolution: a practical approach to global optimization. Springer, Berlin
53. Hua X, Fenga G, Duanb S, Liu L (2015) Multilayer RTD-memristor-based cellular neural networks for color image processing. Neurocomputing 162:150–162
54. Douglas DH, Peucker TK (1973) Algorithms for the reduction of the number of points required to represent a digitized line or its caricature. Can Cartogr 10(2):112–122
55. Mikolajczyk K, Tuytelaars T, Schmid C, Zisserman A, Matas J, Schaffalitzky F, Kadir T, Van Gool L (2005) A comparison of affine region detectors. Int J Comput Vis 65(1–2):43–72
56. Schmid C, Mohr R, Bauckhage C (2000) Evaluation of interest point detectors. Int J Comput Vis 37(2):151–172
57. Bowyer K, Kranenburg C, Dougherty S (1999) Edge detector evaluation using empirical ROC curves. In: IEEE conference on computer vision and pattern recognition, 1, pp 354—359
58. Bilotta E, Pantano P, Vena S (2017) Speeding up cellular neural network processing ability by embodying memristors. IEEE Trans Neural Netw Learn Systs 28(5):1228–1232
59. Zarandy A (1999) The art of CNN template design. Int J Circ Theory Appl 27(1):5–23

Chapter 5
Blood Vessel Segmentation Using Differential Evolution Algorithm

In the last years, medical image processing has become an important tool for health care. In particular, the analysis of retinal vessel images has become crucial for achieving a more accurate diagnosis and treatment for several cardiovascular and ophthalmological deceases. However, this task is extremely hard and time-consuming, often requiring human supervised segmentation of fundus images as well as some degree of professional skills. An automatic yet accurate procedure for retinal vessel segmentation is essential to assist ophthalmologists with illness detection. Several retinal vessel segmentation methods have been developed with satisfactory results. Nevertheless, the image preprocessing techniques implemented in this kind of procedure is known to have poor performance, mainly due to the complex nature of retinal vessel imaging. To improve the image preprocessing stage, a fast and accurate approach for retinal vessel segmentation is presented. This approach aims to enhance the contrast between retinal vessels and the image's background by implementing a natural inspired technique called Lateral Inhibition (LI). Additionally, a cross-entropy minimization procedure based on Differential Evolution (DE) is applied to find the appropriate threshold and then used to define whether an image pixel is a vessel or non-vessel. The described method has been tested by considering two well-known image databases: DRIVE and STARE. The obtained results are compared against those of other related methods in order to prove the accuracy, effectiveness, and robustness of the presented approach.

5.1 Introduction

An important source of information for an eye care expert is the retinal vessel pattern. Some pattern alterations such as length, width, angles, and branching structure, may suggest the presence of several diseases including diabetes, hypertension, arteriosclerosis, choroidal neovascularization, stroke, and cardiovascular issues [1–4].

© The Author(s), under exclusive license to Springer Nature Switzerland AG 2021

E. Cuevas et al., *Recent Metaheuristic Computation Schemes in Engineering*,
Studies in Computational Intelligence 948,
https://doi.org/10.1007/978-3-030-66007-9_5

The manifestation of microaneurysms (the distortion of blood vessels width due to capillary pressurizations) is the main signal of Diabetic Retinopathy, which represents one of the main causes of loss of sight in the world [5]. Under such conditions, accurate image segmentation of retinal vessels has become a relevant task for assisting on the diagnostic of a possible retinal disease. Manual segmentation represents hard and exhausting labor. Therefore alternatives have been considered to automatize it. However, automatic retinal vessel segmentation is a challenging task mainly due to the presence of other eye components such as the optic disk, fovea, macula, etc. Another challenging problem represents the fact that retinal vessels manifest a wide variety of widths, with some being too narrow to be appropriately distinguished from the eye's background.

With the purpose of achieving an automated blood vessel segmentation, different methods have been proposed. These techniques are divided into several categories depending on the approach on which they are based, such a pattern recognition, supervised and unsupervised machine learning methods, tracking-based, model-based, mathematical morphology, matched filtering, and multiscale approach [6, 7]. Some examples include the methods described in [1] and [8], where the authors introduced a supervised method based on a neural network. In [9], convolutional neural networks are used as a feature extractor, and then both are combined with the random forest method to classify the pixels. In [8], the authors proposed the use of Gabor filters and moment invariants-based features to describe vessel and non-vessel information with the purpose of training a neural network. In [10], the authors propose to extract the green and red color channels from an RGB image in order to enhance the non-uniform illumination, followed by a matched filter used to improve the contrast between background and blood vessels. Finally, a spatially weighted fuzzy c-means clustering is employed to segment the vessels so that they maintain a tree structure. In [11], self-organized maps are used to create an input set of training samples, and then a K-means clustering is implemented to separate the regions comprising the background and blood vessels. In [12], the authors proposed a probabilistic tracking method based on a Bayesian classifier. Similarly, in [13], a probabilistic tracking method that combines a multiscale line detection scheme is presented. In [14], it is proposed to apply orthogonal projections to represent the texture of the vessels' structure. In [15], the authors propose to divide the segmentation process into two steps: First, an image enhancement operation implemented to remove noise, low contrast, and non-uniform illumination, and second, morphological processing step used to define retinal blood vessels in the binarized image. In [16], Mathematical morphology and K-means clustering are combined to segment retinal vessels. In [17], the authors use a technique called Contrast Limited Adaptive Histogram Equalization (CLAHE) to enhance the eye's image, while a segmentation process is done by combining Gaussian and Laplacian of Gaussians (LoG) filters. Similarly, [18] implements a matched filter based on a Gaussian function with zero-mean to detect the vessels, while a classification threshold is modified depending on the first-order derivate of Gaussian's image response. In [19], the authors presented an algorithm that applies an anisotropic diffusion filter to remove noise and reconnect vessel lines, and then, multiscale line tracking is used to identify vessels with similar size. While many of these retinal vessel segmentation methodologies have to yield to notable results, there still exist some critical elements that need be addressed, including the presence of false negatives, connectivity loss in retinal vessels, the presence of noise, and others.

Recently, metaheuristic optimization algorithms have been extensively studied in order to solve a wide variety of different computer vision problems such as circle detection [20], segmentation of blood cell and brain images [21, 22], template matching [23], and others [24, 25]. Unlike traditional optimization methods, Metaheuristic techniques are not conditioned by continuity, differentiability, or initial conditions, which have favored their implementations in many real problems [26–30]. In particular, the Differential Evolution (DE) approach is widely known as one of the most successful metaheuristics, with thousands of applications being published every year [31–34]. In this chapter, a retinal vessel segmentation method based on DE is presented to find the threshold that minimizes the cross-entropy for determining a pixel as a vessel and non-vessel element, which is a critical step in the described approach.

The methodology consists of three stages: First, a preprocessing stage that implements LI to enhance the eye image is applied. Secondly, DE is used to minimize the cross-entropy score and to provide an appropriate segmentation threshold. Finally, morphological operations are applied to eliminate noise and artifacts in the output image.

5.2 Methodology

The objective of the method is to improve the vessel segmentation process. To provide that enhancement, the LI technique is included in order to achieve better results in the treatment of fundus images. In the processing stage, the cross-entropy is minimized by using the DE algorithm to find a threshold that allows us to classify pixels into a vessel or non-vessel element. Finally, in the postprocessing step, several morphological operations are applied to eliminate the undesired artifacts present in the image (see Fig. 5.1). In the following, the whole methodology is described in detail.

5.2.1 Preprocessing

It is well known that image processing is more efficient when it is performed on a grayscale rather than a color due to the higher complexity of color images. Besides, the computational cost decreases significantly. Furthermore, since the result of the segmentation process is a binary image with information of interest, the color characteristic is irrelevant to this procedure. Thus, working on an image transformed from color to grayscale improves the process.

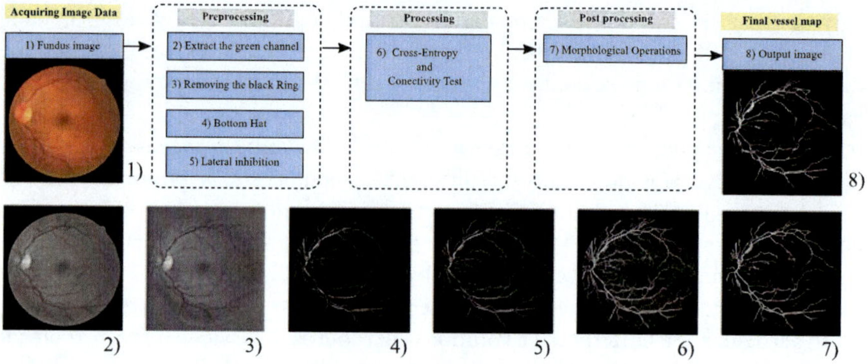

Fig. 5.1 The process diagram of the methodology

Retinal vessel images usually show a weak contrast between thin vessels and the background making them appear blurred [35]. Besides, thick vessels may appear thinner than they are. These effects and low quality in the fundus images are due to several factors such as the misalignment in the camera, poor focus, lighting problems, eye movement, and low pupil dilation [36, 37]. Then, the preprocessing stage is crucial to improve image quality before the retinal vessel detection process.

The first step in the preprocessing of fundus images is the conversion of the RGB image to grayscale by extracting the green channel. Green channel provides better contrast between vessels and the background [38, 39]. In addition, the human eye perceives the green channel stronger than any of the other two. Thus, the green channel is chosen for RGB to grayscale conversion.

Secondly, in preprocessing, the black background around the retina is removed from the grayscale image to improve the uniformity of the image background. A background with a high level of homogeneity is necessary to obtain better results when it is applying the bot-hat filter to the image [4]. The process of removing the black background means to replace the background with an intensity level average from the retinal. The average intensity value is obtained by taking three random samples of retinal intensity levels and calculating the mean value from those samples. Each sample consists of an array of 50×50 pixels.

After removing the black background around the retina, a third step of preprocessing is carried out. In this step, the bot-hat filter is applied to the image. The bot-hat filter, also known as Bottom-Hat, is used to adjust the intensities to improve contrast on grayscale and binary images by using morphological operations [37]. The filter applies the closing operation on an image, and then, the resulting image of this transformation is subtracted from the original fundus [40]. The bot-hat filter is defined by:

$$I_{bh} = \left(I_g \cdot S\right) - I_g \tag{5.1}$$

where the I_{bh} is the image result after applying the filter, I_g is the grayscale image coming from the previous step, $\cdot$ is the closing operation and S is a structural element.

The application of the bot-hat filter removes the noisy areas and information not corresponding to the retinal vessel network such as the optic disc and macula.

At this point, the image presents a certain grade of contrast between the fundus and venules. However, it is insufficient considering that the image will be the input of the following processes. Therefore, to improve the fundus-venules variation, the image is processed by the LI, which is detailed in the following section.

5.2.1.1 Lateral Inhibition

The mechanism of lateral inhibition was discovered and corroborated by Hart and Ratliff when they perform, in 1932, a series of electrophysiology experiments on the Limulus' vision in [41]. In the experiment, every individual microphthalmia of Limulus' ommateum is considered as a receptor, which is inhibited by its contiguous receptors. The inhibitory effect is mutual and spatially additive [42]. The operation of LI is in the time inhibited by its contiguous receptors. Therefore, it also inhibits its contiguous receptors at the same time.

The closer proximity to certain receptors among them, the more intensive the inhibited effect will be. In the context of the retinal image, the excited receptors in a highly illuminated area inhibit to those receptors in the weak illuminated area [43]. An interesting demonstration is shown in Fig. 5.2 [44]. Under such conditions, the light and shadow contrast are improved. Thus, in this method, the LI principle is implemented as a preprocessing step to improve contrast and avoiding the distortion of sensory information.

Specifically, in the LI model [45], an original grayscale image is denoted as I_g and the enhanced image is denoted as I_R. LI is implemented according to Eq. (5.2):

Fig. 5.2 An example of the contrast effect due to lateral inhibition in this visual illusion. Observe two objects that are identical; nonetheless, the left one is perceived brighter when placed in darker backgrounds

$$I_R(x, y) = I_g(x, y) + \sum_{i=-M}^{M} \sum_{j=-N}^{N} \delta_{ij} \cdot I_g(x + i, y + j) \tag{5.2}$$

where $I_g(x, y)$ is related to the original image in gray level value at the coordinate (x, y), $I_R(x, y)$ refers to the enhanced pixel gray level, δ_{ij} is the LI weight parameter, M, and N specify the inhibition scale of the receptive neighborhood, which is $M \times N$ in this method is the coefficient matrix size is 5×5. A grid schema is presented as representative of the condition of $M = 2$ and $N = 2$ in Fig. 5.3. Then the coefficients employed is as follow:

$$I_R(x, y) = \delta_0 \times I_g(m, n) + \delta_1 \left[\sum_{i=-1}^{j=1} \sum_{i=-1}^{i=1} I_g(m + i, n + j) - I_g(m, n) \right]$$

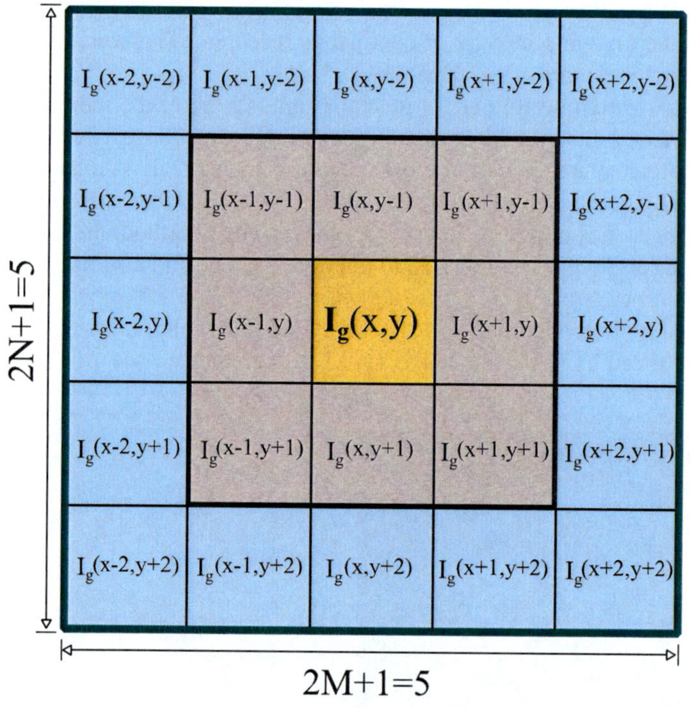

Fig. 5.3 The schematic grid related to the lateral inhibition model when $\boldsymbol{I_R}(x, y)$ is processed under the condition when the parameters are $M = 2$ and $N = 2$. The central position is related to the original gray level centered on the pixel (x, y) on $\boldsymbol{I_g}(x, y)$ and all the contiguous pixels modify the outcome of the computation of $\boldsymbol{I_R}(x, y)$.

$$+\delta_2\left[\sum_{i=-2}^{j=2}\sum_{i=-2}^{i=2} I_g(m+i,n+j) - \sum_{i=-1}^{j=1}\sum_{i=-1}^{i=1} I_g(m+i,n+j)\right] \quad (5.3)$$

It is necessary to satisfy a certain requirement in the way of selecting δ_{ij} as expressed below:

$$\sum_{i=-M}^{M}\sum_{j=-N}^{N}\delta_{ij} = 0 \quad (5.4)$$

Which implies balanced inhibition energy. In this case, the inhibition weight matrix is $\left[\delta_{ij}\right]_{5\times 5}$, while $\delta_0 = 1$, $\delta_1 = -0.025$, $\delta_2 = -0.075$ as [46] to form the following matrix presented below:

$$\delta_{ij} = \begin{bmatrix} -0.025\ -0.025\ -0.025 - 0.025 - 0.025 \\ -0.025\ -0.075\ -0.075 - 0.075 - 0.025 \\ -0.025\ -0.075\ 1 - 0.075 - 0.025 \\ -0.025\ -0.075\ -0.075 - 0.075 - 0.025 \\ -0.025\ -0.025\ -0.025 - 0.025 - 0.025 \end{bmatrix} \quad (5.5)$$

5.2.2 *Processing*

Once the image has been operated under the preprocessing stage, the processing is performed for vessel segmentation. In this procedure, cross-entropy is minimized to obtain the threshold that defines a pixel as a vessel or non-vessel.

5.2.2.1 Cross-Entropy

The cross-entropy is proposed by [47] as a distance D between two probability distributions U and V. Let $\boldsymbol{U} = \{u_1, u_2, \ldots, u_N\}$ and $\boldsymbol{V} = \{v_1, v_2, \ldots, v_N\}$ the cross entropy is defined as follow:

$$D(U,V) = \sum_{i=1}^{N} u_i \log\frac{u_i}{v_i} \quad (5.6)$$

Let the original image I_g and its corresponding histogram $h^g(i)$, $i = 1, 2, \ldots, L$, with L gray intensity levels. The segmented image I_{th}, by:

$$I_{th} = \begin{cases} \mu(1, t) & I_g(x, y) < t \\ \mu(t, L+1) & I_g(x, y) \geq t \end{cases} \tag{5.7}$$

where the threshold that divides a pixel as fundus and object is t, and the μ is given in Eq. (5.8)

$$\mu(a, b) = \sum_{i=a}^{b-1} i h^g(i) / \sum_{i=a}^{b-1} h^g(i) \tag{5.8}$$

Then as proposed in [48], the Eq. (5.7) is redefined as cost function:

$$D_{CE}(t) = -\sum_{i=1}^{t-1} i h^g(i) \log\left(\frac{1}{\mu(1, t)}\right) + \sum_{i=t}^{t-1} i h^g(i) \log\left(\frac{1}{\mu(t, L+1)}\right) \tag{5.9}$$

where the optimal threshold $\hat{t}$ is achieved by minimizing the Minimum Cross-Entropy Thresholding (MCET):

$$\hat{t} = \arg\min_t D_{CE}(t) \tag{5.10}$$

5.2.2.2 Differential Evolution Algorithm

The Differential Evolution (DE) is a vector-based evolutionary algorithm approach introduced by Storn and Price in 1996 [49] and, the Genetic Algorithms (GA) [50] is one of the most simple and powerful optimization methods inspired in the evolution phenomena.

Analogous to GA, at the actual generation s, DE applies a series of crossover, selection, and mutation operators to allow a population to evolve toward an optimal solution. This population represents the solutions $\mathbf{X} = \{\mathbf{x}_1, \mathbf{x}_2, \ldots \mathbf{x}_M\}$. For the mutation operation, new candidates or mutant solutions $\mathbf{m}_j^{s+1} = \left[m_{j,1}^{s+1}, m_{j,2}^{s+1}, \ldots, m_{j,d}^{s+1}\right]$ are generated for each individual $\mathbf{x}_j$ by adding the weighted difference between $\mathbf{x}_{ra_1}$ and $\mathbf{x}_{ra_2}$ to a third candidate solution $\mathbf{x}_{r_{a3}}$, all of those are randomly chosen solutions as the operation is illustrated as follows:

$$\mathbf{m}_j^{s+1} = \mathbf{x}_{ra_3}^s + \delta\left(\mathbf{x}_{r_{a1}}^s - \mathbf{x}_{ra_2}^s\right) \tag{5.11}$$

where $ra_1, ra_2, ra_3 \in \{1, 2, \ldots, M\}$ and with the condition $ra_1 \neq ra_2 \neq ra_3 \neq j$, all denote a randomly chosen solution's index, while the parameter $\delta \in [0, 1]$ stand for the named differential weight $\left(\mathbf{x}_{r_1}^s - \mathbf{x}_{r_2}^s\right)$. This magnitude is used to control the differential variation.

Additionally, for the crossover operation, DE generates a testing or trial solution vector $\mathbf{u}_j^{s+1} = \left[u_{j,1}^{s+1}, u_{j,2}^{s+1}, \ldots, u_{j,d}^{s+1}\right]$ corresponding to all population members' J'. In this trial vector $u_{j,n}^{s+1}$ are involved in the following elements such, the mutant solution $\mathbf{m}_j^{s+1}$ and the candidate solution $\mathbf{x}_j^s$, as follows:

$$u_{j,n}^{s+1} = \begin{cases} m_{j,n}^{s+1} \; if \; (\text{rand}(0, 1) \leq CR) \\ x_{j,n}^{s} \quad if \; (\text{rand}(0, 1) > CR) \, otherwise \end{cases} \quad \text{for } n = 1, 2, \ldots, d \tag{5.12}$$

where $n \in \{1, 2, \ldots, d\}$ stands for a randomly chosen dimension index, the rand(0, 1) is designed for a random number from within the interval [0, 1]. Additionally, the parameter $CR \in [0, 1]$ represents the crossover rate into the DE, which is used to regulate the probability of an element $u_{j,n}^{s+1}$ is being given by either a dimensional part of the mutant solution $\mathbf{m}_j^{s+1}$ ($m_{j,n}^{s+1}$) or a dimensional part from the candidate solution $\mathbf{x}_j^s$ ($x_{j,n}^s$). This way to carry out the crossover is known as the "Binomial scheme" [51].

Finally, for DE's selection process, each testing or trail solution $\mathbf{u}_j^{s+1}$ is compared against its respective candidate solution $\mathbf{x}_j^s$ in terms of his respective fitness value, and a greedy criterion is applied. This regard the follow actions if the trial solution $\mathbf{u}_j^{s+1}$ yields to a better fitness value than $\mathbf{x}_j^s$, then the value of the candidate solution for the next generation '$s+1$' takes the value of $\mathbf{u}_j^{s+1}$. Otherwise, it remains unchanged. as shown:

$$\mathbf{x}_j^{s+1} = \begin{cases} \mathbf{u}_j^{s+1} \; if \; f\left(\mathbf{u}_j^{s+1}\right) > if \; f\left(\mathbf{x}_j^{s}\right) \\ \mathbf{x}_j^{s} \quad otherwise \end{cases} \tag{5.13}$$

For the experiments, DE is implemented with a population of 25 individuals. All individuals from the population are involved in modification and replacement. The crossover probability is 0.5, and the differential factor varies from 0.3 to 1 with step 0.1.

5.2.2.3 Cross-Entropy—Differential Evolution

Cross entropy thresholding is applied to the retinal vessel image to partition it into two classes by determining the threshold value. A vessel or non-vessel classification depends on the threshold selection. Experimentally define an appropriate threshold value is a hard and tedious task that generates inefficient results. In order to make a properly threshold selection, the DE algorithm is used to minimize the cross-entropy among classes.

To find an acceptable threshold value, the DE algorithm is executed. A set of candidate solutions are given in every generation where a candidate solution represents a threshold value. The quality of the solution is evaluated in an objective function that uses cross-entropy. According to the DE operators and the value of the objective function, new candidate solutions are generated along the process. As the method evolves, the quality of the solutions improves.

The objective function of cross-entropy can be defined as:

$$\min f_{cross}(\boldsymbol{t})$$

$$\boldsymbol{t} \in \mathbf{X}$$

$$\mathbf{X} = \left\{ t \in \mathbb{R}^n | 0 \leq t_i \leq 255, \quad i = 1, 2, \ldots, n \right\} \tag{5.14}$$

where $f_{cross}(\boldsymbol{t})$ is the cross-entropy function given by Eq. (5.9), and $\boldsymbol{t}$ is a candidate solution.

5.2.3 *Postprocessing*

Although the connectivity test process removes most of the noise contain, it is not completely eradicated from the image. Under such conditions, it is necessary to eliminate the remaining noisy artifacts. Therefore, morphological operations, such as closing and opening, are applied. These operations are based on the objects' structure features and aim the vessel refinement to improve its definition and reconstruction.

5.3 Experiments

To evaluate the effectiveness of the methodology, it is tested in two public data image sets, DRIVE (Digital Retinal Images for Vessel Extraction) [52] and STARE (Structured Analysis of the Retina) [53]. The DRIVE data set is composed of forty retinal images (565 × 584 pixels 8 bits per color channel) captured by a Canon CR5 nonmydriatic 3CCD camera with a 45° field of view. The data set is subdivided into the training and test group, each of twenty images. The images in the training group were manually segmented once, while the test case images were twice. The segmentations were performed by three different human observers previously trained by an ophthalmologist. The sets X and Y resulting from manual segmentation of the test case are used in this method as ground truth.

The STARE is a database with twenty images (605 × 700 pixels 8 bits per color channel) for blood vessel segmentation digitalized by a Topcon TVR-50 fundus camera with a 35° field of view. This data set was manually segmented by two human observers. Where the first segmented the 10.4% as a vessel pixel while the second 14.9%. The reported results consider the segmentation of both observers as the ground truth.

In the retinal vessel segmentation task, the resultant is a pixel-based classification where a pixel belongs to a vessel or no vessel. In order to evaluate the performance of correct vessel classification, four measurements are used; Sensitivity (*Se*), which reflects the capability of the algorithm to detect vessel pixels, Specificity (*Sp*) ability to detect the no vessel pixels, Accuracy (*Acc*) measure the confidence of the algorithm. The measurements are defined as follows:

$$Sensitivity\,(Se) = \frac{T_P}{T_P + F_N} \tag{5.15}$$

$$Specificity\,(Sp) = \frac{T_N}{T_N + F_P} \tag{5.16}$$

$$Accuracy\,(Acc) = \frac{T_P + T_N}{T_P + T_N + F_P + F_N} \tag{5.17}$$

where the T_P (True Positive) indicates a pixel classification as a vessel in the segmented image and the ground truth, T_N (True Negative) shows the number of pixels classified into a non-vessel in the ground truth and segmented image. On the other hand, the F_P (False Positive) quantify the pixel pointed as a vessel in the segmented image and no vessel in the ground truth, F_N (False Negative) classify a pixel as no vessel in the segmented image and vessel in the ground truth.

Table 5.1 shows the results of the presented methodology for both observers of the database DRIVE, while Table 5.2 presents the results of the STARE database.

Figure 5.4 illustrated six segmented images by the presented method in comparison with the observers' ground truth. The first four are from the DRIVE database and the other two from the STARE database.

Table 5.3 the numerical comparison between the presented methodology results and other related methods is illustrated. Table 5.3 shows the accuracy of the presented method to segment the retinal. In addition to good performance, the methodology does not require a training stage or a sophisticated technique.

5.4 Summary

In this chapter, an approach for retinal vessel segmentation has been presented. The method enhances the image in the preprocessing stage by introducing the LI technique, which improves retinal vessel contrast to achieve higher accuracy.

Through the procedure, the green channel is extracted from the fundus image, the black background is eliminated, the bot-hat filter is applied for grayscale redistribution to enhance quality, and LI increases the contrast between retinal vessels and background. Subsequently, the minimization of cross-entropy is performed by the

Table 5.1 Results of the DRIVE database

Image	Drive observer 1			Drive observer 2		
	Se	Sp	Acc	Se	Sp	Acc
1	0.8316	0.9730	0.9621	0.8359	0.9753	0.9645
2	0.8918	0.9606	0.9555	0.9235	0.9649	0.9618
3	0.8360	0.9564	0.9479	0.8306	0.9675	0.9578
4	0.8793	0.9613	0.9561	0.8952	0.9671	0.9625
5	0.8714	0.9596	0.9540	0.8770	0.9733	0.9671
6	0.8378	0.9574	0.9492	0.8588	0.9634	0.9562
7	0.8516	0.9594	0.9527	0.8004	0.9771	0.9660
8	0.8121	0.9596	0.9509	0.7442	0.9760	0.9623
9	0.8532	0.9655	0.9591	0.8766	0.9671	0.9620
10	0.8250	0.9684	0.9592	0.8077	0.9785	0.9676
11	0.8677	0.9632	0.9572	0.8922	0.9716	0.9666
12	0.8003	0.9680	0.9561	0.8052	0.9748	0.9628
13	0.8457	0.9563	0.9489	0.8789	0.9551	0.9499
14	0.8030	0.9708	0.9596	0.8018	0.9773	0.9656
15	0.7713	0.9758	0.9629	0.7885	0.9738	0.9620
16	0.8812	0.9671	0.9613	0.8722	0.9720	0.9652
17	0.8540	0.9676	0.9604	0.8246	0.9768	0.9672
18	0.8019	0.9750	0.9629	0.8857	0.9677	0.9620
19	0.8510	0.9792	0.9696	0.8766	0.9637	0.9572
20	0.7704	0.9737	0.9608	0.8531	0.9589	0.9522
Mean	0.8368	0.9659	0.9573	0.8464	0.9701	0.9619

DE algorithm to define a threshold for vessel or non-vessel classification. In addition, a connectivity test is performed to remove non-vessel and noise from the image. Finally, morphological operations are applied for vessel refinement.

This method aims to provide an automatic system for medical image processing to assist experts in diagnosing diseases such as glaucoma and diabetic retinopathy.

This approach has been tested in public databases STARE and DRIVE. Furthermore, the performance of the presented method has been analyzed by comparing it with other related approaches. To evaluate this performance, Sensitivity, Specificity, and Accuracy measurements were used. The experimental results demonstrate that this technique is fast, accurate, and robustness over its contenders.

Table 5.2 Results of the STARE database

Image	Stare observer 1			Stare observer 2		
	Se	Sp	Acc	Se	Sp	Acc
1	0.8488	0.9575	0.9524	0.8426	0.9439	0.9392
2	0.7595	0.9528	0.9475	0.6646	0.9615	0.9534
3	0.7998	0.9770	0.9686	0.6821	0.9735	0.9596
4	0.8224	0.9466	0.9432	0.7019	0.9452	0.9386
5	0.8944	0.9574	0.9539	0.8936	0.9418	0.9391
6	0.8368	0.9453	0.9433	0.9392	0.9019	0.9026
7	0.7782	0.9669	0.9550	0.9322	0.9242	0.9247
8	0.6877	0.9624	0.9468	0.8830	0.9077	0.9063
9	0.8737	0.9753	0.9688	0.9606	0.9343	0.9360
10	0.7948	0.9581	0.9497	0.9541	0.8904	0.8936
11	0.8439	0.9793	0.9710	0.9641	0.9411	0.9425
12	0.8860	0.9761	0.9705	0.9852	0.9415	0.9442
13	0.8765	0.9611	0.9561	0.9814	0.9124	0.9165
14	0.8814	0.9594	0.9548	0.9560	0.9193	0.9215
15	0.8459	0.9566	0.9507	0.9562	0.9217	0.9235
16	0.8212	0.9293	0.9248	0.9204	0.8810	0.8827
17	0.9140	0.9617	0.9589	0.9761	0.9213	0.9245
18	0.9050	0.9767	0.9745	0.9503	0.9666	0.9661
19	0.8142	0.9819	0.9766	0.8931	0.9641	0.9619
20	0.7777	0.9567	0.9510	0.8753	0.9259	0.9243
Mean	0.8331	0.9619	0.9559	0.8956	0.9310	0.9300

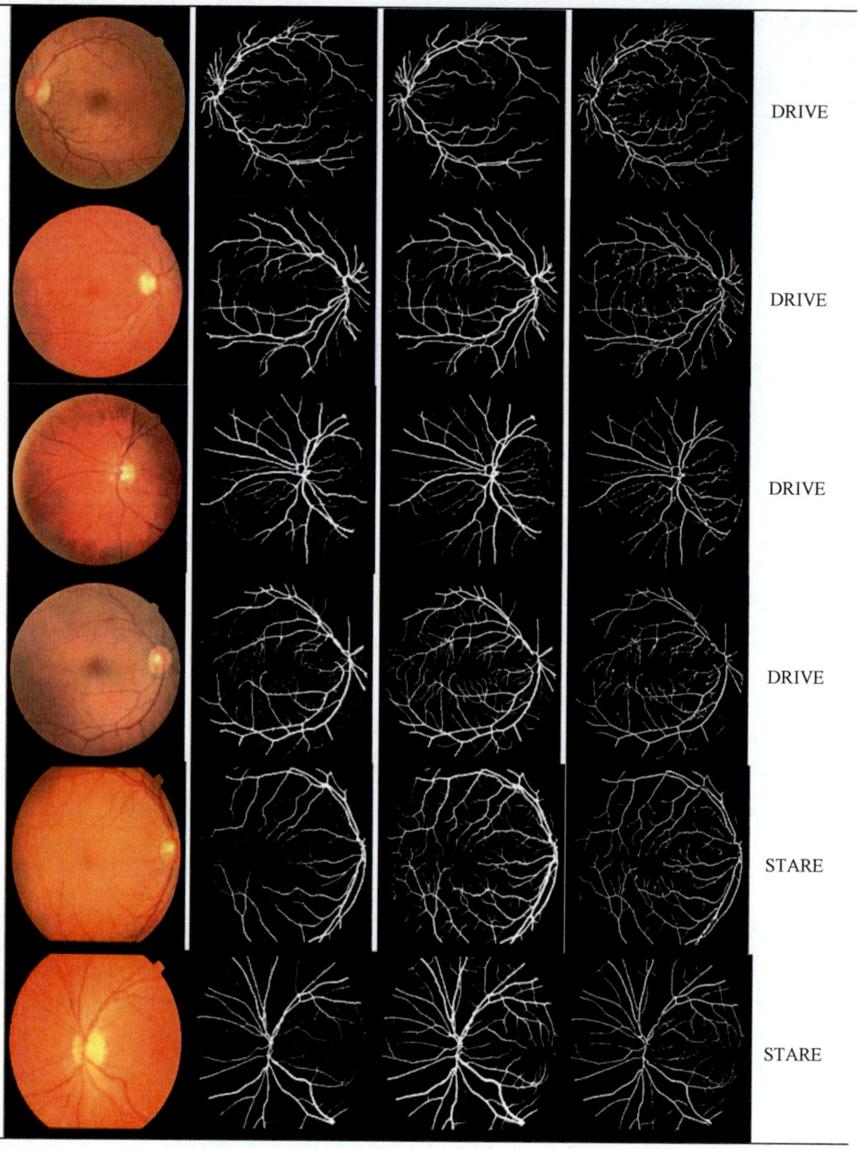

Fig. 5.4 Segmented images by presented methodology

Table 5.3 Comparison results with related methods

Methods	Drive			Stare		
	Se	Sp	Acc	Se	Sp	Acc
Jiang et al. [54]	–	–	0.9212	–	–	–
Zhang et. Al [18]	0.7120	0.9724	0.9382	0.7171	0.9753	0.9483
Staal et al. [52]	0.7194	0.9773	0.9442	0.6970	0.9810	0.9516
Qian et al. [1]	0.7354	0.9789	0.9477	0.7187	0.9767	0.9509
Câmara Neto et al. [55]	0.7942	0.9632	–	0.7695	0.9537	0.8616
Rezaee et al. [56]	0.7189	0.9793	0.9463	0.7202	0.9741	0.9521
Zhao [1]	0.7354	0.9789	0.9477	0.7187	0.9767	0.9509
Rodrigues et al. [57]	0.7165	0.9801	0.9465	–	–	–
Marin et al. [58]	0.7067	0.9801	0.9452	0.6944	0.9819	0.9526
Presented method	0.8464	0.9701	0.9619	0.8331	0.9619	0.9559

References

1. Qian Zhao Y, Hong Wang X, Fang Wang X, Shih FY (2014) Retinal vessels segmentation based on level set and region growing. Pattern Recognit 47(7):2437–2446s
2. Stanton AV et al (1995) Vascular network changes in the retina with age and hypertension. J Hypertens 13(12 Pt 2):1724–1728
3. Skovborg F, Nielsen AV, Lauritzen E, Hartkopp O (1969) Diameters of the retinal vessels in diabetic and normal subjects. Diabetes 18(5):292–298
4. Martinez-Perez ME, Hughes AD, Thom SA, Bharath AA, Parker KH (2007) Segmentation of blood vessels from red-free and fluorescein retinal images. Med Image Anal 11(1):47–61
5. Lázár I, Hajdu A (2015) Segmentation of retinal vessels by means of directional response vector similarity and region growing. Comput Biol Med 66:209–221
6. Fraz MM et al (2012) Blood vessel segmentation methodologies in retinal images—a survey. Comput Methods Progr Biomed 108(1):407–433
7. Kirbas C, Quek F (2003) A review of vessel extraction techniques and algorithmss
8. Franklin SW, Rajan SE (2014) Retinal vessel segmentation employing ANN technique by Gabor and moment invariants-based features. Appl Soft Comput 22:94–100
9. Wang S, Yin Y, Cao G, Wei B, Zheng Y, Yang G (2015) Hierarchical retinal blood vessel segmentation based on feature and ensemble learning. Neurocomputing 149:708–717
10. Kande GB, Subbaiah PV, Savithri TS (2010) Unsupervised fuzzy based vessel segmentation in pathological digital fundus images. J Med Syst 34(5):849–858
11. Lupaşcu CA, Tegolo D (2011) Automatic unsupervised segmentation of retinal vessels using self-organizing maps and K-means clustering. Springer, Berlin, Heidelberg, pp 263–274
12. Yin Y, Adel M, Bourennane S (2012) Retinal vessel segmentation using a probabilistic tracking method. Pattern Recognit 45(4):1235–1244
13. Zhang J, Li H, Nie Q, Cheng L (2014) A retinal vessel boundary tracking method based on Bayesian theory and multiscale line detection. Comput Med Imaging Graph 38(6):517–525
14. Zhang Y, Hsu W, Lee ML (2009) Detection of retinal blood vessels based on nonlinear projections. J Sig Process Syst 55(1–3):103–112
15. Khdhair N, Abbadi E, Hamood E, Saadi A (2013) Blood vessels extraction using mathematical morphology. J Comput Sci Publ Online 9(910):1389–1395
16. Hassan G, El-Bendary N, Hassanien AE, Fahmy A, Abullah SM, Snasel V (2015) Retinal blood vessel segmentation approach based on mathematical morphology. Procedia Comput Sci 65:612–622

17. Kumar D, Pramanik A, Kar SS, Maity SP (2016) Retinal blood vessel segmentation using matched filter and Laplacian of Gaussian. In: International conference on signal processing and communications (SPCOM), pp 1–5
18. Zhang B, Zhang L, Zhang L, Karray F (2010) Retinal vessel extraction by matched filter with first-order derivative of Gaussian. Comput Biol Med 40(4):438–445
19. Ben Abdallah M et al (2015) Automatic extraction of blood vessels in the retinal vascular tree using multiscale medialness. Int J Biomed Imaging 2015:519024
20. Cuevas E, Sención-Echauri F, Zaldivar D, Pérez-Cisneros M (2012) Multi-circle detection on images using artificial bee colony (ABC) optimization. Soft Comput 16(2):281–296
21. Oliva D, Cuevas E (2017) A medical application: blood cell segmentation by circle detection. Springer, Cham, pp 135–157
22. Oliva D, Hinojosa S, Cuevas E, Pajares G, Avalos O, Gálvez J (2017) Cross entropy based thresholding for magnetic resonance brain images using crow search algorithm. Expert Syst Appl 79:164–180
23. González A, Cuevas E, Fausto F, Valdivia A, Rojas R (2017) A template matching approach based on the behavior of swarms of locust. Appl Intell, pp 1–12
24. Díaz P, Pérez-Cisneros M, Cuevas E, Hinojosa S, Zaldivar D (2018) An im-proved crow search algorithm applied to energy problems. Energies 11(3):571
25. Cuevas E, Gálvez J, Hinojosa S, Zaldívar D (2014) Pérez-Cisneros, M, A com-parison of evolutionary computation techniques for IIR model identification. Journal of Applied Mathematics 2014:827206
26. Valdivia-Gonzalez A, Zaldívar D, Fausto F, Camarena O, Cuevas E, Perez-Cisneros M (2017) A states of matter search-based approach for solving the problem of intelligent power allocation in plug-in hybrid electric vehicles. Energies 10(1):92
27. Yang Y, Wang Z, Yang B, Jing Z, Kang Y (2017) Multiobjective optimization for fixture locating layout of sheet metal part using SVR and NSGA-II. Math Probl Eng 2017:1–10
28. Zhang H, Dai Z, Zhang W, Zhang S, Wang Y, Liu R (2017) A new energy-aware flexible job shop scheduling method using modified biogeography-based optimization. Math Probl Eng 2017:1–12
29. Pang C, Huang S, Zhao Y, Wei D, Liu J (2017) Sensor network disposition facing the task of multisensor cross cueing. Math Probl Eng 2017:1–8
30. Kóczy LT, Földesi P, Tüű-Szabó B (2017) An effective discrete bacterial memetic evolutionary algorithm for the traveling salesman problem. Int J Intell Syst 32(8):862–876
31. Céspedes-Mota A et al (2016) Optimization of the distribution and localization of wireless sensor networks based on differential evolution approach. Math Probl Eng 2016:1–12
32. Lai L, Ji Y-D, Zhong S-C, Zhang L (2017) Sequential parameter identification of fractional-order duffing system based on differential evolution algorithm. Math Probl Eng 2017:1–13
33. Bhattacharyya S, Konar A, Tibarewala DN (2014) A differential evolution based energy trajectory planner for artificial limb control using motor imagery EEG signal. Biomed Sig Process Control 11(1):107–113
34. Elsayed S, Sarker R (2016) Differential evolution framework for big data optimization. Memetic Comput 8(1):17–33
35. Rahebi J, Hardalaç F (2014) Retinal blood vessel segmentation with neural network by using gray-level co-occurrence matrix-based features. J Med Syst 38(8):85
36. Zheng Y, Kwong MT, Maccormick IJC, Beare NAV, Harding SP (2014) A comprehensive texture segmentation framework for segmentation of capillary non-perfusion regions in fundus fluorescein angiograms. PLoS One 9(4)
37. Bai X, Zhou F, Xue B (2012) Image enhancement using multi scale image features extracted by top-hat transform. Opt Laser Technol 44(2):328–336
38. Salazar-Gonzalez A, Kaba D, Li Y, Liu X (2014) Segmentation of the blood vessels and optic disk in retinal images. IEEE J Biomed Heal Inform 18(6):1874–1886
39. Soares JVB, Leandro JJG, Cesar RM, Jelinek HF, Cree MJ (2006) Retinal vessel segmentation using the 2-D Gabor wavelet and supervised classification. IEEE Trans Med Imaging 25(9):1214–1222

40. Bai X, Zhou F (2010) Multi structuring element top-hat transform to detect linear features. In: IEEE 10th international conference on signal processing proceedings, pp 877–880
41. Hartline HK (1938) The response of single optic nerve fibers of the vertebrate eye to illumination of the retina. Am J Physiol Leg Content 121(2)
42. Li B, Li Y, Cao H, Salimi H (2016) Image enhancement via lateral inhibition: an analysis under illumination changes. Optik (Stuttg) 127:5078–5083
43. Fang Z, Dawei Z, Ke Z (2007) Image pre-processing algorithm based on lateral inhibition. In: 2007 8th international conference on electronic measurement and instruments, 2007, pp 2-701-2–705s
44. Coren JS, Girgus S (1978) Seeing is deceiving: the psychology of visual illusions. Lawrence Erlbaum, Hillsdale. References—Scientific Research Publish
45. Liu F, Duan H, Deng Y (2012) A chaotic quantum-behaved particle swarm optimization based on lateral inhibition for image matching. Opt Int J Light Electron Opt 123(21):1955–1960
46. Wang X, Duan H, Luo D (2013) Cauchy biogeography-based optimization based on lateral inhibition for image matching. Optik (Stuttg) 124(22):5447–5453
47. Kullback S (1968) Information theory and statistics. Dover Publications
48. Li CH, Lee CK (1993) Minimum cross-entropy thresholding. Pattern Recognit 26(4):617–625
49. Storn R, Price K (1997) Differential evolution—a simple and efficient heuristic for global optimization over continuous spaces. J Glob Optim 11:341–359
50. Schmitt LM (2001) Theory of genetic algorithms. Theor Comput Sci 259(1):1–61
51. Yang X-S (2014) Nature-inspired optimization algorithms. In: Nature-inspired optimization algorithms
52. Staal J, Abramoff MD, Niemeijer M, Viergever MA, van Ginneken B (2004) Ridge-based vessel segmentation in color images of the retina. IEEE Trans Med Imaging 23(4):501–509
53. Hoover AD, Kouznetsova V, Goldbaum M (2000) Locating blood vessels in retinal images by piecewise threshold probing of a matched filter response. IEEE Trans Med Imaging 19(3):203–210
54. Jiang X, Mojon D (2003) Adaptive local thresholding by verification-based multithreshold probing with application to vessel detection in retinal images. IEEE Trans Pattern Anal Mach Intell 25(1):131–137
55. Câmara Neto L, Ramalho GLB, Rocha Neto JFS, Veras RMS, Medeiros FNS (2017) An unsupervised coarse-to-fine algorithm for blood vessel segmentation in fundus images. Expert Syst Appl 78(C):182–192s
56. Rezaee K, Haddadnia J, Tashk A (2017) Optimized clinical segmentation of retinal blood vessels by using combination of adaptive filtering, fuzzy entropy and skeletonization. Appl Soft Comput J 52:937–951
57. Rodrigues LC, Marengoni M (2017) Segmentation of optic disc and blood vessels in retinal images using wavelets, mathematical morphology and Hessian-based multiscale filtering. Biomed Sig Process Control 36:39–49
58. Marín D, Aquino A, Gegundez-Arias ME, Bravo JM (2011) A new supervised method for blood vessel segmentation in retinal images by using gray-level and moment invariants-based features. IEEE Trans Med Imaging 30(1):146–158

Chapter 6
Clustering Model Based on the Human Visual System

Clustering involves the process of dividing a collection of abstract objects into a number of groups, which are integrated with similar elements. Several clustering methods have been introduced and developed in the literature with different performance levels. Among such approaches, density algorithms present the best advantages, since they are able to find clusters from a dataset under different scales, shapes, and densities without requiring the number of groups as input. On the other hand, there are processes that humans perform much better than deterministic approaches or computers. In fact, humans can visually cluster data exceptionally well without the necessity of any training. Under this unique capability, clustering, instantaneous, and effortless for humans, represents a fundamental challenge for artificial intelligence. In this chapter, a simple clustering model inspired by the way in which the human visual system associates patterns spatially is presented. The model, at some abstraction level, can be characterized as a density grouping strategy. The approach is based on Cellular Neural Networks (CNNs), which have demonstrated to be the best models for emulating the human visual system. In the method, similar to the biological model, CNN is used to build especially groups while an automatic mechanism tries different resolution scales to find the best possible data categorization. Different datasets have been adopted to evaluate the performance of the algorithm. Their results are also compared with popular density clustering techniques from the literature. Computational results demonstrate that the CNN approach presents competitive results in comparison with other algorithms regarding accuracy and robustness.

6.1 Introduction

The objective of Extracting knowledge techniques is to automatically detect patterns in data with the idea of using these uncovered patterns in the decision-making process.

© The Author(s), under exclusive license to Springer Nature Switzerland AG 2021

E. Cuevas et al., *Recent Metaheuristic Computation Schemes in Engineering*,
Studies in Computational Intelligence 948,
https://doi.org/10.1007/978-3-030-66007-9_6

Clustering [1] is one of the most popular methods in machine learning for extracting information from data. Clustering involves the process of dividing a set of abstract elements into groups such that the elements in the same cluster maintain a high similarity but are significantly different from the members of other groups. Clustering has been extensively used in several domains, which include applications in communications [2], dimension reduction [3], ATM covering [4], medical diagnosis [5], image processing [6], information retrieval [7], and bioinformatics [8].

Several clustering methods have been suggested and developed in the literature. Essentially, such approaches are divided into four classes: hierarchical, partitional, grid-based clustering, and density-based. Some representative examples of these categories can be found in [9–22]. Hierarchical, partitional, and grid-based clustering techniques use as similarity criterion the Euclidean distance among cluster elements. However, there exist various scenarios where the Euclidian distance cannot be used to separate the elements into groups appropriately. Under such conditions, those methods cannot be directly applied. On the other hand, density-based algorithms employ the local concentration of data to build groups instead of the euclidian distance. Consequently, they can find clusters from a dataset under different scales, shapes, and densities without requiring the number of groups as input. Some representative density-based algorithms involve the classical Density-based spatial clustering of applications with noise (DBSCAN) [17] and the recent Gaussian Density Distance (GDD) [23] method.

Recognizing visual patterns and associating them with a particular category seems so easy for humans. We can effortlessly detect and classify groups of objects in spite of their configuration complexity [24]. Such capabilities are mainly the product of the human vision system [25]. From an operative perspective, the human visual system represents the most important sensor since our survival depends on the precise and fast classification of object characteristics extracted from visual patterns [26].

The human visual system presents two relevant characteristics according to its capacities to associate spatially patterns: receptive cells [27] and varying spatial resolution [28, 29]. The human visual system is composed of a large number of receptive cells that operate in parallel. These units are spatially organized in such a way that units maintain connections with only other neighbor elements [30]. Under the process of these units, several complex operations take place, such as low-pass filtering, shape detection, etc. [31]. Humans and other species also obtain visual knowledge through the modification of the spatial resolution across the visual field [28]. With the reduction of the spatial resolution, the visual mechanism continually tests whether it is possible to delete more image details or no [29]. The objective of this reduction is to eliminate unimportant image details which do not put at risk its general interpretation.

The benefit of this mechanism is to decrease brain operation and metabolic costs [32].

Cellular neural networks (CNNs) [33, 34] are a class of parallel information-processing paradigm which integrates a group of similar operative elements. Each unit emulates a nonlinear dynamic system that is locally interconnected with its neighbor elements. CNN presents two fundamental features: parallel processing and local connection [35]. Consequently, CNN can work at high speed by using different digital processors, through array computations. Due to these characteristics, CNNs have been extensively used in a wide range of applications defined by their spatial dynamics. Some examples of such applications include temperature prediction [36], modeling atmospheric dispersion [37], desertification disaster prediction [38], communications [39], fuzzy systems [40], image processing [41], to mention a few. On the other hand, the distinct similarities between the human visual system and the general structure of the CNN permit the development of CNN models that accurately simulate the human visual system in certain aspects [27, 42–44]. Such approaches have demonstrated to be the best models for emulating the characteristics and properties of the human visual system in comparison with other proposed paradigms [27, 44].

From biological models, several approaches have been considered to solve engineering problems. In this chapter, a simple clustering model inspired by the way in which the human visual system associates patterns spatially is presented. The model, at some abstraction level, can be characterized as a density grouping strategy. The approach is mainly based on Cellular Neural Networks (CNNs). In the method, similar to the biological model, CNN is used to build groups spatially through operations of the locally interconnected units. During the clustering process, an automatic mechanism tries different scales to find the best possible data categorization. To evaluate the performance of the algorithm, different datasets have been adopted. The experimental clustering results are also compared with popular density techniques from the literature. Computational results demonstrate that the CNN approach presents competitive results in comparison with other algorithms regarding accuracy and robustness.

6.2 Cellular-Nonlinear Neural Network

A CNN involves an arrangement of $M \times N$ locally-connected processing units. In a CNN, each unit $C(i,j)$ represents a nonlinear dynamical system, which is modeled by the state equation given in Eq. 6.1.

$$\dot{x}_{i,j} = -x_{i,j} + \sum_{k=-1}^{1} \sum_{l=1}^{1} a_{k,l} y_{i+k,j+l} + \sum_{k=-1}^{1} \sum_{l=1}^{1} b_{k,l} u_{i+k,j+l} + z_{i,j}, \tag{6.1}$$

where $x_{i,j}$ represent the state of $C(i,j)$. The output $y_{i,j}$ of the unit is functionally related to the state $x_{i,j}$ by the nonlinearity defined as follows:

$$y_{i,j} = f(x_{i,j}) = 0.5 \cdot |x_{i,j} + 1| - 0.5 \cdot |x_{i,j} - 1|, \tag{6.2}$$

The parameters that determine the behavior of each unit $C(i,j)$ are symbolized by the set of elements (**A**, **B**, z). From these parameters $\mathbf{A} = \left[a_{k,l}\right]$ and $\mathbf{B} = \left[b_{k,l}\right]$ represent the feedback and control templates, respectively, where $k, l \in \{-1, 0, 1\}$ for the case of a 3×3 neighborhood. The threshold value is z. Therefore, each unit $C(i,j)$ involves a feed-forward connection $a_{k,l} \cdot u_{i+k,j+l}$ that corresponds to the input effect. In the processing, $C(i,j)$ also considers a feedback link $b_{k,l} \cdot y_{i+k,j+l}$ that represents the output impact produced by the neighbor units. Figure 6.1 exhibits the schematic representation of a CNN (a) and its processing model (b).

For solving Eq. 6.1, the initial state $x_{i,j}(0)$, along with the boundary conditions, must be defined. One particular case of CNNs is the uncouple CNNs. In this scheme, the elements of the feedback template **A** are zero except the central coefficient $a_{0,0}$. Under such conditions, the Eq. 6.1 can be reduced as follows:

$$\dot{x}_{i,j} = h_{i,j}(x_{i,j}, w_{i,j}) = g_{i,j}(x_{i,j}) + w_{i,j}, \tag{6.3}$$

where $g_{i,j}(x_{i,j})$ represents the driving-point (D.P.) element defined as:

$$g_{i,j}(x_{i,j}) = -x_{i,j} + 0.5 \cdot a_{00} \cdot \left|x_{i,j} + 1\right| - 0.5 \cdot a_{00} \cdot \left|x_{i,j} - 1\right| \tag{6.4}$$

The member $w_{i,j}$ corresponds to the offset level formulated as follows:

$$w_{i,j} = \sum_{k=-1}^{1} \sum_{l=-1}^{1} b_{k,l} \cdot u_{i+k,j+1} + z_{i,j} \tag{6.5}$$

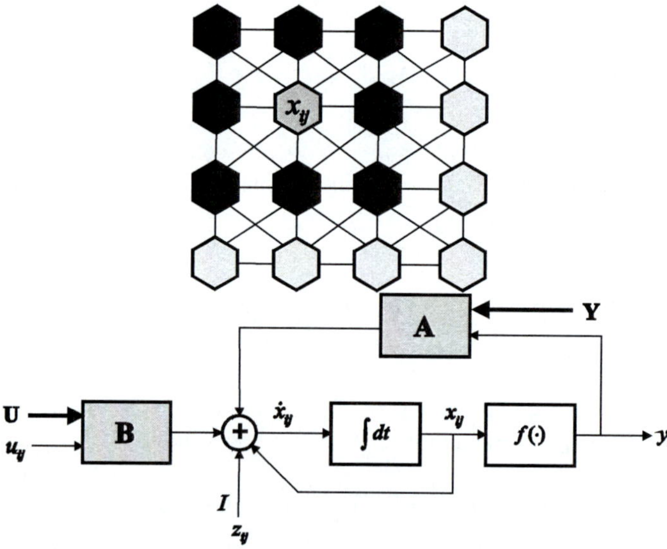

Fig. 6.1 Schematic representation of a CNN **a** and its processing model **b**

Assuming binary elements $u_{i,j} \in \{-1, 1\}$ as inputs, the steady-state output $y_{i,j}(\infty)$ of each unit $C(i,j)$ can be directly calculated considering the following cases [45]:

Case 1. With $a_{0,0} > 1$:

$$y_{i,j}(\infty) = sign\left[(a_{00} - 1)x_{i,j}(0) + w_{i,j}\right] \tag{6.6}$$

Case 2. If $a_{0,0} = 1$:

$$y_{i,j}(\infty) = \begin{cases} sign(w_{i,j}) & \text{if } w_{i,j} \neq 0 \\ x_{i,j}(0) & \text{if } w_{i,j} = 0 \end{cases} \tag{6.7}$$

Case 3. If $a_{0,0} < 1a_{0,0} < 1$:

$$y_{i,j}(\infty) = \begin{cases} sign(w_{i,j}) & \text{if } \left|w_{i,j}\right| \geq 1 - a_{0,0} \\ w_{i,j}(1 - a_{0,0})^{-1} & \text{if } \left|w_{i,j}\right| < 1 - a_{0,0} \end{cases} \tag{6.8}$$

In all cases, $a_{0,0} > 1$, $a_{0,0} = 1$ and $a_{0,0} < 1$, it is assumed that the initial state value $x_{i,j}(0)$ is within the interval $\{-1, 1\}$.

For estimating the CNN parameters **A**, **B**, and z, one of the most straightforward design methods is the case 1 [50]. Under this methodology, the elements of the feedback template **A** are zero except the central coefficient $a_{0,0}$, which must be fixed to a value more than one. Then, the values of the control template **B** are set to a value less than one, according to the local rules that the CNN must emulate. Likewise, the value of z is set to zero. Finally, the initial states $x_{i,j}(0)$ are determined from Eq. 6.6.

6.3 Human Visual Models

The human visual system can effortlessly combine different types of information to detect and classify groups of objects from visual stimuli. As an obscure structured system, the visual system captures and processes the information from an image projected by the optical system of the eye. This mechanism is remarkably complex. As a consequence, to model, analyze, and even simulate simple human visual schemes represent an extremely challenging process.

In this chapter, a simple clustering model inspired by the way in which the human visual system associates patterns spatially is presented. The human visual system presents two relevant characteristics according to its capacities to associate spatially patterns: receptive cells [27] and varying spatial resolution [28, 29]. In this section, the schemes and models employed to represent both biological mechanisms are discussed.

6.3.1 Receptive Cells

The human visual system is composed of a large number of receptive cells that operate in parallel. These units are spatially organized in such a way that units maintain connections with only other neighbor elements [30]. Under the process of these units, several complex operations take place, such as low-pass filtering, shape detection, etc. [31].

Among the models used for emulating the receptive field of the visual system, the Cellular neural networks are the most accurate. CNN scheme represents an integrated approach that accurately models the spatial–temporal properties of the receptive cells in the visual system.

Under the scheme of CNN, each receptive field of the visual system corresponds to a unit $C(i,j)$. Therefore, the complete retinal processing structure is simulated by a CNN of $M \times N$ units. Figure 6.2a presents a simple representation of the CNN located in the ocular. The model of each receptive field $C(i,j)$ can be visualized as a three interconnected layer arrangement. The first layer **I** collects the input visual stimulus $\mathbf{u} = \{u_{1,1}, \ldots, u_{M,N}\}$. The second layer **X** represents the state of each receptive field $\mathbf{x} = \{x_{1,1}, \ldots, x_{M,N}\}$. Finally, the third layer **Y** symbolizes the output of each receptive field $\mathbf{y} = \{y_{1,1}, \ldots, y_{M,N}\}$. Figure 6.2b illustrates the model for a receptive field corresponding to the unit $C(i,j)$.

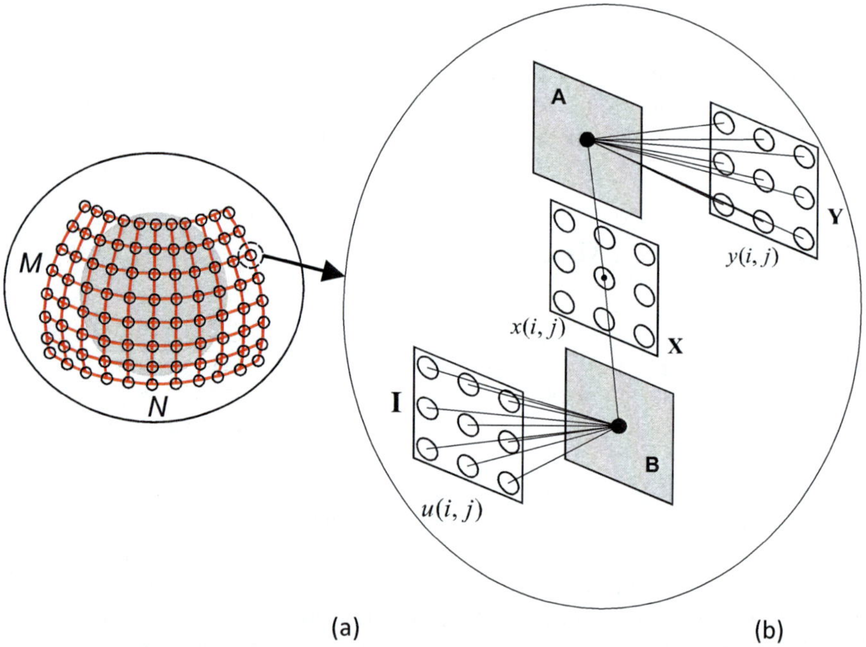

Fig. 6.2 **a** A simple representation of the CNN located in the ocular and **b** the model for a receptive field corresponding to the unit $C(i,j)$

The potential of a CNN originates from the nonlinear behavior of its dynamics and its architecture. Considering different templates **A** and **B**, the set of receptive fields can perform a particular behavior in order to emulate a specific visual mechanism. Several processes of the human visual system have been accurately simulated by using CNN as a net of receptive fields. Some examples [46] include the triadic synapse function, directional selectivity, the Muller-Lyer illusion, Fovea behavior [47], etc.

6.3.2 Modification of the Spatial Resolution

Humans and other species also obtain visual knowledge through the modification of the spatial resolution across the visual field [28]. If the visual system always used a high spatial resolution, it would drastically expand the brain operation, increasing the metabolic cost. The benefit of this varying resolution architecture is to decrease the number of required operations and to improve the accuracy detection [32]. The capacity of humans to favorably modify the spatial resolution of the visual system relies on different mechanisms such as the microsaccades or visual cortex [48].

The process of modifying the spatial resolution in visual perception has been a profoundly discussed topic that is still currently unsolved [49]. There exist several theories that explain this phenomenon considering distinct mechanisms. In general, the idea is to reduce the spatial resolution gradually (the number of active receptive fields) until the critical visual information remains still clear [28]. The content of several image details increases brain operation and metabolic cost. The objective of this reduction is to eliminate unimportant image details which do not put at risk its general interpretation. With the reduction of the spatial resolution, the visual mechanism constantly tests whether it is possible to delete more image details or no [29]. In order to illustrate this process, Fig. 6.3 shows the effect of reducing spatial resolution gradually in two images. The original images (a) and (d) contain a set of artifacts as noisy information. According to the Figure, the resolution can be decreased by eliminating the set of undesired information, but without losing critical information for their interpretation.

6.4 Clustering Method

The objective of a clustering technique is to separate a collection **D** of n objects ($\mathbf{D} = \{\mathbf{d}_1, \ldots, \mathbf{d}_n\}$) into subgroups according to their distribution. Hierarchical, partitional, and grid-based clustering techniques use as clustering criterion the Euclidean distance among the elements in the groups. However, there exist several scenarios where the Euclidian distance cannot be used to separate the elements in groups appropriately. Figure 6.4 illustrates two different cases. In the first case (Fig. 6.4a), the data distributions maintain convex shapes. Therefore, the Euclidian distance can be

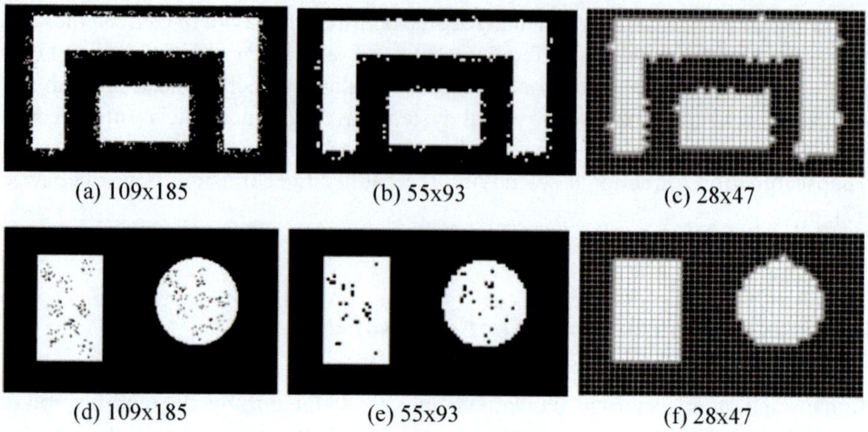

Fig. 6.3 Effect of reducing the spatial resolution gradually **a** and **d** Original images, **b**, **c**, **e**, and **f** images with lower resolution

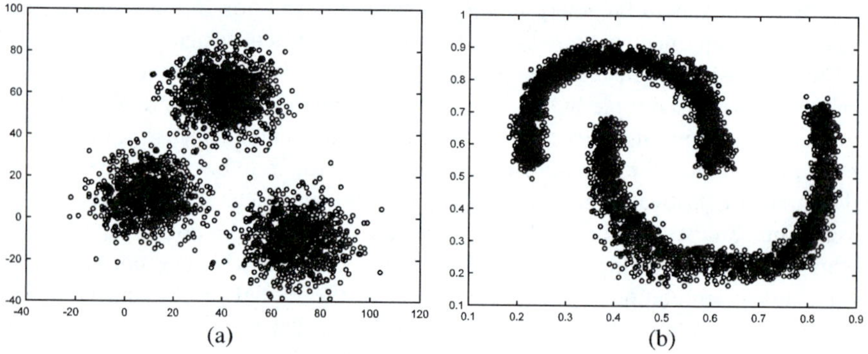

Fig. 6.4 Data distributions: **a** convex shapes that can use the Euclidian distance as clustering criterion and **b** arbitrary shapes which cannot employ the distance as similarity index

considered as a similarity criterion, so that the elements in the same cluster maintain a small distance (high similarity) but are significantly different (large distance) from the members of other groups. On the other hand, Fig. 6.4b shows data distributions with arbitrary shapes in which are impossible to use the Euclidian distance for building groups. In these configurations, elements that belong to a certain group maintain small distances to elements which, in fact, are affiliated with another group. Under such conditions, methods based on Euclidian distance cannot be directly applied. As an alternative to these methods, density-based algorithms employ the local concentration of data to build groups instead of only distance. Consequently, they can find clusters from a dataset under different scales, shapes, and densities without requiring the number of groups as input.

In this chapter, a simple clustering model inspired by the way in which the human visual system associates patterns spatially is presented. The model, at some abstraction level, can be characterized as a density grouping strategy. To build the clustering model, the approach combines two different biological mechanisms, such as receptive cells and modification of the spatial resolution. In this section, the modeling of these two mechanisms and their implementation in a clustering algorithm are discussed.

6.4.1 Representation of Data Distribution as a Binary Image

As the described method is inspired by the biological visual system, the first operation is to transform the data **D** into a binary image. Therefore, the set of data is composed of elements in two dimensions, such as $\mathbf{d}_i = (v_i^1, v_i^2)\ i \in 1, \ldots, n$. Under this process, the set of observations **D** is mapped onto a discrete space $\mathbb{Z}^2$. Considering the maximal and minimal values, each dimension from **D** is divided into S exclusive intervals. With this discretization, a set of $S \times S$ squares is defined. These squares represent a lattice **Q** in the data space containing the sampling points of **D**. Each square is identified by two integers q_1 and q_2 which define the relative position of the square in the arrangement **Q** $(q_1, q_1 \in 1, \ldots S)$. The selection and modification of the parameter S that determines the resolution of the lattice **Q** are discussed in Sect. 6.3.

Then, the sampling of the data observations **D** on the arrangement **Q** of $S \times S$ points is obtained by just summing the number of data elements of **D** that fall inside each square of **Q**. This process, which is similar to the computation of a two-dimensional histogram, is considered in order to use the robust and powerful functions implemented in almost all programming languages. Finally, the binary image **BQ** is produced by thresholding the arrangement **Q**. Therefore, cells from **Q** that include at least one data observation are set to one in **BQ** In contrast, the empty cells from **Q** are set to -1 in **BQ** Fig. 6.5 illustrates the process of representing data distribution. Figures 6.5a, b show **Q** and **BQ**, respectively, considering the data distribution of Fig. 6.4b.

6.4.2 Receptive Cells

Once obtained the image **BQ** from the data distribution **D**, the next step is to process it by a CNN, which models the effect of the receptive cells over the visual stimulus representing for **BQ** To associate patterns, the receptive fields build groups spatially through the integration of visual information. This local integration can be emulated by eliminating the white values (-1 elements of **BQ**) inside a certain neighborhood. Under this process, areas with a low concentration of data (black points) are rebuilt to produce solid visual objects. To implement this behavior in a CNN, its parameters

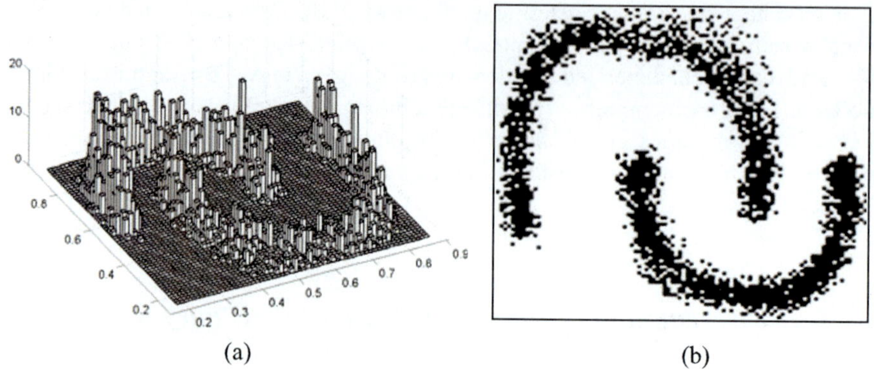

Fig. 6.5 The process of representing data distribution D from 4 **b**, **a Q**, and **b BQ**

A, **B**, and z must be defined. The estimation of such parameters is typically based on two steps: (I) the specification of the local rules and (II) the determination of the parameters that correspond to the behavior of the local rules according to the CNN properties.

Local rules establish the set of possible scenarios and the way in which the CNN has to operate. Therefore, to rebuild low data concentrations, a local rule must impose that all the white cells (outputs equal to -1) surrounded by at least one black cell (outputs equal to 1) become black. Figure 6.6 shows a simplified representation of the distinct scenarios presented by the local rules in which the central unit will be set from -1 to 1. In the Figure, the x value represents the don't care condition, that is, it can be whether 1 or -1 without changing the analysis. Under such circumstances, the only case in which the central unit is set to -1 (white) represents the configuration when all elements inside the 3×3 neighborhood are -1.

Once defined the local rules, the next step is to calculate the CNN parameters **A**, **B**, and z. To estimate such values, the simplest method is to use the CNN properties defined in Cases 1–3. Considering particularly the case 1, the elements of the feedback template **A** are zero except the central coefficient $a_{0,0}$, which must be more than one. In the design, this value is fixed to 2. In the analysis, the value of z is set to zero. Then, the control template **B** is set with the following values:

$$\mathbf{B} = \begin{bmatrix} 1/9 & 1/9 & 1/9 \\ 1/9 & 1/9 & 1/9 \\ 1/9 & 1/9 & 1/9 \end{bmatrix} \tag{6.9}$$

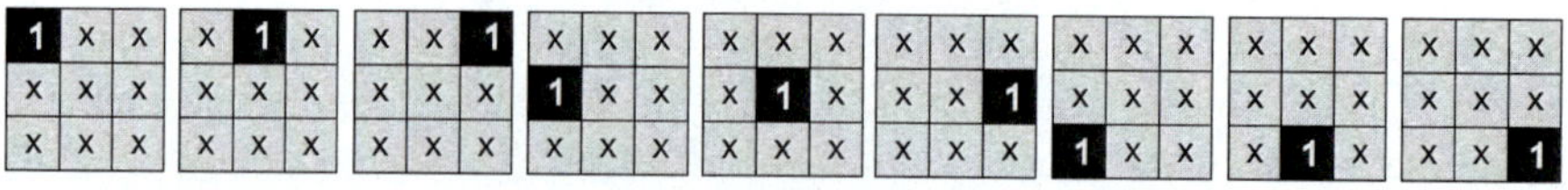

Fig. 6.6 A simplified representation of the distinct scenarios presented by the local rules

Such values are identical and less than one. Finally, the initial states $x_{i,j}(0)$ are determined according to Eq. 6.6. Considering the local rules in Fig. 6.6, a unit will produce one if at least one element in its 3×3 neighborhood is 1. On the other hand, the unit will produce -1 only when all the elements in its 3×3 neighborhood are -1. Therefore, if all elements of the neighborhood are -1 except one that maintains a value of 1, the result of $w_{i,j}$(Eq. 6.5) is $-7/9$. Consequently, the unit produces one under the following condition:

$$(a_{00} - 1)x_{i,j}(0) + w_{i,j} > 0 \tag{6.10}$$

Solving Eq. 6.10, it is obtained $x_{i,j}(0) = 8/9$. With this value, the CNN behaves as receptive fields that are able to rebuild areas with a low concentration of data. The design of the CNN considers an architecture of $S \times S$ nonlinear dynamical units.

6.4.3 Modification of the Spatial Resolution

In the beginning, the data distribution **D** to be grouped is transformed into a binary image **BQ** of 160×160 bins ($S \times S$). Then, the operation on the CNN is executed over **BQ** producing the image **R**. Then, the binary image **R** is analyzed through a connected-component labeling (CCL) method [51]. The CCL algorithm is used to detect connected regions in binary digital images. After its execution, it delivers the number of clusters c contained in the image and the number of bins nb_i that are included in them ($i \in 1, \ldots, c$). With this information, the number of elements nd_i of **D** contained in each cluster i is calculated. Then, the ideal number of elements (*Ine*) contained in a cluster is computed as follows:

$$Ine = \text{Int}\left(\frac{n \cdot 0.2}{c}\right), \tag{6.11}$$

Where $\text{Int}(p)$ delivers the integer part of p. The index *Ine* represents the minimal number of elements that a cluster should contain. Therefore, if in the current iteration k, there exists at least one cluster i such that $nd_i < Ine$, a new iteration $k+1$ is tried with a resolution $(S-10) \times (S-10)$. This process is repeated until all the c clusters contain more than *Ine* elements. If this condition is not reached, the clustering process is also ended in a resolution of 40×40.

6.4.4 Computational Clustering Model

In the approach, the data clustering process is an iterative method that starts with an initial resolution $S \times S$. At each iteration, the quality of the groups produced by the

CNN operation is analyzed. If a certain criterion is not fulfilled, a new iteration is executed by using a lower resolution. This process is continued until the criterion, or a minimum allowed resolution has been reached.

The algorithm considers several steps in its execution. Algorithm 6.1 summarizes the complete process in the form of pseudo-code. The grouping model does not use any parameter that needs to be calibrated before its execution.

Therefore, the first step ($k = 1$) is to transform the data **D** into the binary image **BQ** (Line 5). The initial resolution is 160 × 160 bins. Then, the CNN is applied to **BQ** in order to associate spatially data elements producing the binary image **R** (Line 6). The processed image **R** is analyzed by using a connected-component labeling (CCL) method (Line 7). After its execution, the CCL technique delivers the number of clusters c contained in the image. With this information, the number of elements nd_i of **D** contained in each cluster i is calculated (Line 8). Afterward, the index *Ine* is computed (Line 9). Finally, if there exists at least one cluster i such that $nd_i < Ine$, a new iteration $k + 1$ is executed with a lower resolution so that $(S - 10) \times (S - 10)$ (Line 10). This process is repeated until all the c clusters contain more than *Ine* elements. If this condition is not reached, the clustering process is also ended in a resolution of 40 × 40 (Line 10). The final result corresponds to the number of clusters c and their elements nd_i (Line 11).

Algorithm 6.1. Pseudo-code for the clustering method

(1) **Input:** Data distribution **D**
(2) $K = 0; S = 170;$
(3) do {
(4) $k = k + 1; S = S - 10;$
(5) **BQ** ←← **TransformData** (**D**, $S \times SS \times S$);
(6) **R** ←← **CNN**(**BQ**);
(7) c ←← **CCL**(**R**);
(8) $nd_i nd_i$ ←← **ElementInCluster** (c, **R**); ($i \in 1, \ldots, c$)
(9) *Ine* ←← **CalculateIne** (c, n);}
(10) **While** (($\exists i{:}nd_i < Ine \exists i{:}nd_i < Ine$) or ($S > 40$));
(11) **Output:** **R**, $c, nd_i nd_i$

Algorithm 6.1. Summarized processes of the clustering method.

Figure 6.7. shows the evolution of the method during its execution. Considering the image 7**a** as de original data distribution **D**, the images 7**b**, **c**, d, e, **f** and **g** represent the progress of the binary image **R** in the iterations $k = 1$, $k = 7$, $k = 9$, $k = 11$, $k = 12$ and $k = 13$, respectively. Finally, image 7**h** exhibits the clustering results over data set **D**.

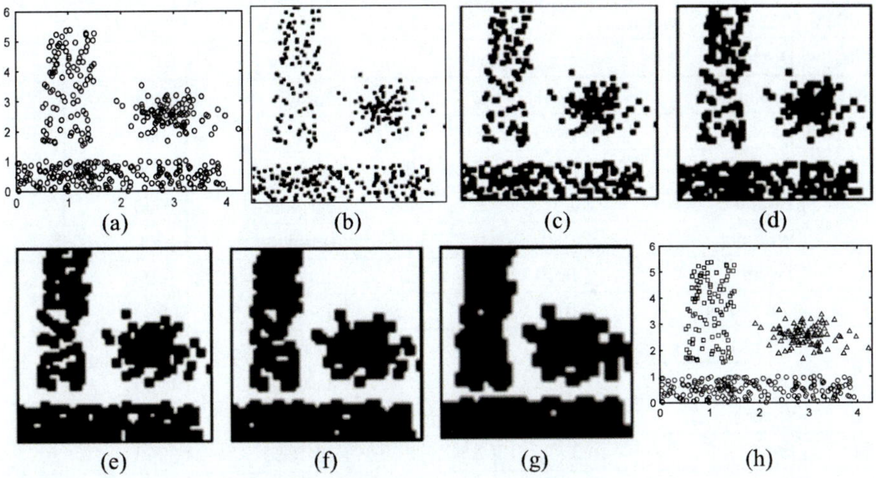

Fig. 6.7 Evolution of the presented method during its execution **a** Original data and binary image of **R** in **b** $k = 1$, **c** $k = 7$, **d** $k = 9$, **e** $k = 11$, **f** $k = 12$, **g** $k = 13$. The final clustering results are shown in **h**

6.5 Experiments

In this section, the performance of the presented method is numerically analyzed against two well-known density-based clustering methods considering a collection of 28 data sets with different complexity levels. The datasets have been collected from several sources [52]. They represent the most popular data distribution employed to evaluate the performance of new clustering schemes. All datasets include the data observations and their actual classification, according to a human expert. Table 6.1 shows the most important properties of each dataset. In the Table, the data size of the distribution is represented by n and the number of clusters by c.

In this chapter, a simple clustering model inspired by the way in which the human visual system associates spatially patterns is presented. In this section, the clustering method is called henceforth CNNCA. The numerical results obtained by the CNNCA are compared with the numeric results of two representative density-based algorithms, such as the classical Density-based spatial clustering of applications with noise (DBSCAN) [17] and the recent Gaussian Density Distance (GDD) [23] method.

In the comparisons, the following configurations have been adopted. DBSCAN has two parameters: the minimum number of points *minPts* and the threshold *epsilon*. Several tests have been necessary to obtain their best possible values. Such parameters are reported in Table 6.2. In this Table, it is remarkable that dataset 28 has no *minPts* and epsilon *values*. This fact is because the DBSCAN algorithm could no perform the clustering process due to the large number of observations contained in dataset 28. On the other hand, the GDD and the CNNCA method do not present any parameter to be configured.

Table 6.1 Dataset Properties

No	Dataset	n	c	No	Dataset	n	c
1	Banana	4811	2	15	2d-4c	1261	4
2	Cassini	1000	3	16	curves1	1000	2
3	Cure-t0-2000n-2D	2000	3	17	curves2	1000	2
4	Target	770	6	18	dartboard1	1000	4
5	Wingnut	1016	2	19	spiral	1000	2
6	Twenty	1000	20	20	donut1	1000	2
7	Gaussians1	100	2	21	donut3	999	3
8	Zelnik1	299	3	22	triangle1	1000	4
9	Lsun	400	3	23	dartboard2	1000	4
10	Shapes	1000	4	24	donutcurves	1000	4
11	Jain	373	2	25	2d-10c	2990	9
12	3MC	400	3	26	golfball	4002	1
13	Aggregation	788	7	27	long1	1000	2
14	2sp2glob	2000	4	28	birch-rg2	100,000	1

Table 6.2 Parameter settings of the DBSCAN algorithm

No	Dataset	*minPts*	*epsilon*	No	Dataset	*minPts*	*epsilon*
1	Banana	10	0.05	15	2d-4c	10	5.00
2	Cassini	10	0.20	16	curves1	10	0.05
3	cure-t0-2000n-2D	10	0.40	17	curves2	10	0.005
4	target	10	0.40	18	dartboard1	10	0.07
5	wingnut	10	0.29	19	spiral	10	1.00
6	twenty	10	0.90	20	donut1	10	0.05
7	gaussians1	10	0.09	21	donut3	10	0.02
8	zelnik1	10	0.05	22	triangle1	10	3.50
9	lsun	10	0.58	23	dartboard2	10	0.01
10	shapes	10	0.58	24	donutcurves	10	0.013
11	jain	10	2.80	25	2d-10c	10	6.00
12	3MC	10	1.00	26	golfball	10	0.50
13	aggregation	10	2.00	27	long1	10	0.39
14	2sp2glob	10	2.50	28	birch-rg2	–	–

To evaluate the performance results, different criteria have been considering, such as the clustering Accuracy ACC, the Error Rate $E.R.$, and the Purity P of the clustering process. Additionally, another important index is the number of detected clusters $F.C.$ for each clustering approach. From the data size n, the number of data samples that have been correctly classified is symbolized by *TCD*. On the other hand, the number of data samples that have been misclassified is referred to as *MCD*.

ACC measures the accuracy of the clustering algorithm. An *ACC* value of 1 specifies that the algorithm has been able to classify correctly every observation $\mathbf{d}_i$ from a dataset $\mathbf{D}$ into its actual cluster $G_j(j \in 1, \ldots, c)$, while an *ACC* value different to 1 means that not all data has been assigned correctly to its corresponding class. Under such conditions, as the *ACC* value gets closer to 0, the performance of the clustering algorithm is worse. The *ACC* is calculated as follows:

$$ACC = \frac{TCD}{n} \tag{6.12}$$

The *TCD* index is given by:

$$TCD = \sum_{j=1}^{c} \sum_{i=1}^{n} \{ f(\mathbf{d}_i) \big| \mathbf{d}_i \in g_j \}, \tag{6.13}$$

where

$$f(\mathbf{d}_i) = \begin{cases} 1 & \mathbf{d}_i \in G_j \\ 0 & \text{otherwise} \end{cases}, \tag{6.14}$$

Here, g_j is the class assigned to the observation $\mathbf{d}_i$ by the clustering algorithm while G_j is the actual class $\mathbf{d}_i$. The *E.R.* indicator exhibits the error in classifying the data into its actual class, and it can be calculated as:

$$ER = \frac{MCD}{n}, \tag{6.15}$$

where

$$MCD = n - TCD, \tag{6.16}$$

The parameter *P* indicates the purity of the clustering process, considering the number of clusters that have been correctly classified. If a clustering algorithm assigns more data in a certain cluster than it really has, it is considered not pure. The same occurs if the algorithm designates fewer data to a certain group than it actually has.

A *P* value of 1 means that for every cluster g_j, all data contained in the cluster g_j actually belong to the class G_j. A *P* value closer to 0 implies that some data are not entirely classified. The parameter *P* is an important measure to consider since the accuracy is not enough to test the performance of a clustering algorithm completely. The purity of the clustering process could be critical for some applications. Therefore, the purity of the clustering process is defined as:

$$P = \frac{q}{c}, \tag{6.17}$$

where q is the number of pure clusters defined as follows:

$$q = \sum_{j=1}^{c} \prod_{i=1}^{n} \{ f(\mathbf{d}_i) | \mathbf{d}_i \in G_j \} \quad f(\mathbf{d}_i) = \begin{cases} 1 & \mathbf{d}_i \in g_j \\ 0 & \text{otherwise} \end{cases} \tag{6.18}$$

In Table 6.3, the experimental results are listed for quantitative analysis. In the Table, all evaluation parameters for each algorithm in all datasets are shown. For every dataset, the method that better results produce is highlighted in boldface.

In DBSCAN, all data that cannot be correctly classified are considered noisy information, which is reported as *MCD* in the experiments. From Table 6.3, it can be observed that the CNNCA outperforms its competitors in datasets 5, 23, 24, 27, and 28. Dataset 28 could not be grouped by GDD and DBSCAN algorithms due to its large amount of observations, while the CNNCA algorithm manages to classify it. The CNNCA clustering result of dataset 28 can be observed in Fig. 6.8.

In dataset 14, the CNNCA and the DBSCAN algorithms identify three from four clusters. Both methods are not able to detect the spiral shape as two different clusters instead of only one. On the contrary, the GDD method has the capacity to find separately two groups from the spiral shape. However, the GDD algorithm finds more clusters exceeding the actual number. It detects 14 groups considering that there exist only four groups, which represents a significant difference.

The CNNCA and the GDD algorithm obtain similar results for datasets 4, 8, and 11. For datasets 1, 6, 7, 9, 10, 15, 20, 21, 22, and 25, the CNNCA and the DBSCAN maintain a comparable performance while for datasets 2, 3, 12, 13, 16, 17, 18, 19, and 26 the three algorithms reach similar results. In general, the GDD algorithm achieves high accuracy but low purity in the clustering process. Besides, the number of detected clusters exceeds the actual amount of groups.

The DBSCAN algorithm reports the data elements in *MCD* as noise instead of classifying them in any other group. This fact could be convenient in the sense that reporting *MCD* as undesired information reduces the number of false detected groups, avoiding to exceed the actual number of clusters. On the other hand, the CNNCA conducts the clustering process by using a simple model of the biological visual system. For this reason, the CNNCA reaches high purity and, consequently, high accuracy in the clustering process compared to GDD and DBSCAN.

From Table 6.3, it can be concluded that the CNNCA algorithm outperforms its counterparts, reaching 24 of 28 datasets with $ACC = 1$ and $P = 1$ while the DBSCAN algorithm achieves 17 of 28 with $ACC = 1$ and $P = 1$. Finally, the GDD attains 11 of 28 with the same ACC and P values. It is remarkable that although the CNNCA and the DBSCAN achieved comparable results in some tests, the DBSCAN requires parameter configurations for every dataset, which are difficult to estimate. The CNNCA is a non-parametric algorithm that automatically adjusts to the density and the shape of the data. The GDD algorithm is also a non-parametric

Table 6.3 Experimental Results of GDD, CNNCA and DBSCAN clustering algorithms

No	Dataset	Algorithm	ACC	ER	P	FC/c	n	TCD	MCD
		GDD	0.9981	0.0019	0.0000	10/2	4811	4802	9
1	Banana	**CNNCA**	1.0000	0.0000	1.0000	2/2	4811	4811	0
		DBSCAN	1.0000	0.0000	1.0000	2/2	4811	4811	0
		GDD	1.0000	0.0000	1.0000	3/3	1000	1000	0
2	Cassini	**CNNCA**	1.0000	0.0000	1.0000	3/3	1000	1000	0
		DBSCAN	1.0000	0.0000	1.0000	3/3	1000	1000	0
		GDD	1.0000	0.0000	1.0000	3/3	2000	2000	0
3	Cure-t0-2000n-2D	**CNNCA**	1.0000	0.0000	1.0000	3/3	2000	2000	0
		DBSCAN	1.0000	0.0000	1.0000	3/3	2000	2000	0
		GDD	1.0000	0.0000	1.0000	6/6	770	770	0
4	Target	**CNNCA**	1.0000	0.0000	1.0000	6/6	770	770	0
		DBSCAN	0.9844	0.0156	0.3333	2/6	770	758	12
		GDD	0.9902	0.0098	0.0000	6/2	1016	1006	10
5	Wingnut	**CNNCA**	1.0000	0.0000	1.0000	2/2	1016	1016	0
		DBSCAN	0.9803	0.0197	0.0000	2/2	1016	996	20
		GDD	0.9960	0.0040	0.8000	24/20	1000	996	4
6	Twenty	**CNNCA**	1.0000	0.0000	1.0000	20/20	1000	1000	0
		DBSCAN	1.0000	0.0000	1.0000	20/20	1000	1000	0
		GDD	0.9700	0.0300	0.0000	3/2	100	97	3
7	Gaussians1	**CNNCA**	1.0000	0.0000	1.0000	2/2	100	100	0
		DBSCAN	1.0000	0.0000	1.0000	2/2	100	100	0
		GDD	1.0000	0.0000	1.0000	3/3	299	299	0
8	Zelnik1	**CNNCA**	1.0000	0.0000	1.0000	3/3	299	299	0
		DBSCAN	0.6689	0.3311	0.6667	2/3	299	200	99
		GDD	0.9800	0.0200	0.6667	4/3	400	392	8
9	Isun	**CNNCA**	1.0000	0.0000	1.0000	3/3	400	400	0
		DBSCAN	1.0000	0.0000	1.0000	3/3	400	400	0
		GDD	0.9560	0.0440	0.5000	9/4	1000	956	44
10	Shapes	**CNNCA**	1.0000	0.0000	1.0000	4/4	1000	1000	0
		DBSCAN	1.0000	0.0000	1.0000	4/4	1000	1000	0
		GDD	1.0000	0.0000	1.0000	2/2	373	373	0
11	Jain	**CNNCA**	1.0000	0.0000	1.0000	2/2	373	373	0
		DBSCAN	0.9088	0.0912	0.5000	3/2	373	339	34
		GDD	1.0000	0.0000	1.0000	3/3	400	400	0
12	3MC	**CNNCA**	1.0000	0.0000	1.0000	3/3	400	400	0
		DBSCAN	1.0000	0.0000	1.0000	3/3	400	400	0
		GDD	0.7919	0.2081	0.4286	5/7	788	624	164

(continued)

Table 6.3 (continued)

No	Dataset	Algorithm	ACC	ER	P	FC/c	n	TCD	MCD
13	Aggregation	**CNNCA**	0.7919	0.2081	0.4286	5/7	788	624	164
		DBSCAN	0.7919	0.2081	0.4286	5/7	788	624	164
		GDD	0.9895	0.0105	0.5000	14/4	2000	1979	21
14	2sp2glob	CNNCA	0.7500	0.2500	0.5000	3/4	2000	1500	500
		DBSCAN	0.7500	0.2500	0.5000	3/4	2000	1500	500
		GDD	0.9976	0.0024	0.7500	7/4	1261	1258	3
15	2d-4c	**CNNCA**	1.0000	0.0000	1.0000	4/4	1261	1261	0
		DBSCAN	1.0000	0.0000	1.0000	4/4	1261	1261	0
		GDD	1.0000	0.0000	1.0000	2/2	1000	1000	0
16	Curves1	**CNNCA**	1.0000	0.0000	1.0000	2/2	1000	1000	0
		DBSCAN	1.0000	0.0000	1.0000	2/2	1000	1000	0
		GDD	1.0000	0.0000	1.0000	2/2	1000	1000	0
17	Curves2	**CNNCA**	1.0000	0.0000	1.0000	2/2	1000	1000	0
		DBSCAN	1.0000	0.0000	1.0000	2/2	1000	1000	0
		GDD	1.0000	0.0000	1.0000	4/4	1000	1000	0
18	Dartboard1	**CNNCA**	1.0000	0.0000	1.0000	4/4	1000	1000	0
		DBSCAN	1.0000	0.0000	1.0000	4/4	1000	1000	0
		GDD	1.0000	0.0000	1.0000	2/2	1000	1000	0
19	Spiral	**CNNCA**	1.0000	0.0000	1.0000	2/2	1000	1000	0
		DBSCAN	1.0000	0.0000	1.0000	2/2	1000	1000	0
		GDD	0.9750	0.0250	0.5000	10/2	1000	975	25
20	Donut1	**CNNCA**	1.0000	0.0000	1.0000	2/2	1000	1000	0
		DBSCAN	1.0000	0.0000	1.0000	2/2	1000	1000	0
		GDD	0.9680	0.0320	0.3333	14/3	999	967	32
21	Donut3	**CNNCA**	1.0000	0.0000	1.0000	3/3	999	999	0
		DBSCAN	1.0000	0.0000	1.0000	3/3	999	999	0
		GDD	0.9880	0.0120	0.0000	10/4	1000	988	12
22	Triangle1	**CNNCA**	1.0000	0.0000	1.0000	4/4	1000	1000	0
		DBSCAN	1.0000	0.0000	1.0000	4/4	1000	1000	0
		GDD	0.7500	0.2500	0.5000	3/4	1000	750	250
23	Dartboard2	**CNNCA**	1.0000	0.0000	1.0000	4/4	1000	1000	0
		DBSCAN	0.5000	0.5000	0.5000	2/4	1000	500	500
		GDD	0.9990	0.0010	0.7500	5/4	1000	999	1
24	Donutcurves	**CNNCA**	1.0000	0.0000	1.0000	4/4	1000	1000	0
		DBSCAN	0.9990	0.0010	0.7500	4/4	1000	999	1
		GDD	0.8839	0.1161	0.4444	16/9	2990	2643	347
25	2d-10c	**CNNCA**	0.8870	0.1130	0.7778	8/9	2990	2652	338

(continued)

Table 6.3 (continued)

No	Dataset	Algorithm	*ACC*	*ER*	*P*	*FC/c*	*n*	*TCD*	*MCD*
		DBSCAN	0.8870	0.1130	0.7778	8/9	2990	2652	338
		GDD	1.0000	0.0000	1.0000	1/1	4002	4002	0
26	Golfball	**CNNCA**	1.0000	0.0000	1.0000	1/1	4002	4002	0
		DBSCAN	1.0000	0.0000	1.0000	1/1	4002	4002	0
		GDD	0.8830	0.1170	0.0000	12/2	1000	883	117
27	Long1	**CNNCA**	0.9990	0.0010	0.5000	3/2	1000	999	1
		DBSCAN	0.9950	0.0050	0.0000	2/2	1000	995	5
		GDD	–	–	–	–	–	–	–
28	Birch-rg2	**CNNCA**	1.00	0.00	1.00	1/1	100,000	100,000	0
		DBSCAN	–	–	–	–	–	–	–

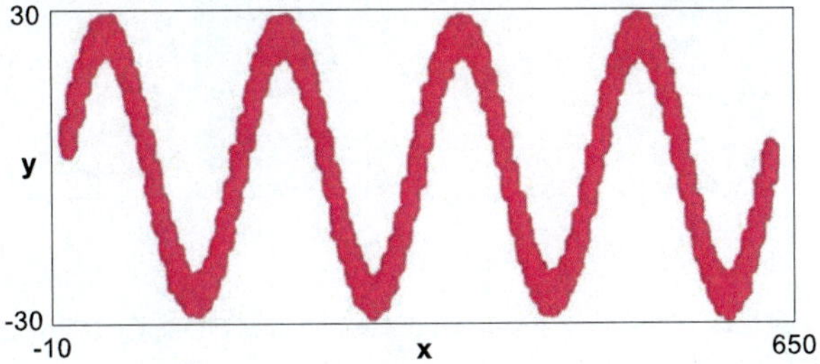

Fig. 6.8 CNNCA clustering result of dataset 28

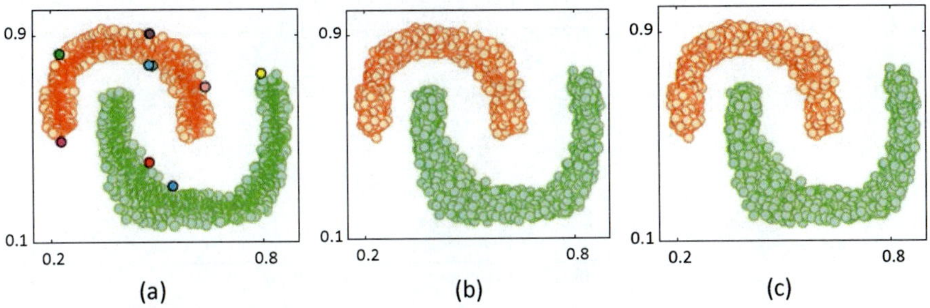

Fig. 6.9 Clustering results of dataset 1 using GDD, CNNCA, and DBSCAN algorithms. **a** GDD, **b** CNNCA, **c** DBSCAN

and self-adjustable algorithm. However, the performance of the CNNCA is superior compared to the GDD algorithm.

Figures from 6.9, 6.10, 6.11, 6.12, 6.13, 6.14, 6.15 and 6.16 show the clustering results of representative datasets (1, 2, 4, 5, 11, 14, 23, 24 and 27), which, according to their complexity, are considered difficult to detect. In these figures, the noise reported by the DBSCAN algorithm can be recognized by circles in gray color. Clusters can be identified by different colors.

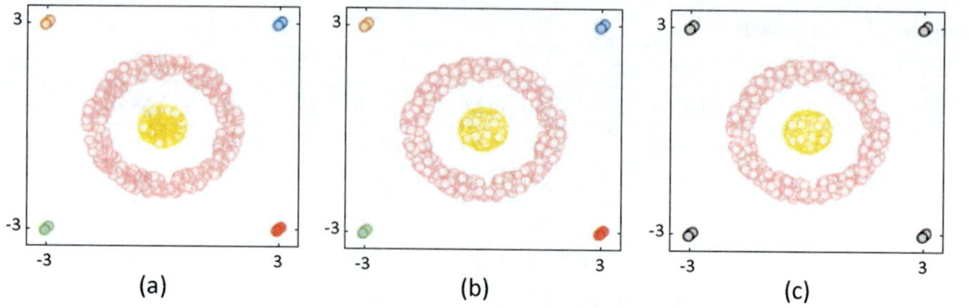

Fig. 6.10 Clustering results of dataset 4 using GDD, CNNCA, and DBSCAN algorithms. **a** GDD, **b** CNNCA, **c** DBSCAN

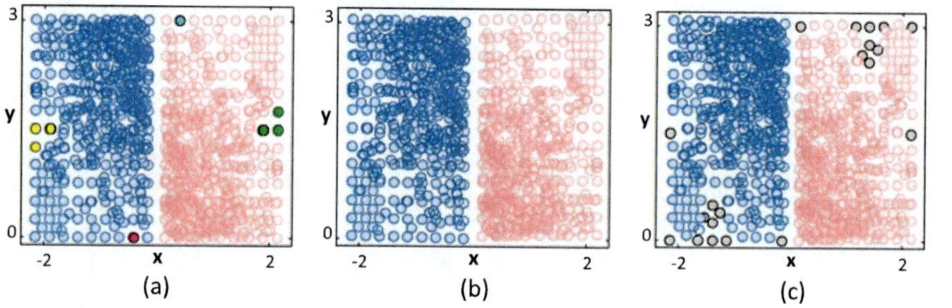

Fig. 6.11 Clustering results of dataset 5 using GDD, CNNCA, and DBSCAN algorithms. **a** GDD, **b** CNNCA, **c** DBSCAN

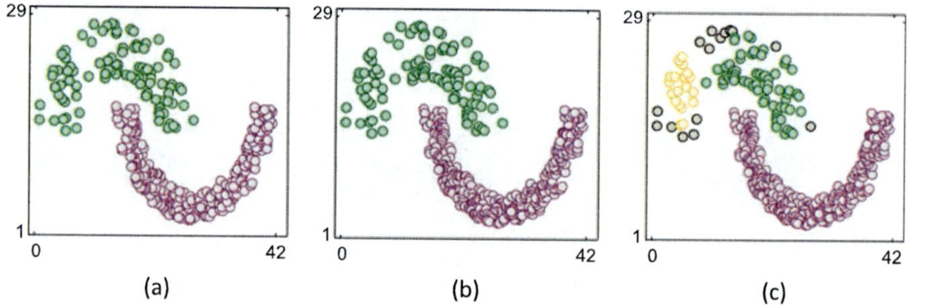

Fig. 6.12 Clustering results of dataset 11 using GDD, CNNCA, and DBSCAN algorithms. **a** GDD, **b** CNNCA, **c** DBSCAN

6.6 Summary

In this chapter, a simple clustering model inspired by the way in which the human visual system associates patterns spatially is presented. The model, at some abstraction level, can be characterized as a density grouping strategy. To build the clustering

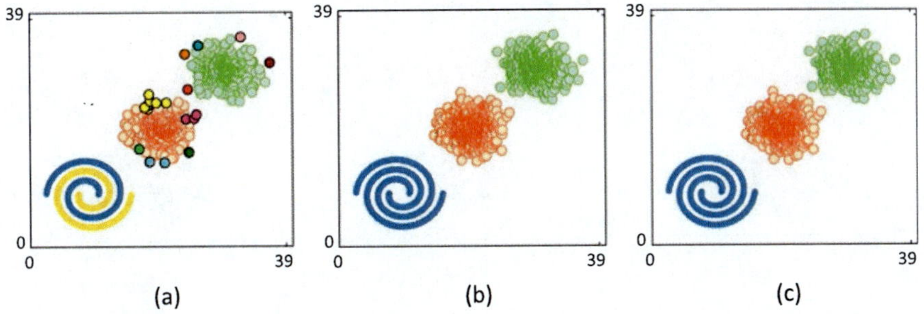

Fig. 6.13 Clustering results of dataset 14 using GDD, CNNCA, and DBSCAN algorithms. **a** GDD, **b** CNNCA, **c** DBSCAN

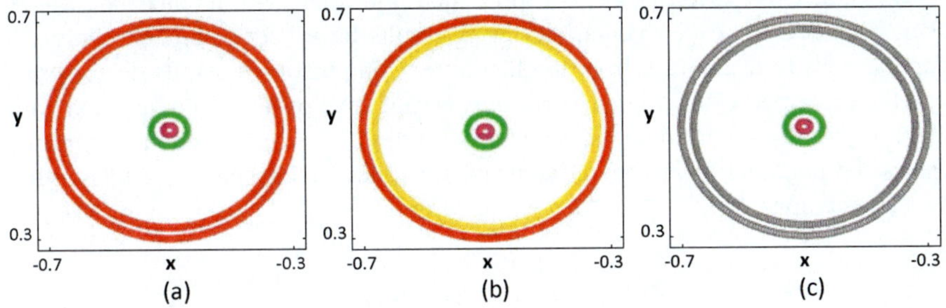

Fig. 6.14 Clustering results of dataset 23 using GDD, CNNCA, and DBSCAN algorithms. **a** GDD, **b** CNNCA, **c** DBSCAN

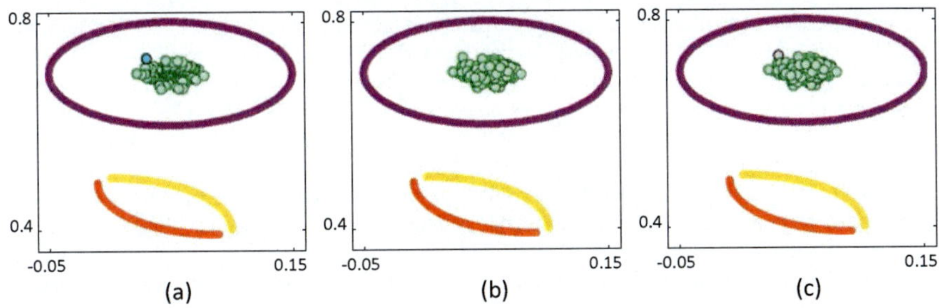

Fig. 6.15 Clustering results of dataset 24 using GDD, CNNCA, and DBSCAN algorithms. **a** GDD, **b** CNNCA, **c** DBSCAN

model, the approach combines two different biological mechanisms, such as receptive cells and modification of the spatial resolution. The approach is mainly based on Cellular Neural Networks (CNNs). In the method, similar to the biological model, the CNN is used to build groups spatially through operations of the locally interconnected units. During the clustering process, an automatic mechanism tries different scales to find the best possible data categorization.

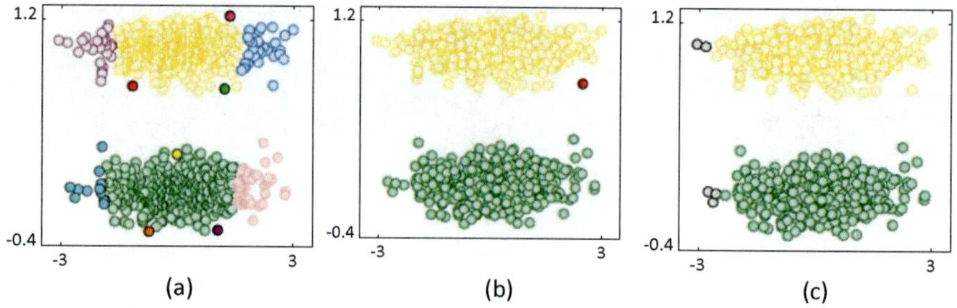

Fig. 6.16 Clustering results of dataset 27 using GDD, CNNCA, and DBSCAN algorithms. **a** GDD, **b** CNNCA, **c** DBSCAN

To evaluate the performance of the presented algorithm, 28 different datasets have been adopted. The experimental clustering results have been compared with two popular density techniques from the literature. Computational results demonstrate that the presented CNN approach presents competitive results in comparison with other algorithms regarding accuracy and robustness. This remarkable performance shows the potential of biological models over traditional schemes for identifying data associations.

References

1. Han J, Kamber M, Pei J (2011) Data mining: concepts and techniques, Elsevier
2. He R, Li Q, Ai B, Geng YLA, Molisch AF, Kristem V, Zhong Z, Yu J (2017) A kernel-power-density-based algorithm for channel multipath components clustering. IEEE Trans Wireless Commun 16(11):7138–7151. https://doi.org/10.1109/TWC.2017.2740206
3. Laohakiat S, Phimoltares S, Lursinsap C (2016) A clustering algorithm for stream data with lda-based unsupervised localized dimension reduction. Inf. Sci. 381:104–123
4. Kisore NR, Koteswaraiah CB (2016) Improving atm coverage area using density based clustering algorithm and voronoi diagrams. Inf Sci 376:1–20
5. Nguyen TT, Le HS (2015) HIFCF: an effective hybrid model between picture fuzzy clustering and intuitionistic fuzzy recommender systems for medical diagnosis. Expert Syst Appl 42(7):3682–3701
6. Jiao Y, Jianshe Wu, Jiao L (2018) An image segmentation method based on network clustering model. Phys A 490:1532–1542
7. Youcef D, Asma B, Philippe F-V, Jerry C-W, Lin (2018) Fast and effective cluster-based information retrieval using frequent closed itemsets. Inf Sci 453:154-167
8. Iván G, Grolmusz V (2014) On dimension reduction of clustering results in structural bioinformatics, Biochimica et Biophysica Acta (BBA)—proteins and proteomics 1844(12): 2277–2283
9. MacQueen JB (1967) Some methods for classification and analysis of multivariate observations. In: Proceedings of 5th Berkeley symposium on mathematical statistics and probability, Berkeley, vol 1, University of California Press, pp 281–297
10. Bezdek JC (1981) Pattern recognition with fuzzy objective function algorithms. Plenum Press, New York
11. Dempster AP, Laird NM, Rubin DB (1977) Maximum likelihood from incomplete data via the em algorithm. J R Stat Soc 39(1):1–38

12. Kaufman L, Rousseeuw RJ (1990) Finding groups in data: an introduction to cluster analysis, Wiley
13. Camastra F, Verri A (2005) A novel kernel method for clustering. IEEE Trans Pattern Anal Mach Intell 27(5):801–805
14. Zhang T, Ramakrishnan R, Livny M (1996) BIRCH: an efficient data clustering method for very large databases. In: SIGMOD Conference, pp 103–114
15. Guha S, Rastogi R, Shim K (1998) CURE: an efficient clustering algorithm for clustering large databases. In: Proceedings of the symposium on management of data (SIGMOD), pp 73–84
16. Karypis G, Han EH, Kumar V (1999) CHAMELEON: a hierarchical clustering algorithm using dynamic modeling. IEEE Comput 32(8):68–75
17. Ester M, Kriegel H, Sander J, Xu X (1996) A density-based algorithm for discovering clusters in large spatial databases with noise. In: Simoudis E, Han J, Fayyad U (eds) Proceedings of the 2nd international conference on knowledge discovery and data mining (KDD-96), AAAI Press, pp 226–231
18. Ankerst M, Breunig M, Kriegel HP (1999) OPTICS: ordering points to identify the clustering structure. In: Proceedings of international conference on management of data (SIGMOD99), Philadelphia, PA, pp 49–60
19. Hinneburg A, Keim D (1998) An efficient approach to clustering in large multimedia databases with noise. In: Proceedings of the fourth international conference on knowledge discovery and data mining, New York, pp 58–65
20. Wang W, Yang J, Muntz R (1997) STING: a statistical information grid approach to spatial data mining. In: Proceedings of the 23rd international conference on very large data bases, Athens, Morgan Kaufmann, pp 18–195
21. Sheikholeslami G, Chatterjee S, Zhang AD (1998) Wavecluster: a multi-resolution clustering approach for very large spatial databases. In Gupta A, Shmueli O, Widom J (eds) Proceedings of the 24th international conference on very large data bases, Morgan Kaufmann, New York, pp 428–439
22. Andrew Y, Ng MI, Jordan Y (2001) On spectral clustering: analysis and an algorithm. Adv Neural Inf Process Syst 14:849–856
23. Güngör E, Özmen A (2017) Distance, and density based clustering algorithm using Gaussian kernel. Expert Syst Appl 69:10–20
24. Mark P (2014) Mattson, Superior pattern processing is the essence of the evolved human brain, Front. Neurosciences 8:265–278
25. Aguirre GK, Farah MJ (1998) Human visual object recognition: what have we learned from neuroimaging? Psychobiology 26(4):322–332
26. Wu BW, Fang YC (2015) Human vision model in relation to characteristics of shapes for the Mach band effect. Appl Opt 54(28):181–188
27. Wilson JG, Mitchell RJ (2000) Object detecting artificial retina. Kybernetes 29(1):31–52
28. Akbas E, Eckstein MP (2017) Object detection through search with a foveated visual system. PLoS Comput Biol 13(10):1–28
29. Zanker JM, Harris JP (2002) On temporal hyperacuity in the human visual system. Vision Res 42:2499–2508
30. Lindeberg T (2013) A computational theory of visual receptive fields. Biol Cybern 107(6):589–635
31. Yang X, Li Y (2015) Contour extraction based on human visual system. In: Zha H, Chen X, Wang L, Miao Q (eds) Communications in computer and information science computer vision. CCCV 2015 (21015), vol 547. Springer, Berlin, Heidelberg
32. Carrasco M, Barbot A (2014) How attention affects spatial resolution. Cold Spring Harb Symp Quant Biol 79:149–160
33. Chua LO, Yang L (1988) Cellular neural networks: theory. IEEE Trans Circuits Syst 35:1257–1272
34. Chua LO, Yang L (1988) Cellular neural networks: applications. IEEE Trans Circuits Syst 35:1273–1290

35. Li H, Liao X, Li C, Huang H, Li C (2011) Edge detection of noisy images based on cellular neural networks. Commun Nonlinear Sci Numer Simulat 16:3746–3759
36. Starkov SO, Lavrenkov YN (2017) Prediction of the moderator temperature field in a heavy water reactor based on a cellular neural network. Nuclear Energy Technol 3(2):133–140
37. Lauret P, Heymes F, Aprin L, Johannet A (2016) Atmospheric dispersion modeling using Artificial Neural Network based cellular automata. Environ Modell Softw 85:56–69
38. Fuxin Z, Guodong L, Wenxia X (2016) Xinjiang desertification disaster prediction research based on cellular neural networks. In: 2016 international conference on smart city and systems engineering (ICSCSE), pp 545–548
39. Shen S, Chang CH, Wang LC (2009) A cellular neural network and utility-based radio resource scheduler for multimedia CDMA communication systems. IEEE Trans Wireless Commun 8(11):5508–5519
40. Hou YY, Liao TL, Yan JJ (2007) Stability analysis of Takagi–Sugeno Fuzzy cellular neural networks with time-varying delays. IEEE Trans Syst Man Cybern Part B (Cybernetics) 37(3):720–726
41. Hu X, Feng G, Duan S, Liu L (2017) A memristive multilayer cellular neural network with applications to image processin. IEEE Trans Neural Netw Learning Syst 28(8):1889–1901
42. Huang CH, Lin CT (2007) Bio-Inspired computer fovea model based on hexagonal-type cellular neural network. IEEE Trans Circuits Syst I 54(1):35–47
43. Leon FWTR, Chua O (1995) The analogic cellular neural network as a bionic eye. Circuit Theory Appl 23(6):541–569
44. Gál V, Hámori J, Roska T, Bálya D, Borostyánkői ZS, Brendel M, Lotz K, Négyessy L, Orzó L, Petrás I, Rekeczky CS, Takács J, Venetiáner P, Vidnyánszky Z, Zarándy Á (2004) Receptive field atlas and related cnn models. Int J Bifurcation Chaos 14(2):551–570
45. Chua LO, Roska T (2002) Cellular neural networks and visual computing, Cambridge University Press, Cambridge
46. Roska T, Hamori J, Labos E, Lotz K, Orzo L, Takacs J, Venetianer PL, Vidnyanszky Z, Zarandy A (1993) The use of CNN models in the subcortical visual pathway. IEEE Trans Circuits Syst I Fund Theory Appl 40(3):182–195
47. Huang CH, Lin CT (2007) Bio-inspired computer fovea model based on hexagonal-type cellular neural network. IEEE Trans Circuits Syst I: Regular Papers 54(1): 35–47 (2007)
48. Rolfs M (2009) Microsaccades: Small steps on a long way. Vision Res 49(20):2415–2456
49. Dimigen O, Valsecchi M, Sommer W, Kliegl R (2009) Human microsaccade-related visual brain responses. J Neurosci 29(39):12321–12331
50. Gilli M, Corinto F, Civalleri PP (2003) Design and synthesis methods for cellular neural networks, In: Proceedings of the international joint conference on neural networks
51. Samet H, Tamminen M, (1988) Efficient component labeling of images of arbitrary dimension represented by linear bintrees. IEEE Trans Pattern Anal Mach Intell IEEE 10(4):579–589
52. https://github.com/deric/clustering-benchmark/tree/master/src/main/resources/datasets/artificial

Chapter 7
Metaheuristic Algorithms for Wireless Sensor Networks

This chapter presents the main concepts of metaheuristic schemes for Wireless Sensor Networks (WSNs). WSNs are multi-functional, low-cost, and low-power networks and rely on communications among devices, from sensor nodes to one or more sink nodes. Sink nodes, sometimes called coordinator nodes or root nodes, may be more robust and have larger processing capacity than the other nodes. Sensor networks can be widely used in various environments, sometimes hostile. Some of the many applications of WSNs are in the medical field, agriculture, monitoring and detection, automation, and data mining. The objective of this chapter is to introduce the main approaches and algorithms applied in WSNs. An important propose of this chapter is understand the usefulness of optimization of metaheuristic algorithms for wireless networks and their use for the reliable delivery of information packets.

7.1 Introduction

WSN is considered as the most critical element in the Internet of Things (IoT) model. In the context of the IoT, they play an essential role in increasing the ubiquity of networks [1] since wireless technology is the fundamental way in which "intelligent objects" communicate with each other and the Internet. In this sense, WSNs allow IoT scalability and provide enough functionality to support its integration with the current Internet architecture. Moreover, it is essential to study the scalability and adaptation methods of the network in the face of packet transmission failures and topology changes [2].

Most of the nodes in a sensor network have a limited power supply and do not have the ability to generate their own energy. Therefore, the design of efficient energy protocols is critical for the longevity of the network. A protocol for sensor networks must be configured in such a way that its operation does not require human attention. Since a direct link between any node in the network and the coordinating node

© The Author(s), under exclusive license to Springer Nature Switzerland AG 2021

E. Cuevas et al., *Recent Metaheuristic Computation Schemes in Engineering*,
Studies in Computational Intelligence 948,
https://doi.org/10.1007/978-3-030-66007-9_7

cannot necessarily be established, multi-hop network topology and an algorithm are required to determine the route the messages follow. These are dynamic topologies with nodes that can stop operating due to physical failures or lack of batteries, with bandwidth restrictions, links with variable capacities, and equipment that can operate with energy restrictions. All these factors result in reconfigurations or unpredictable changes in the topology that handles the routing protocol [3]. On the basis that many routes communicate to a node with the base station or coordinating node, the objective of an energy-aware algorithm is the selection of those routes that maximize the lifetime of the network. In consequence, those routes composed of nodes that have greater autonomy are marked as preferred.

Routing or communication protocols work under algorithms that have steps and rules for traffic packets (information) to arrive satisfactorily to the destination. Therefore, the exchange of so many packets congests the network and, it is a problem in communications. Metaheuristic algorithms provide a general framework for creating new hybrid algorithms, combining different concepts derived from artificial intelligence, biological evolution, and statistical mechanisms. Metaheuristic methods provide a general framework to establish new hybrid algorithms, combining concepts from various fields such as biology, artificial intelligence, mathematics, physics, and neurology, among others. They are defined as solution methods that articulate interactions between some improvements in local heuristics and high-level strategies, aimed at escaping from local optimum in a solution space, aimed at a global optimum. Conventional optimizing algorithms in routing protocols for sensor networks only find a solution in a process. Nowadays, heuristic algorithms are not able to optimize whether the universe of solutions is highly dynamic and under variable traffic [4]. Heuristic allows finding solutions that are not necessarily optimal but, they are good solutions and not only one solution but several in the same algorithmic process, giving robustness to this type of algorithms.

This chapter describes some of the most representative metaheuristic algorithms used in communication protocols for wireless sensor networks. These algorithms optimize performance metrics for a successful delivery of information in a network. We define a metaheuristic as an iterative generation process, which guides a subordinate heuristic intelligently combining different concepts to explore the search space. Strategies learned are used to structure information, to find near-optimal solutions efficiently [5].

The metaheuristic algorithms in three main classes, as is shown in Fig. 7.1:

1. Evolutionary algorithms are based on the evolution of nature.
2. Physics-based techniques are based on the laws of physics.
3. Collective intelligence and social behavior of herds, swarms, and herds in nature.

The main feature of those algorithms are:

- Proximity: the population must be able to perform simple calculations of space and time.
- Quality: the population must be able to respond to quality factors in the environment.

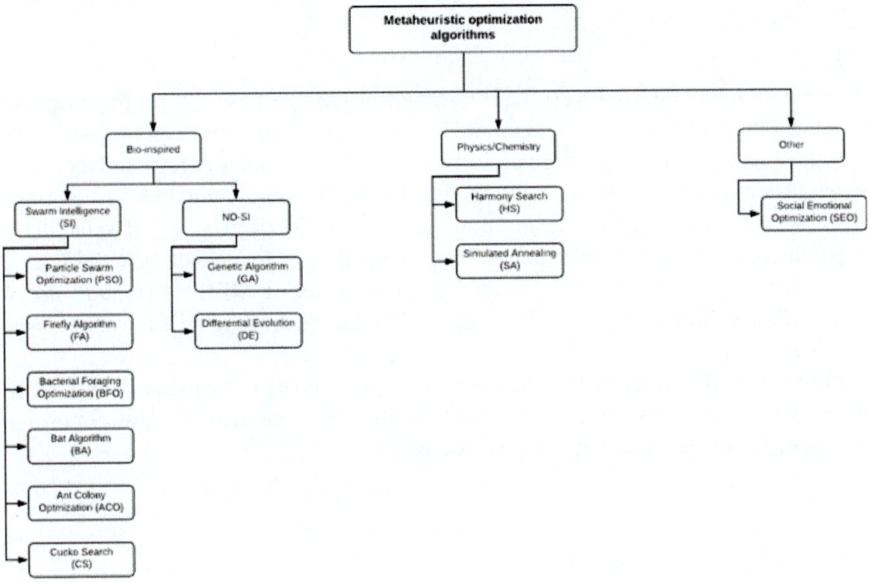

Fig. 7.1 Metaheuristics algorithms

- Diverse Response: the population manages its activities through multiple and wide channels.
- Stability: the population does not change their behaviors as the environment also changes.
- Adaptability: the population could change their behavior when it enriches the computational benefit.

In order to choose the right optimization technique, it is essential to consider the following [6]:

1. The efficiency of the algorithm.
2. Identify the appropriate algorithm for the problem.
3. The numerical effectiveness.

A vital issue for the application of metaheuristics in real problems is the possibility of exploiting the parallelism in its application. The metaheuristics based on local searches can be parallelized following three types of strategies. The first possibility is to parallelize the treatment of the environments by distributing different parts of them among the different processors. This would be the way to approach the standard procedures of local search as the voracious strategy. Another possibility to take advantage of parallelism is to replicate the metaheuristics in each of the processors, which perform independent searches. This corresponds to the natural way of parallelizing a hybrid between the multiple start strategy and the metaheuristic in question. The third intermediate alternative way consists in involving in the parallelism some of the proper components of metaheuristics.

7.2 Fast Energy-Aware OLSR

Optimized Link State Routing (OLSR) [7] is a well-known proactive routing protocol based on the optimization of the standard link-state protocols, based on link status. To keep track, packets are exchanged periodically, so the network topology and the status of neighboring nodes are learned. OLSR uses the MPR (Multi-Point Relay) technique [8]; then the number of retransmissions is reduced. MPR consists of selecting a minimum set of neighboring nodes at a hop away; they can reach all the neighboring nodes that are two hops away. A node selects its set of MPR nodes, and it can only exchange control messages with them. This avoids sending broadcast messages massively.

In the work described in [9], the authors propose the use of a metaheuristic technique, differential evolution, or DE (Differential Evolution) to obtain an optimal configuration of the OLSR routing protocol.

7.2.1 Differential Evolution

DE is a stochastic population algorithm designed to solve problems of continuous optimization [10], which is characterized by being simple, efficient, and fast. Like any evolutionary algorithm, it consists of two general stages: initialization and main cycle. In the first stage, the population is pseudo-randomly initialized in the search space of NP vectors that are potential solutions to the optimization problem to be solved, while in the main cycle, new individuals (candidate solutions) are generated by crossing and mutation. This algorithm is adapted in the solution of several inverse problems of parameter estimation in engineering [11].

The process followed by DE to solve an optimization problem is characterized by iterating over a population of vectors to evolve candidate solutions with respect to a fitness function called fitness. The generation of new individuals is made through operators such as mutation and differential crossing. A mutated individual, represented by ω_{g+1}^{i}, is generated by the following Eq. 7.1:

$$\omega_{g+1}^{i} = \upsilon_{g}^{r1} + \mu\left(\upsilon_{g}^{r2} - \upsilon_{g}^{r3}\right) \tag{7.1}$$

where, $r1, r2, r3 \in \{1, 2, 3 \ldots, i-1, i+1, \ldots, N\}$ without repetition. The adoption of values below the unit for these constants implies an attenuation in their impact. On the other hand, if values higher than the unit are considered, an increase in the impact on fitness values will be achieved. The mutation constant or scale factor, $\mu > 0$, introduces diversity among individuals υ_{g}^{r2} and υ_{g}^{r3}, preventing stagnation in the search process.

To further increase diversity among individuals in the population, each mutated individual performs a crossover operation with the target individual υ_{g}^{i}, then an intermediate individual υ_{g+1}^{i} is generated. Therefore, a random position of the mutated individual is obtained to prevent the intermediate individual replicates the target individual. This process aims to add genetic diversity to the population. Depending on the probability of crossing, P_c, an element of the target individual or the intermediate individual will be selected. If the new individual is better then, it replaces the individual that gave rise to it. This is shown in Eq. 7.2.

$$u_{g+1}^{i}(j) = \begin{cases} \omega_{g+1}^{i}(j) \; if \;\; r(j) \leq P_c \;\; or \;\; j = j_r \\ \upsilon_{g}^{i}(j) \quad in \; other \; case \end{cases} \tag{7.2}$$

Then, the intermediate individual will be accepted for the next generation if and only if it is better than the current individual is. Here, we are using a fitness function, called f. This is described in Eq. 7.3.

$$\upsilon_{(g+1)}^{i} = \begin{cases} u_{g+1}^{i} \;\; if \;\; f\left(u_{g+1}^{i}\right) \leq f(\upsilon_{g}^{i}) \\ \upsilon_{g}^{i}(j) \; in \; other \; case \end{cases} \tag{7.3}$$

The optimization method for the OLSR protocol, used to simulate VANET networks (Vehicular Ad-Hoc Network), is an evolutionary process driven by the DE algorithm. When DE wants to evaluate a solution, it invokes the simulation process of OLSR configuration on a scenario defined for VANETs.

We obtain a solution vector whose values are continuous within the defined range. To evaluate the performance of the different OLSR configurations (solutions), we analyze three performance metrics as an example: PDR-Packet delivery ratio, NRL-Normalized routing load, and E2ED-Average End-to-End delay of a data packet. Once the values of the metrics are obtained, the solution fitness is calculated, as shown in Eq. 7.4.

$$fit = \omega_1(-PDR) + \omega_2(-NLR) + \omega_3(-E2ED)C \tag{7.4}$$

The authors use an aggregate minimization function, where the PDR is formulated with a negative sign. The factors ω_1, ω_2 and ω_3 are used to introduce a weight to the influence of each metric on the fitness function. The PDR is clearly a priority over the other two, as the effectiveness of the routing protocol is prioritized. The constant C normalizes the value of E2ED, so that its range is of the same order of magnitude as that of the other two metrics.

7.3 Ant Colony Optimization (ACO) for Ad Hoc Mobile Networks

ACO [12] is a probabilistic algorithm inspired by ant colonies adapting solutions of combinatorial optimization characterized by their dynamism, decentralization, and computational complexity. It solves problems of searching for optimal or quasi-optimal paths in graphs. In this type of algorithm, ants choose a route and build the hop-by-hop path in a probabilistic way using the pheromone information. The use of pheromone information allows us to build on the experience previously acquired by ants. The algorithm appropriately defines the meaning of pheromone marks for decision-making. The best trajectories will attract a greater number of ants that, in turn will strengthen by depositing a new amount of pheromone, thus giving rise to positive feedback. The fact that ants build their paths in a probabilistic way allows the exploration of multiple routes. This makes the algorithm adapt to changes in the network, increasing robustness through the availability of backup routes and network throughput. Another advantage is that the random forwarding of data packets is based on pheromone information. This is related to the use of multiple routes. In this process, the use of pheromone ensures that the data is routed along the best routes. If the pheromone is kept up to date by the use of enough ants, load balancing automatically follows changes in the network. One of the disadvantages in which we pay special attention is the fact that ants always show complete routes between origin and destination, and this leads to increased overload.

The basic parameters of the algorithm are the initial pheromone trail, the number of ants in the colony, the weights that define the proportion of heuristic information in the probabilistic transition rule. It is established as the point of origin, n, the ant colony nest, and as the destination point f, the food source. There are multiple routes that link the origin n and the destination f, with a different distance. Ants arrive at bifurcations where they must choose between two or more alternative routes to reach their destination. When an ant leaves nest n in search of food source f, it will first track if there is a previous path with pheromone remains and uses some prior knowledge stored in its memory. If it detects pheromones, it will most likely select the path in which the greatest amount of pheromones has been deposited. However, if there is no pheromone path in its environment to follow, a path is established completely randomly until it finds the food source f. Once f is reached, the ant returns to the nest n again selecting more likely that path with more pheromones. If the ant was the first to reach the destination f, its way back to the nest n will be the same way back in the opposite direction, since it is the only pheromone path that exists, as long as it is not short enough to having evaporated, then, the ant will take a random path again. Ants that choose the shortest path take less time than the others to travel their way from the origin to the destination, this generates a greater accumulation of pheromones in the most used path. Other ants will leave the nest n towards the food source f and all will continue using the same mechanics of following and depositing the pheromone trajectories or moving randomly. Over a period, several paths will be used to join nest n and source of food f, but over time those shortest paths will be the most chosen. Based on the pheromone trail, ants follow the route whose probability of selection is from highest to lowest. A constructive algorithm is designed, where in each iteration, each ant constructs a solution to the problem by crossing a graph that represents the problem instances, whose edges represent possible steps that an ant can take to move from one node to another. There are two reasons that the colony converges to the optimal solution: (1) being shorter routes, the deposited pheromone suffers less evaporation than the longer routes, so the level of pheromone is higher. (2) The more ants choose the same trajectory; the higher the level of pheromones will accumulate the same trajectory, avoiding depositing pheromones on the rest of the routes that will suffer, even greater evaporation over time [13].

ACO obtains a minimum feasible cost path on a graph, $G = \{C, L, W\}$. The feasibility of the path will depend on the restrictions imposed, Ω. The behavior of ants has two modes of work: forward and backward. $C = \{c_1, c_2, \ldots, c_n\}$ is the finite set of components and $L = \{l_{c_i c_j}, |c_i, c_j \in C\}$ is the set of possible connections between the elements of C. W is the set of weights associated with the components of C, with links L or both, and $\Omega(C, L)$ is a finite set of restrictions on elements of C and L which may vary over time. To make this probabilistic choice, the ant relies on the pheromone trail deposited in the graph previously by other backward ants. Forward ants do not deposit pheromones when they move. When an ant in forward mode reaches its destination, f, it changes to backward mode and begins its journey back to the source, n.

Once the graph representation is obtained, a population of agents similar to ants travels through the different links exploring the search space and collecting information. Information collected is stored in the pheromone path τ_{ij} associated with the connection l_{ij}, coding the long-term memory of the entire search process. The pheromone intensity trail is proportional to utility, that the ants estimate, that this arc lends to build new solutions. The movement between each link is made based on a stochastic policy with respect to the local pheromone values τ, which represent the ability to select a node. Combining both collection and selection functions, ants incrementally build a solution with each of displacements.

In addition to the generation of ants, ACO includes two procedures: a process of evaporation of pheromones and demon actions. The evaporation process decreases the intensity of the pheromone paths τ over time, which prevents the algorithm from rapidly converging to possible sub-optimal solutions and favors exploration of new areas of the search space. Demon actions are an optional process that allow actions to be performed centrally while the actions of the ant population are executed in parallel.

The ACO algorithm application to the rules of a routing protocol in a network is associated with a cost or, in the case of ACO, a pheromone value is associated with each pair of nodes. Therefore, we have a directed graph $G = \{C, L, W\}$, in which $C = \{c_1, c_2, \ldots, c_n\}$ represents the set of network nodes, $L = \{l_{c_i c_j}, |c_i, c_j \in C\}$ the set of directed links between nodes and, W, the set costs for links, which will depend on the physical characteristics of the link (bandwidth and delay) or traffic they support. Ants are generated and sent from a source node s to a destination node, d, trying to find a feasible path between the two by selecting adjacent nodes. Pheromone paths accumulate in nodes like routing tables, structured in as many rows as neighbors the current node k has, and in as many columns as destinations the network has, in a manner that τ_k represents the ability to select neighbor n to reach destination d [14].

Pheromone evaporates locally and globally. Local evaporation of pheromone is modeled according to Eq. 7.5.

$$\tau_{ij} = (1 - \varepsilon)\tau_{ij} - \varepsilon\tau_0 \tag{7.5}$$

where τ_0 is the initial value of the global pheromone trail and ε with $(0 < \varepsilon < 1)$ is the local pheromone evaporation rate. Then, the global evaporation is modeled by Eq. 7.6.

$$\tau_{ij} = (1 - \rho)\tau_{ij} + \rho\Delta\tau_{ij}^{bs} \tag{7.6}$$

where $\Delta\tau_{ij}^{bs}$ is the amount of pheromone deposited by the best ant and ρ is the global evaporation rate.

The advantage of using ACO is that it allows to obtain a routing table adapting to network traffic changes. ACO has been applied to find approximate solutions to various optimization problems, simulating it through software. In addition, it allows redistributing the load between the different nodes of the network. Some applications in networks for the use of ACO are also based on parameters that regulate the processing, using metrics such as packet size, simulation time, relative importance of heuristics in the pheromone and the processing time of the packet in each node [15]. An aspect of benefit to the algorithm is to implement activities from a global point of view, which cannot be carried out by ants, such as determining the quality of the solutions generated, depositing new amounts of pheromone in the paths of some solution that we want to maximize.

This metaheuristic has some disadvantages, one of them is the difficulty of adjusting several parameters that do not have a theoretical basis for their values. This requires investment of time by users to obtain tests and tests on the parameters given for a particular problem; the update of the routes demarcated by the pheromones with each network and path must be achieved by increasing the difficulty of coding and calculation times for roads.

7.4 Greedy Randomized Adaptive Search Procedure (GRASP)

It is a metaheuristic that combines constructive heuristics with local search. This algorithm can be useful for route searches in node routing tables. GRASP [16] is an iterative procedure, composed of two phases: first, the construction of a solution (construction stage) and then, an improvement process (local search stage). The improved solution is the result of the search process. The mechanism for construction solutions is a random constructive heuristic. It is said that an algorithm is randomized if its response is determined not only by the input data but also by the values produced by a random number generator. It adds, step by step, different components c to the partial solution, s^p, which begins with an empty solution initially. This procedure is repeated several times until a stop criterion determines the extent to which iteration is continued and the best solution found is assumed as a result. Instead of always choosing the best candidate, components that are added in each step are chosen randomly from a restricted list of candidates, (RCL), Restricted Candidate List. This list is a subset of $N(s^p)$, the set of components allowed for partial solution s^p. The term adaptive refers to the fact that benefits associated with each element are updated in each iteration of the construction phase to reflect the changes produced by previous selections. To generate this list, the solution components in $N(s^p)$ are sorted according to some problem dependent function (η) [17].

The local search algorithm phase improves the solution given by the construction phase, replacing it with a better one in the neighborhood. Then, a stop criterion is established. The (RCL) list is composed of the best α components of that set. In the extreme case of $\alpha = 1$, the best component found in a deterministic way is always added. At the other end, with $\alpha = |N(s^p)|$ the component to be added is chosen completely randomly from all the available ones. Therefore, α is a key parameter that influences how the search space will be sampled, also called threshold, where $\alpha \in [0, 1]$. If $\alpha = 0$ corresponds to a pure greedy algorithm and if $\alpha = 1$ corresponds to a random construct. The second algorithm phase consists in applying a local search method to improve the generated solution.

One way to restrict the list of candidates is based on their quality. The items to be included in the list, in addition to preserving the feasibility of the solution, should also have a higher quality in the cost function than the threshold value α. This is described in Eq. 7.7.

$$N(s^p) \in \left[N^{min}, N^{min} + \alpha\left(N^{max} - N^{min}\right)\right] \tag{7.7}$$

where N^{min} and N^{max} represent the minimum and maximum increase in cost respectively.

The quality of the solution depends on: the neighborhood structure used in the local search phase, the rapid evaluation of the cost functions and the main solution built. If the benefit of good choices exceeds the cost of bad choices, a random selection of good and bad choices may eventually produce good results [18]. As an example, the problem of scheduling tasks is solved in the construction phase; that of the variable process times, in the improvement phase, while the problem of variable assignment is treated in both phases.

7.5 Gray Wolf Optimizer (GWO)

The metaheuristic of GWO [19] is inspired by the behavior of the gray wolf herds (Canis Lupus) and their social organization for prey hunting. GWO mimics the leadership hierarchy and the mechanism of gray wolf hunting in nature. Gray wolves prefer to live in groups of between 5 and 12 members on average. The herd consists of two leaders called alpha, α, (one male and one female), and the rest of the herd. The group of alpha wolves are those who lead the herd and influence the search space in more forceful way. In this social structure, alpha members are responsible for making decisions about hunting, where to sleep, time to wake up, and the rest of the members must follow their orders. The alpha member of a herd should not necessarily be the strongest, but the best in terms of group management and the ability to make decisions. The second hierarchy level of the group is the beta members, β, (subordinates of alphas), who must collaborate with the decision-making and compliance with the orders of the alphas. These members are the candidates to become alpha when any of these dies or ages. There is another classification that is below in the hierarchy of gray wolf herds and is called omega, ω, these must be submitted to the alphas and betas obeying them, in many cases usually seen in this type to members who officiate "babysitters", onside the herd [20].

This methodology is widely used in different fields of study, due to its flexibility and simplicity compared to traditional optimization methods. The algorithm presents the following processes, taken from the natural behavior of the gray wolf hunt:

- Encircling prey.
- Hunting.
- Attacking prey (exploitation).
- Search for prey (exploration).

The mathematical model of the social hierarchy of the gray wolf herds is described based on the three main actions related to hunting activity: follow, surround, and attack.

In order to mathematically model the social wolves hierarchy, it is considered that the most suitable solution for the problem to be solved is α. Consequently, the second and third-best solutions are called β and δ, respectively. This last level is made up of scout wolves and sentries. The rest of the candidate solutions are supposed to be ω. In the GWO algorithm, hunting activity is translated as an optimization process, and it is guided by α, β, and δ, while wolves ω are at the last level of the social pyramid and must submit to the orders of the upper wolves.

Gray wolves surround the prey during the hunt. Mathematically this behavior is modeled in Eq. 7.8.

$$D = \left|CX_p(t) - X(t)\right| \tag{7.8}$$

Also,

$$X(t+1) = X_p(t) - AD \tag{7.9}$$

where t indicates the current iteration, A and C are vectors of coefficients, X_p is the vector of prey position, and X indicates the vector position of a certain gray wolf. On the other hand, vectors A and C are calculated in Eqs. 7.10 and 7.11.

$$A = 2ar_1 - a \tag{7.10}$$

$$C = 2r_2 \tag{7.11}$$

Vector A components decrease linearly from 2 to 0, over the course of iterations and r_1, r_2 are random vectors in the range [0, 1]. The mathematical model of GWO describes the approach operation in function of which a gray wolf in a certain position will be able to update its position with respect to the prey, moving wolf's location by adjusting the value of vectors A and C.

In nature, the alpha member usually guides hunting, although beta and delta members may eventually participate in hunt. In order to mathematically simulate hunting behavior of gray wolves, we assume that the alpha member (best candidate solution), beta and delta have a better understanding of the possible location of the prey. Therefore, we save the first three best solutions obtained so far and force the other search agents (including omegas) to update their positions according to the position of the best search agents.

$$D_\alpha = |C_1 X_\alpha - X|,\ D_\beta = \left|C_2 X_\beta - X\right|,\ D_\delta = |C_3 X_\delta - X| \tag{7.12}$$

$$X_1 = X_\alpha - A_1(D_\alpha),\ X_2 = X_\beta - A_2\left(D_\beta\right),\ X_3 = X_\delta - A_3(D_\delta) \tag{7.13}$$

$$X(t+1) = (X_1 + X_2 + X_3)/3 \tag{7.14}$$

where, D_α, D_β, D_δ are the distances of the wolves; C_1, C_2, C_3 are vectors that vary according to the random value r; X_α, X_β, X_δ are the positions of the three best search agents values; X is the update parameter in each iteration for the following searches. X_1, X_2, X_3 are equations to calculate the positions of the three best wolves.

Depending on the fact that each agent updates its position according to: alpha, beta and delta in a search space, we note that alpha's, beta's and delta's role is to estimate the position of the prey and the other wolves update their positions at random around it.

Wolves approach the prey and attack it when it stops moving. The mathematical model of such behavior of approaching the prey is translated as the decrease in the value of vector A (coefficient). The fluctuation range of vector A is also reduced by the value of "a" (Eq. 7.10). The wolves finish attacking the prey when it stops moving. To simulate this procedure, remember that A is a function of a. The value of a decreases from 2 to 0 as the iterations pass. Then the value of A will be in a range $[-2a, 2a]$, in the course of the iterations the value decreases from $|2a|$ up to zero. When $|A| < 1$, GWO forces the wolves to attack the prey. GWO is prone to stagnation in local solutions with these operators and it needs more operators to emphasize exploration.

Wolves have different positions regarding the prey, but converge to it to attack. The mathematical model of this divergence is reflected in the use of random values for vector A (greater than 1 and less than -1) to ensure divergence and the search encompasses the search space in the best way.

In Fig. 7.2 we describe the GWO functionality.

Fig. 7.2 Flow char of GWO

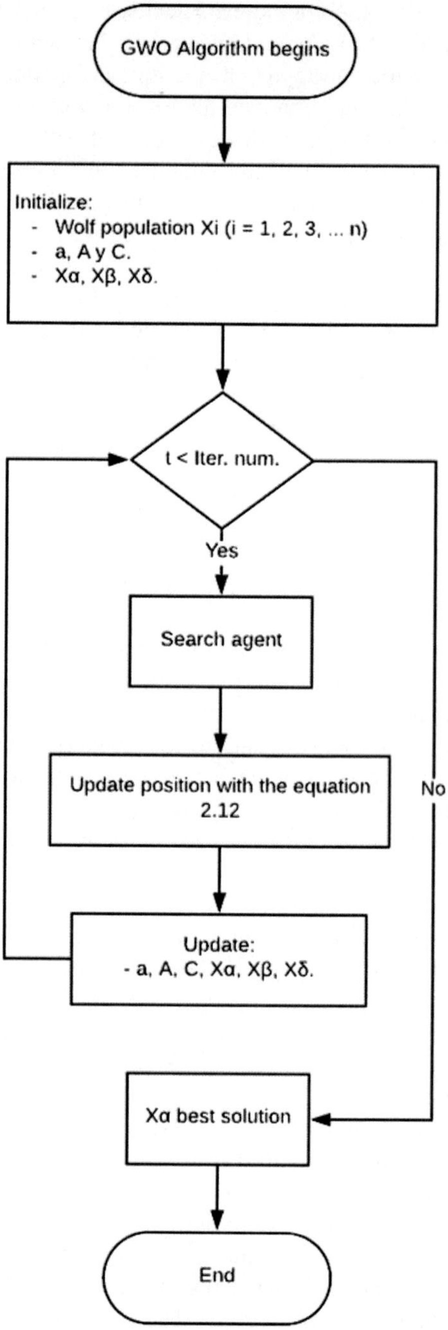

7.5.1 Application to a Network Model for Energy Optimization

GWO is based on hierarchical order and having a local and a global search that helps to have a rapid convergence. Taking as an example of the previous model, in a network model [21], the following assumptions are made: (1) The coordinator node is located in the center of the sensing area and is externally powered. (2) All nodes are randomly distributed, and once deployed, the nodes are not changing their positions. (3) All sensors nodes are homogeneous and have the function of data fusion. (4) After deployment, the coordinator node knows all information of all sensor nodes.

To transmit a $l - bit$ long data packet over the distance d, the required energy is:

$$E_{TX}(l,d) = \begin{cases} lE_{elec} + l\varepsilon_{fs}d^2, & if \ \ d < d_0 \\ lE_{elec} + l\varepsilon_{mp}d^4, & if \ \ d > d_0 \end{cases} \tag{7.15}$$

E_{TX} is the transmitted energy,
E_{elec} is the energy dissipated per bit in the transmitter or receiver circuit,
ε_{fs} and ε_{mp} depend on the transmitter amplifier model.

If the distance between the transmitter and the receiver is less than a threshold d_0, we use the free space model; otherwise, we use the multi-path model.

$$d_0 = \sqrt{\frac{\varepsilon_{fs}}{\varepsilon_{mp}}} \tag{7.16}$$

The energy consumption for the receiver to receive a $l - bit$ long packet is calculated by the Eq. 7.17.

$$E_{TX}(l) = lE_{elec} \tag{7.17}$$

Now, the FIGWO (Fitness value based Improved GWO) [21] defines rounds and each round is divided into cluster construction phase and data transmission phase. A cluster head (CH) is responsible for collecting data from member nodes inside the cluster. It is also responsible for aggregating and delivering data to base station (BS).

We select initial clusters by the following targets. The set of alive sensor nodes is partitioned into k equal subsets according to the fitness value. In each set, the sensor node that is the nearest to the middle point is taken as the initial cluster head. Then, we calculate the node's fitness value according to its distance to the BS, and its residual energy is described in Eq. 7.18.

$$F = \begin{cases} a\frac{E_r}{E_i} + (1-a)\left(\frac{d_{MAX}-d}{d_{MAX}-d_{MIN}}\right), & if \ \ E_r > 0 \\ 0, & if \ \ E_r \leq 0 \end{cases} \tag{7.18}$$

a is a coefficient which indicates the contribution between E_r and d in the fitness function F;

E_r describes the residual energy for alive node;

E_i is the initial energy of a node;

d is the distance from the node to the BS;

d_{MAX} and d_{MIN} are the maximum and the minimum distance between a sensor node and the BS, respectively.

The FIGWO algorithm is used to select CHs in initial clusters. Based on GWO and Eq. 7.18, the fitness value is calculated and used as weights to determine the final position of the optimal solution, which fully considers the node's current state. The new position of prey is computed as follows:

$$X_p^{t+1} = F_{w\alpha} X_\alpha^{t+1} + F_{w\beta} X_\beta^{t+1} + F_{w\delta} X_\delta^{t+1} \tag{7.19}$$

$$F_{w\alpha} = F_\alpha / \left(F_\alpha + F_\beta + F_\delta\right) \tag{7.20}$$

$$F_{w\beta} = F_\beta / \left(F_\alpha + F_\beta + F_\delta\right) \tag{7.21}$$

$$F_{w\delta} = F_\delta / \left(F_\alpha + F_\beta + F_\delta\right) \tag{7.22}$$

where, $F_{w\alpha}$, $F_{w\beta}$ and $F_{w\delta}$ represent the new weights of α wolf, β wolf and δ wolf, respectively, which are calculated by Eq. 7.20, 7.21 and 7.22. F_α, F_β and F_δ are the best three fitness values.

Regarding the algorithm performance, we can analyze some metrics:

(a) Residual energy: This includes the average residual energy of each node and the energy difference between the node with the most energy and the node with the least energy.
(b) Stability period: The time duration from the network operation until the first node is turned off.
(c) Packet Delivery Ratio: The number of data packets received by BS.

7.6 Intelligent Water Drops (IWD)

This constructive metaheuristic [22] is inspired by natural rivers and how to find optimal paths to a destination. These routes follow the actions and reactions that occur between the drops of water with their channels. In this algorithm, several artificial water drops cooperate to change their environment in such a way that the optimal route is revealed as the one with the lowest ground in their links. Soil is the amount that is carried by each drop of artificial water in the algorithm. In this method, we consider that each drop moves in discrete length steps, through a certain number of points, called nodes, starting randomly in one of them and then to their properties, decides which is the next node to visit, fulfilling their route through all nodes that correspond to a particular problem. Keep in mind that the same node is not visited twice by the same drop.

A drop of water in its path always acquires a certain amount of land, which increases its weight and therefore its speed, also causes the amount of soil between the nodes to decrease, causing the next drops of water to choose the routes that have less amount of land, thus facilitating its movement. This phenomenon is modeled by a uniform random distribution, which assigns a determined value to the amount of soil between links of each pair of nodes.

The intelligent virtual water drops algorithm has two properties obtained from a drop of natural water:

- The amount of soil it carries, denoted by $soil(IWD)$.
- The speed, denoted by $velocity(IWD)$.

An IWD algorithm is made up of two parts: a graph that plays the role of distributed memory in which soils of different edges are preserved and the moving part of the DIM algorithm, which is a small number of smart water droplets. These smart water drops (IWDs) both compete and cooperate to find better solutions and by changing the floors of the graph, the paths to the best solutions become more accessible. It is mentioned that DIM based algorithms need at least two IWDs to work.

The IWD algorithm has two types of parameters: static and dynamic ones. Static parameters are constant during the IWD algorithm process. The dynamic parameters are reset after each iteration of the IWD algorithm. The pseudo-code of an algorithm based on the IWD can be described in the following steps [22]:

1. Static parameter initialization
    (a) Problem representation in a graph
    (b) Adjustment of values for static parameters
2. Dynamic parameter initialization: the ground and the speed of IWDs
3. Distribution of IWDs in the graph of the problem
4. Solution construction by IWDs along with ground and update speed
    (a) The local update of the soil in the graph
    (b) Soil and update speed in IWDs
5. Local search on the solution of each DIM (optional).
6. The global update of the soil.
7. Total-best upgrade solution.
8. Go to step 2 unless the termination condition is satisfied.

Assuming that a drop of intelligent water moves in discrete length steps, from one point to another point, its velocity is increased by an amount defined by Δvel^{IWD}, which is inversely proportional to the ground that exists between the two nodes, that is to say that If the drop of water reduces the soil of the riverbed by taking a certain amount of land, it causes a change in its speed. This behavior is modeled by the Eq. 7.23.

$$\Delta vel^{IWD}(t) = a_{\upsilon}/(b_{\upsilon} + c_{\upsilon} soil(i, j)) \tag{7.23}$$

a_{υ}, b_{υ}, and c_{υ} are positive parameters selected by the user;
$soil(i, j)$ is the amount of land between the nodes i and j.

The amount of land increases non-linearly due to the increase of its speed, and it is related to the time it takes to go from one node to another, we see this behavior in Eq. 7.24.

$$soil(i, j) = a_s/\left(b_s + c_s\ time\left(i, j, vel^{IWD}\right)\right) \tag{7.24}$$

a_s, b_s, and c_s are positive numbers, defined by the user.

Since velocity is linear, time is inversely proportional to the velocity of the water drop and proportional to the distance. Therefore, time taken for a drop of water to travel between the two nodes is determined by the Eq. 7.25.

$$time\left(i, j, vel^{IWD}\right) = HUD(i, j)/(\max(\varepsilon_{\upsilon}, vel^{IWD})) \tag{7.25}$$

$HUD(i, j)$ is a local heuristic function, defined depending on the particular problem, which measures the undesirability of a drop of water to move from one node to another. The term $\max(\varepsilon_{\upsilon}, vel^{IWD})$ means that the maximum velocity of the drop must be selected.

The amount of soil between two nodes is affected by the action performed by the drop to the path, therefore the following expression is found in Eq. 7.26.

$$soil(i, j) = \rho_0 soil(i, j) - \rho_n \Delta soil(i, j) \tag{7.26}$$

ρ_0 and ρ_n are numbers between 0 and 1. $soil(i, j)$ is the amount of land that loses the path, but is earned by the drop of water. The amount of soil the water drop earns is expressed by the Eq. 7.27.

$$soil^{IWD} = soil^{IWD} + \Delta soil^{IWD} \tag{7.27}$$

Another modeled behavior is the preference of the drop of water on roads with fewer obstacles. It is implemented by a uniform distribution, which establishes the probability that a drop will choose one path or another. This probability is inversely proportional to the amount of soil between the two nodes. The smaller the amount of soil on a road, the more possibility this road has to be selected. The probability function must fulfill the following properties:

$$f(x) \geq 0, \quad for\ all\ \ x.$$

$$\sum f(x) = 1.$$

Equation 7.28 determines the probability as follows:

$$p(i, j, IWD) = \frac{f(soil(i, j))}{\sum k \notin N(IWD) f(soil(i, k))} \tag{7.28}$$

$N(IWD)$ is the set of nodes that have been visited by the drop. The expression $f(soil(i, j))$ is a function that depends on the amount of soil between two nodes and is expressed by Eq. 7.29.

$$f(soil(i, j)) = \frac{1}{\varepsilon_s + g(soil(i, j))} \tag{7.29}$$

The constant ε_s is a small positive number that prevents a possible indeterminacy of the algorithm. The term $g(soil(i, j))$ is used to shift the amount of soil between and, and is expressed in Eq. 7.30.

$$g(soil(i, j)) = \begin{cases} soil(i, j) & if \ min_{l \notin vc(IWD)}(soil(i, l)) \leq 0 \\ soil(i, j) - min_{l \notin vc(IWD)}(soil(i, l)) & in \ other \ case \end{cases} \tag{7.30}$$

$min_{l \notin vc(IWD)}$ returns the smallest value of the argument, and the algorithm selects $soil(i, j)$ when $min_{l \notin vc(IWD)}(soil(i, l))$ is greater than or equal to zero.

For cases where the argument value is less than zero, the algorithm chooses the option $oil(i, j) - min_{l \notin vc(IWD)}(soil(i, l))$, so this expression aims to return the negative values to positive since the real lengths are positive.

7.7 Particle Swarm Optimization (PSO)

This algorithm [23] is inspired by the social behavior of the congregation of birds or fish. It is a global optimization algorithm to solve problems in which a better solution can be represented as a point or a surface in an n-dimensional space. Hypotheses are represented in this space with an initial speed, as well as with a communication channel between the particles. Particles move through the solution space, and are evaluated according to some criteria after each time step. Over time, the particles are accelerated towards those particles within their communication group that have better fitness values [24]. The main advantage of this approach over other global minimization strategies such as simulated annealing is that the large number of the members that make up the swarm of particles make the technique impressively resistant to the problem of local minima.

It was developed by James Kennedy and Russell C. Eberhart around 1995 [24], and was inspired by the behavior of insect swarms in nature.

Using as an example bees, which at the time they go in search of pollen, look for the region of space where the highest density of flowers is found, because in that region the probability of finding pollen is greater. This idea was where it started to generate the algorithm.

To explain the algorithm, we start with an unknown two-dimensional function $f(x, y)$, where it is randomly evaluated by generating "particles".

Each particle or individual consists of a position p in the search space and a velocity v that determines its movement through space, in the two-dimensional case, p is represented by a vector in the form (x, y) and v with a vector (v_x, v_y).

Nevertheless, since they are particles in a physical real world, they have a quantity of inertia, which makes them keep the same direction in which they move, as well as an acceleration that mainly depends on two characteristics:

- Each particle is "attracted" to the best location that it, individually, has found in its history (BEST LOCAL).
- Each particle is "attracted" to the best location that has been found by the set of particles in the search space (BEST GLOBAL).

The forces to which the particle is subjected and therefore "pushed" depend on two adjustable parameters:

- Attraction to the best staff
- Attraction to the best global

Therefore, at a greater distance from these better ones, the forces are greater. The velocity of the particles is described by Eq. 7.31.

$$v_i(t+1) = v_i(t) + c_1 r_1 \left(p_i^{best} - p_i(t)\right) + c_2 r_2 \left(p_g^{best} - p_i(t)\right) \tag{7.31}$$

where, $v_i(t)$ is the velocity of the particle i in the instant t.

c_1 and c_2 are constants of attraction to the best staff and the best global respectively.

r_1 and r_2 are random numbers between [0, 1].

p_i^{best} is the best personal position.

p_g^{best} is the best global position of the swarm.

Once the speeds are updated the positions are updated and are described by the following Eq. 7.32.

$$p_i(t+1) = p_i(t) + v_i(t) \tag{7.32}$$

7.7.1 Application in Routing in Networks. Minimum Spanning Tree Problem

Many metaheuristic algorithms are applied in network routing; one of the most used is the Ant Colony Optimization (ACO). Besides, PSO is also used to optimize routing

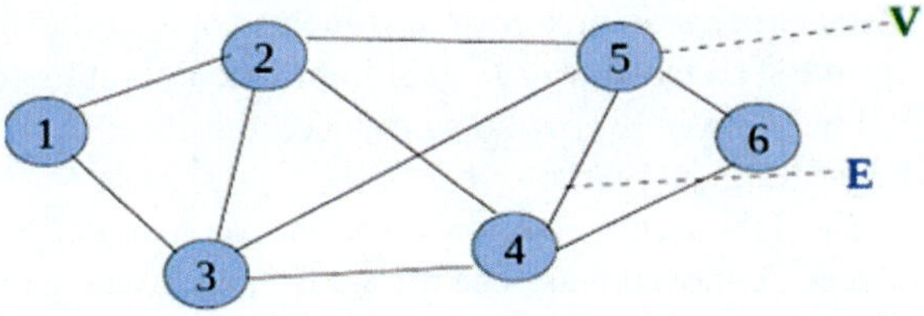

Fig. 7.3 Described network like (V, E).

times, depending on the topological architecture of the network to optimize. The type of protocol to be used is also significant [25].

Algorithms find the ideal route for packets, avoiding collisions or wasting time if we focus these algorithms on the energy consumption problem, they are beneficial in applying them in transmission and receiving tasks because these activities consume the most energy in a sensor device.

If we set an example of an x sensor, where more energy is spent is in the transmission of data. Therefore, if the routing protocol is optimized to be able to send packets more efficiently and in less time, energy consumption will be optimized consequently.

A specific problem that is solved with the PSO algorithm is the Minimum Spanning Tree and Path-Finding, which seeks to find a fast, new and easy way to solve the problem of routing multiple destinations [26].

To understand this problem, we will define what a network is. A network consists of a series of nodes connected to each other. A simple notation for a network is (V, E), where V is the number of vertical nodes; and E is the number of connections or links.

Using Fig. 7.3, as an example, the network would be described as follows in Eqs. 7.33 and 7.34.

$$V = \{1, 2, 3, 4, 5\} \tag{7.33}$$

$$E = \{(1, 2), (1, 3), (2, 3), (2, 4), (2, 5), (3, 4), (3, 5), (4, 5)\} \tag{7.34}$$

A path can be defined as the sequence of different connections or links that join two nodes. A network with n nodes needs $(n - 1)$ links to provide a path between a pair of nodes.

Using the PSO algorithm in this problem, we seek to optimize the distance between nodes, which means finding the shortest path that will be optimal.

7.8 Tabu Search

The Tabu Search [26] is a metaheuristic algorithm that guides a heuristic local search procedure to explore the solution space beyond local optimization. One of the main characteristics of Tabu Search is the use of adaptive memory, which creates a more flexible search behavior in addition to incorporating receptive scanning.

Fred Glover [27], who introduced the metaheuristic term, introduced the term Tabu Search in 1986.

The Tabu Search approach focuses on a strategy of prohibition or penalization of specific movements. The reason for classifying a movement as prohibited that is, tabu, is mainly to prevent cycling.

From an Artificial Intelligence (AI) point of view, the Tabu Search deviates to an extension from intelligent human behavior. Humans are commonly dragged into acting in a hypothesized manner according to some random (probabilistic) elements. The human tendency to act based on hypotheses often leads to events that are classified as errors, but these errors provide valuable information. That receptive exploration enters the Tabu Search, where the assumption that a bad strategic choice can often provide more information than an excellent random choice [28].

The Tabu Search uses a local or neighborhood search procedure to iteratively move from a potential solution x to an improved solution x in the vicinity of x, until some detention criteria have been met. Generally, an attempt limit or a scoring threshold.

Solutions that are supported by the new neighborhood, $N^*(x)$, are determined through the use of memory structures. These memory structures form what is known as a tabu list, a series of prohibited rules and solutions that function as a filter for the solutions that will be admitted in $N^*(x)$. The tabu list consists of a group of solutions that have changed by the process of moving from one solution to another.

Memory structures used in the Tabu Search can be classified into three categories:

- Short term: Recently considered solutions, if a potential solution appears in the tabu list, it cannot be obtained again until an expiration term is met.
- Medium term: Intensification rules aimed at skewing the search towards promising areas of the search space.
- Long-term: Diversification rules that drive the search to new regions (i.e., regarding restarts when the search is stuck in an optimal impasse).

Memory structures of the tabu search work by referring to three main dimensions: being recent, frequency, and quality.

Memories based on the recent and frequent complement each other to strike a balance between intensification and diversification strategies. Diversification seeks to obtain solutions not previously seen, so that previously seen solution is not generated. They are preventing the algorithm from falling into a loop.

On the other hand, the intensification is in charge of looking for good solutions, not only that they solve the problem; nevertheless, they solve it fulfilling a series of characteristics, for instance, reducing the problem time, obtaining a cheap solution, obtaining a quick solution, and so on.

The quality dimension refers to being able to recognize the benefits of a solution, finding the common elements that make a solution a good solution. Quality is when the algorithm learns what kindness and ways there are. They are the ones that usually lead to a right solution and which paths lead to a wrong solution.

7.8.1 Performance of Tabu Search for Location in Wireless Sensor Networks

In WSN environments, an algorithm for location of the sensor nodes is essential; this is because it is essential knowing where nodes are located in a sensor network allows to reach quickly if a failure occurs or some event that requires to be there [29].

In a WSN, sensors are dispersed in large quantities under a specified area. They are distributed from a small environment (such as a house, a park, and so on) to environments where they are further from each other (a factory or a subdivision), along years, algorithms developments capable of locating sensor nodes in a wireless network has become an important issue.

Most of these algorithms share a common characteristic, as they seek to estimate the locations of sensor nodes with initially unknown locations (these locations are known as "target nodes") using prior knowledge of the positions of nodes that will function as points. of reference (anchor nodes) and measurement between sensors.

The following is usually used to determine the distance between two nodes: the strength of the received signal (received signal strength—RSS), time of arrival (TOA), travel time (Round Trip Time—RTT).

Estimating the position of the target nodes is formulated as an optimization problem, where a target function must be minimized by representing the error in locating the target nodes. Mean squared error ranges between the target node and neighboring anchor nodes can be considered as the objective function of this problem.

The Root Mean Square Error (RMSE) measures the amount of error between two data sets. In other words, it compares a predicted value and an observed or known value.

The objective function (Eq. 7.35) is the equation that will be optimized given the given limitations or restrictions and with variables that need to be minimized or maximized using linear or nonlinear programming techniques.

$$f(x, y) = \frac{1}{M} \sum_{i=1}^{M} \left(\sqrt{(x - x_i)^2 + (y - y_i)^2} - \hat{d}_i \right)^2 \tag{7.35}$$

(x_i, y_i) are coordinates of the anchor node number i,

(x, y) are coordinates of the node to be estimated,

$M \geq 3$ is the number of anchor nodes that are in the transmission range of the target node.

We break down the objective function with Eqs. 7.36 and 7.37.

$$\sqrt{(x - x_i)^2 + (y - y_i)^2} - \hat{d}_i \tag{7.36}$$

$$\hat{d}_i = \sqrt{(x - x_i)^2 + (y - y_i)^2} + n_i \tag{7.37}$$

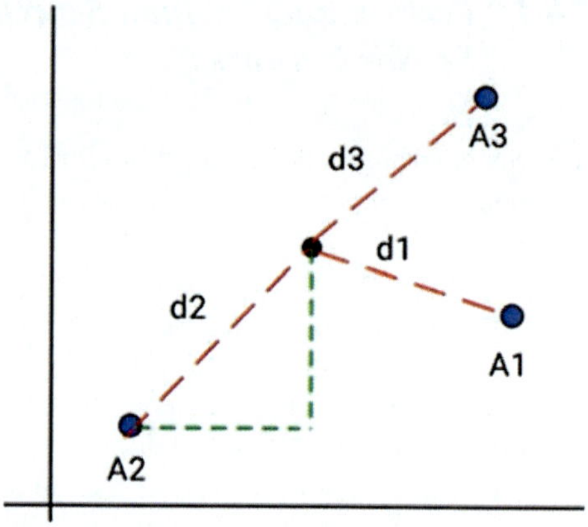

Fig. 7.4 Plane of coordinates

where n_i is the distance measurement error.

The first part of the objective equation deals with the Pythagorean Theorem, using the Pythagorean Theorem you can know the distance between two coordinate points as shown in Fig. 7.4.

In Fig. 7.4, we have the target node (the black node) and the anchor nodes $(A1, A2, A3)$. If we assign to each one a coordinate in the plane, given by the Eq. 7.38.

$$A1 = (x_1, y_1) \quad A2 = (x_2, y_2) \quad A3 = (x_3, y_3) \quad T = (x, y) \tag{7.38}$$

Using the Pythagorean Theorem, the distance between two coordinates can be calculated, where the values calculated are d_1, d_2 and d_3.

$$\begin{aligned} d_1 &= \sqrt{(x_1 - x_i)^2 + (y_1 - y_i)^2} \\ d_2 &= \sqrt{(x_2 - x_i)^2 + (y_2 - y_i)^2} \\ d_3 &= \sqrt{(x_3 - x_i)^2 + (y_3 - y_i)^2} \end{aligned} \tag{7.39}$$

Resulting in Eq. 7.40.

$$f(x, y) = \frac{1}{M} \sum_{i=1}^{M=3} \left(\left(d_1 - \hat{d}_1\right) + \left(d_2 - \hat{d}_2\right) + \left(d_3 - \hat{d}_3\right) \right)^2 \tag{7.40}$$

Once the respective distances between the anchor nodes and the target node are calculated, the term $\hat{d}_i$ is added to the Eq. 7.40. Besides, in that Eq. 7.37, a new term is added, n_i, this term is used as the distance measurement error.

$$(x, y) = \frac{1}{M}\sum_{i=1}^{M=3}((d_1 - d_1 - n_1) + (d_2 - d_2 - n_2) + (d_3 - d_3 - n_3))^2 \quad (7.41)$$

If we look at Eqs. 7.36 and 7.37, by substituting Eq. 7.37 into Eq. 7.36, our result will eliminate the part of the Pythagorean Theorem leaving only the measurement error value. Considering the outer sign, that error is negative.

This is because the mean squared error measures the error that exists between our predictions ($\hat{d}_i$) with the real values.

$$f(x, y) = \frac{1}{M}\sum_{i=1}^{M=3}((-n_1) + (n_2) + (-n_3))^2 = \frac{1}{M}\sum_{i=1}^{M=3}(-N)^2 = \frac{1}{M}(N) \quad (7.42)$$

The resulting value that we obtained in Eq. 7.41 afterwards is squared, but the result of the function cannot be negative, which assures that the result will always be positive.

Finally, the value obtained from the sum of errors is averaged with the number of anchor nodes that were available. Notice that M is the number of anchor nodes that are within the transmission range of the target node.

The result obtained from our equation is interpreted as the greater the quantity, the worse our localization algorithm is. Conversely, the closer to zero, the better the algorithm. If it is zero, it would be perfect.

7.8.2 Location Algorithm for Wireless Sensor Networks

In a Tabu Search-based distributed localization process, each target node performs the following steps [30]:

1. Formation of the tabu list. The tabu list is formed with an initial estimate as its first entry.
2. Formation of the neighborhood complex. A disturbance is inserted into the estimated initial value, which moves it to a new value. Although any number of disturbances is allowed, in this case they are restricted to four movements that take the initial estimate (x_c, y_c) to the neighbors $(x_c \pm \Delta x, y_c \pm \Delta y)$. The objective functions of these neighbors are evaluated.
3. The solution and the tabu list are updated. The current solution is moved to the best available non-tabu neighbor even if it deteriorates the current value of the target function and the tabu list is updated.
4. Steps 2 and 3 are repeated until the value of the objective function falls below a predefined threshold value or the number of iterations crosses its threshold value.

7.9 Firefly Algorithm (FA)

This algorithm is based on the natural behavior of fireflies, which are attracted to the brightest individual, this being the best individual in the population. The algorithm starts with a random population moving in the search space, where each individual is attracted to the best, which in this case, is characterized by the best objective function [31].

In the algorithm, there are two functions that a firefly performs: attracting a partner or finding potential prey. This can be a warning or defense technique for potential predators. The rhythmic flash, the blink rate and the time are parameters of this behavior, the females use a unique pattern to attract the male, but they can also imitate a pattern of another species to obtain food [32].

The intensity of the light I decreases at a particular distance as the distance r increases, due to the law of the inverse square $I \propto \frac{1}{r^2}$, the air absorbs the light which becomes progressively weaker as the distance increases (the characteristics of the transmission medium, such as air, which is the channel, for wireless networks). Because of these factors, fireflies can only be seen up to a certain distance, but it is sufficient for communication between them. The flashing light can be modeled as an objective function.

The algorithm generates three main rules, based on the behavior of fireflies:

(a) All fireflies are gender independent so that the attraction between them is realized regardless of their sex.
(b) The attraction is proportional to the brightness, so for any glowing firefly, the least bright will move towards the brightest. The attractiveness decreased as the distance increased if there is no brighter firefly they move randomly.
(c) The glow of the firefly is conditioned by the landscape of the objective function.

In the FA algorithm, there are two important questions: the variation of light intensity and the modeling of attraction. The attractiveness of the firefly is determined by its brightness, and this is associated with the objective function of the optimization problem. Furthermore, the objective should vary with the degree of absorption.

Starting from the inverse quadratic law of distance, the intensity of light is formulated according to Eq. 7.43, where I_s is the intensity of light at the source.

$$I(r) = \frac{I_s}{r^2} \tag{7.43}$$

If the medium is specified with a fixed light absorption coefficient and the intensity is dependent on the distance, the effect of combining the inverse square law equation with absorption can be modeled with Eq. 7.44 in its Gaussian form.

$$I(r) = I_0 e^{-\gamma r^2} \tag{7.44}$$

where I_0, is the intensity at a distance of zero and γ is the absorption coefficient. The attraction of the firefly is proportional to the intensity of the light perceived by the nearby fireflies, therefore, the attraction can also be defined with Eq. 7.45.

$$\beta(r) = \beta_0 e^{-\gamma r^2} \tag{7.45}$$

For a real implementation, the attraction of a firefly is a monotonously decreasing function, which varies with the distance r between two individuals i and j, where β_0 is the value of the initial attraction at a distance $r = 0$. The location of a firefly is expressed with the Cartesian coordinates, so the distance r between two individuals, in the case of two dimensions, is represented by Eq. 7.46.

$$r_{ij} = \sqrt{(x_i - x_j)^2 + (y_i - y_j)^2} \tag{7.46}$$

With locations x_i and x_j, the Euclidean distance is expressed in Eq. 7.47.

$$r_{ij} = |x_i - x_j| = \sqrt{\sum_{k=1}^{d} (x_{i,k} - x_{j,k})^2} \tag{7.47}$$

Then, the movement of a firefly i, towards another j, is determined by Eq. 7.48.

$$x_i^{t+1} = x_i^t + \beta_0 e^{-\gamma r_{ij}^2} (x_j^t - x_i^t) + \alpha \varepsilon_i^t \tag{7.48}$$

In Eq. 7.48 there are three terms that make up the firefly movement. The first corresponds to the current position, the second term is the attraction and the third is the random component, where α is the parameter that controls the randomness and ε_i, is an array of random numbers drawn from a Gaussian distribution or a distribution uniform.

Because FA is based on attraction and decreases with distance, this automatically subdivides the population into subgroups and each group can swarm around each local optimum, achieving a global optimum. However, through this subdivision, it is possible to find the optimal ones simultaneously; this will depend on the size of the population. Hence its application in multimodal problems. It is described in Algorithm 1.

Algorithm 1. Firefly Algorithm

(continued)

(continued)

Set target function: f (x), x = {x1, ..., xn}
Define algorithm parameters:
Ii = Light intensity; determined by target function
γ = absorption coefficient; α = randomness parameter
Initialize population, create particle matrices with random locations within the search space
t = 1; iter = number of iterations; i = number of particles
while (number of iterations or stopping criteria):
for all (for every firefly):
Evaluate each firefly and its variables, according to the target function
End for
Sort the fireflies and find the globally best
for i = 1: n (all the n fireflies)
for j = 1: n (all the n fireflies)
Calculate the distance using the Eq. 7.34
if Ii s ≤ Ij
Modify the attraction according to distance
Move the i firefly to j with the Eq. 7.35
End if
End for
End for
t = t + 1
End while
Show the collected results and the globally best

7.9.1 *Firefly Meta-Heuristic Algorithm Applied to Artificial Neural Network*

The field of research regarding Artificial Neural Networks (ANR) is one of the most active in the scientific community with multiple recent applications. The Firefly algorithm has been used successfully in ANR pre-training with the aim of avoiding the convergence in local minima of conventional training methods such as the Stochastic Gradient Descent (SGD) algorithm [33]. However, in networks with a considerable number of parameters, pre-training becomes an optimization problem in spaces with high dimensionality, and the application of the Firefly algorithm, as well as any meta-heuristics, presents computational limitations to consider.

The ANR training problem is formulated as an optimization problem. Formally, given a function $f(w, X)$ that measures the network error when evaluating a set of training patterns X, where $w \in \mathbb{R}^d$ is the vector of weights or parameters of an ANR, the optimization problem is defined as follows in Eq. 7.49.

$$\hat{w} = \min_{w \in \mathbb{R}^d} f(w, X) \tag{7.49}$$

where,

$$f(w, X) = \frac{\sum_{i=1}^{|X|} (\hat{y} - y_i)^2}{|X|} \tag{7.50}$$

where, $\hat{y}_i$ and y_i are the expected output and the actual output of the network respectively for the pattern x_i of the set X. The objective function definition is known as the Mean Squared Error (MSE).

The use of global optimization heuristic methods applied to the ANR pre-training problem is currently limited by the expansion of the search space defined by w and by the cost of evaluating the objective function $f(w, X)$. This work studies the effect of reducing the number of training patterns to direct the search carried out using the Firefly Algorithm [32], which has been used previously in high dimensionality problems. One of many hypothesis is that if the Firefly algorithm is used to solve problem in Eq. 7.49 in the training process, it is possible to decrease the number of patterns in X employees and at the same time obtain similar precision.

ANRs, also known as connectionist models, emerged in 1943, introduced by McCulloch and Pitts. A common perspective for its characterization is the idea of Distributed Parallel Processing [34]. In this perspective, ANRs are variations of a distributed parallel processing model characterized by a group of the following aspects:

(a) Processing units (Neurons). Each processing unit does a relatively simple job: receiving input from neighboring units or external sources, and using this input to produce an output signal that then propagates to other processing units or to the network output. The propagation rule corresponds to Eq. 7.51.

$$s_k(t) = \sum_j w_{jk}(t) y_j(t) + \theta_k(t) \tag{7.51}$$

w_{jk} is the weight associated with the input y_j, and θ_k is the bias corresponding to neuron k at an instant of time t. Subsequently, the value s_k is evaluated in an activation function to limit the contribution of the net input in the activation of the neuron. Often, a non-decreasing function such as that described in Eq. 7.52.

$$F(s_k) = \frac{1}{1 + e^{-s_k}} \tag{7.52}$$

(b) Connectivity pattern between processing units. The processing units are connected to each other. How these connections are established determines what the network is capable of representing and learning. Among the most frequent connection architectures are feed-forward, recurrent, and convolutional networks. Feed-forward networks are the simplest and most used. This architecture is based on a group of cascading unit layers. Units located on the same layer do not have connections between them, they receive their input from the units located in the previous layer, and they send their outputs to the units in the next layer. For simplicity, in the following, it will be assumed that an

ANR of the feed-forward type is made up of a layer of input neurons, which does not perform processing, a layer of intermediate or hidden neurons, and a layer of output neurons. This particular configuration is widely recognized as a multilayer perceptron type network.

(c) Learning rule. In order for a network to recognize a certain problem, a procedure is necessary to modify the connectivity patterns based on the experience obtained from the training patterns. This means training the network or modifying the weights w that weight the importance of the inputs of each neuron. For ANR training, the SGD algorithm is traditionally used. This algorithm loops through the parameter space of a network such that the error function $f(w, X)$ is minimized, iteratively following the direction of a calculated error gradient at a random starting point in the parameter space. In each step, small movements are made in the opposite direction of said gradient until a minimum is found. Due to this behavior, the family of descending gradient-based algorithms is sensitive to converging on local minima in the parameter space, and therefore is sensitive to the initial values of the network weights.

If the parameter space of an ANR were made up of only two weights, the error surface defined by the training patterns could be represented as a landscape made up of valleys and hills. An individual at any point in that landscape looking for an elevation would have to decide which direction to go. This corresponds to the fact of generating a group of candidate solutions and, based on some experience, choosing which could be the best. In practice, this experience is modeled on the objective function. The objective function represents a measure of the height or quality of each candidate solution. It is not necessary to obtain a highly accurate measurement of the objective function to compare candidate solutions qualitatively. For example, it is not always a requirement to know the exact number of meters of height of two elevations to estimate that one is higher than the other. In this way, it is possible to reduce the number of patterns that are used to calculate the value of the objective function without affecting the exploration and exploitation capacity of the Firefly algorithm. To reduce the number of standards, the whole is sampled initial training. This sampling is carried out randomly following a uniform distribution [35].

7.10 Scatter Search (SS)

Scatter Search [36–38] is based on maintaining a relatively small set of tentative solutions (called a reference set or RefSet) that is characterized by containing quality and diverse solutions (distant in the search space).

The method generates a reference set from a wide population of solutions. Subsets of the reference set whose solutions are combined are selected to obtain starter solutions for local improvements. The result of these improvements can lead to an update of the reference set and even of the entire population from which to extract the reference set again.

For the complete definition of SS, five components must be specified: the creation of the initial population, generation of the reference set, generation of subsets of solutions, method of combining solutions, and improvement method. For problems in telecommunications, we may use this metaheuristic because it is very efficient in some instances. It should also be noted that for multi-objective formulations, we have designed a new approach, called AbYSS (Archive-based hYbrid Scatter Search) [39], which is part of state of the art for existing test benches in continuous multi-objective optimization and its application to real-world problems such as those discussed here is of great interest. Scattered search is a metaheuristic introduced in the 1970s [40]. Its foundations are based on the strategies of combining decision rules, especially on sequencing problems, as well as on the combination of constraints (such as the method of surrogate constraints). Its operation is based on the use of a small population, known as a reference set (RefSet), whose elements are systematically combined to create new solutions. In addition, these new solutions can go through a phase of improvement, consisting of applying a local search. The reference set is initialized from an initial population, P, made up of the most dispersed random solutions possible, and is updated with the solutions resulting from the improvement phase. Many implementations of sparse search algorithms take as a reference to the template proposed by Glover, which consists of defining five methods:

- Generation of diverse solutions. The method is based on generating P diverse solutions from which we will extract a small subset called RefSet.
- Improvement. This is typically a local search method to improve solutions.

- Reference set update. This method takes care of both the initial generation of the reference set and its subsequent update. The RefSet stores solutions based on two criteria: quality and diversity. Therefore, the RefSet1 and RefSet2 are distinguished, which store high-quality solutions and very diverse solutions, respectively.
- Subset generation. It is a method to generate subsets of solutions from the RefSet to which the combination method will be applied. SS is based on examining exhaustive all possible combinations of the RefSet. The most usual way is to generate pairs of solutions.
- Combination of solutions. It is the method in charge of combining in some way the subsets of solutions resulting from the previous method to generate new solutions.

In general, sparse search avoids the use of random components, and the typical operators of evolutionary algorithms such as crossover and mutation operations are not theoretically adapted. However, there are works in both the mono-objective and the multiobjective [41] fields that demonstrate that the use of stochastic operators within of the SS scheme allows calculating solutions that are more precise. We describe this scheme in Fig. 7.5. This approach will also allow us to use the same operators as the GAs (Genetic Algorithms) presented in the previous section and to be able to compare thus the efficiency and effectiveness of each of the two search engines, ssGA and SS. Given that each problem considered will have its specific operators, we show below how these methods interrelate to define the general sparse search procedure. We describe this procedure in Algorithm 2.

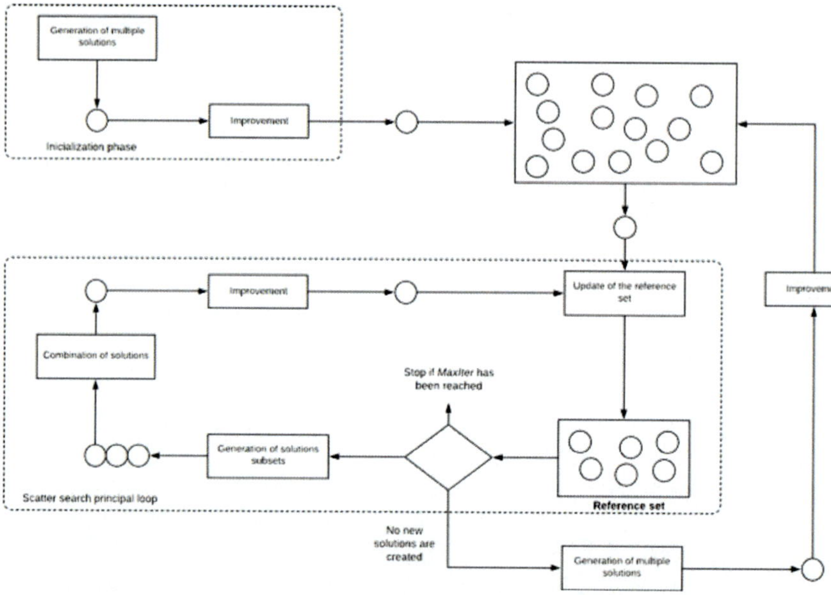

Fig. 7.5 Scatter Search general scheme

Algorithm 2. Scatter Search Algorithm

1. Begin with $P = \emptyset$. Use the generation method to build a solution and the upgrade method to try to improve it; being x the obtained solution. If $x \notin P$ then add x to P. (e.g., $P = P \cup x$), otherwise, reject x. Repeat this phase until P has a preset size
2. Construct the reference set $R = \{x^1, \ldots, x^b\}$ with the best b/2 solutions of P and the most diverse b/2 solutions of P to those already included
3. Evaluate the solutions in R and order them from best to worse with respect to the objective function
4. Do $NewSolution = TRUE$

As long as $(NewSolution)$

5. $NewSolution = FALSE$
6. Generate the subsets of R in which there is at least one new solution. As long as (subsets remain unexamined)
7. Select a subset and label it as examined
8. Apply the combination method to the subset solutions
9. Apply the upgrade method to each solution obtained by combination. Being x the improved solution: $If(f(x) < f(x^b)\ and\ x\ is\ not\ in\ R)$
10. Do $x^b = x$ and reorder R
11. Do $NewSolution = TRUE$

The Scatter Search strategy involves six procedures and three stopping criteria to solve an optimization problem. The procedures are the following [42]:

1. The Initial Population Creation Method. This procedure creates a random initial population P of good and disperse solutions.
2. The Reference Set Generation Method. This procedure selects some of the best representative solutions in the population to be included in the reference set, R.
3. The Subset Generation Method. This procedure generates subsets, which consist of good solutions in the reference set, to apply the combination procedure.
4. The Solution Combination Method. This procedure, which includes parameters used to modulate the intensification and/or diversification, combines the solutions in the previously selected subset to get the new current solution, s.
5. The Improvement Solution Method. This procedure, which includes parameters to modulate the specialization of the method, improves the current solution s to get an improved solution s'.
6. The Reference Set Updating Method. This procedure updates the reference set by deciding when and how the obtained improved solutions are included in the reference set replacing some solutions already in it.

In addition to these six procedures, the metaheuristic involves three stopping procedures that implement the criteria to decide when generating a new reference set, a new population or when stopping the search.

(a) New Reference Set Criterion. The first criterion decides when to generate a new reference set from the population.

(b) New Population Criterion. The second criterion decides when to generate a new population.
(c) Termination Criterion. Finally, it decides when to stop the whole search.

7.10.1 Performance Metrics

When evaluating the performance of multi-objective algorithms, two aspects are usually taken into account: minimizing the distance of the Pareto front obtained by the algorithm to the exact Pareto front of the problem and maximizing the extension of solutions on the front so that the distribution is as uniform as possible. In this sense, the metrics used in the field can be grouped among those that assess the proximity to the exact fronts, those that they measure the diversity of the non-dominated solutions obtained, or both [43]. Here we have used one of each type:

- Generational Distance. This metric was proposed by Van Veldhuizen and Lamont [44] to measure the distance between the set of non-dominated solutions found and the set of Pareto optimal. It is defined in Eq. 7.53 as follows:

$$GD = \frac{\sqrt{\sum_{i=1}^{n} d_i^2}}{n} \tag{7.53}$$

where, n is the number of elements in the set of undominated solutions and d_i is the Euclidean distance (measured in the objective space) between each of these solutions and the closest member of the Pareto front optimum. The solution set needs to be standardized for reliable results.

- Spread. An extension of the Spread metric has been used here. Instead of using the distance between two consecutive solutions, the new metric uses the distance from each point to its closest neighbor.

$$\Delta = \frac{\sum_{i=1}^{m} d(e_i, S) + \sum_{X \in S} |d(X, S) - \overline{d}|}{\sum_{i=1}^{m} d(e_i, S) + |S^*|\overline{d}} \tag{7.54}$$

where S is a set of solutions, S^* is the set of Pareto optimal, $(e_1, \ldots, e_m)$ are the m extreme solutions of S^*, m is the number of objectives and,

$$d(X, S) = \min_{Y \in S, Y \neq X} |F(X) - F(Y)|^2 \tag{7.55}$$

$$\overline{d} = \frac{1}{|S^*|} \sum_{X \in S^*} d(X, S) \tag{7.56}$$

Ideally, $\Delta(X, S) = 0$ for those solution sets with an optimal distribution on the Pareto front. This metric also requires prior normalization for its application.

- Hypervolume. This metric calculates the volume (in the target space) covered by members of a given set, Q, of unmanaged solutions to problems where all targets are to be minimized. Mathematically, for each $i \in Q$ a hypercube v_i is constructed with a reference point W and the solution i that define its diagonal. Point W can be obtained simply with the worst values of the objective functions. So, the union of all hypercubes is what defines the hypervolume (HV) in Eq. 7.57.

$$HV = volume\left(\bigcup_{i=1}^{|Q|} v_i\right) \tag{7.57}$$

Algorithms that achieve higher values for HV are better. As with the previous two metrics, it is necessary to normalize the non-mastered solutions since HV depends on the scaling of the values of the objective function.

AbYSS has been validated using a current and standard methodology within the multi-objective community. Different algorithm configurations have been tested to evaluate their search ability on the ZDT (Zitzler, Deb y Thiele) [45] family of problems. Once I know has successfully resolved these problems, the best AbYSS configuration has been compared to the two state-of-the-art algorithms in the field: NSGA-II [46] and SPEA2 [47]. The results of three different metrics show that, with the benchmark used, AbYSS surpasses these two algorithms in convergence (closeness of the fronts calculated with respect to the Pareto-optimal front) and, especially, in diversity (distribution of solutions not dominated on the Pareto front).

7.11 Greedy Randomized Adaptive Search Procedures (GRASP)

In an optimization problem, we have a set of solutions S and an objective function $f\colon S \to \mathbb{R}$ and we want to find a solution $s^* \in S$ such that $f(s^*) \leq f(s), \forall s \in S$. GRASP is a metaheuristic to obtain approximate solutions to this type of problem. Each GRASP iteration consists of two fundamental phases [48]:

- Construction phase.
- Local search phase.

The best overall solution is saved as a result. Algorithm 3 shows the basic scheme of GRASP.

Algorithm 3. GRASP general algorithm

```
function GRASP
while stop criterion not satisfied
    s ← ConstructGreedyRandomizedSolution()
    LocalSearch(s)
    UpdateSolution(s, bestSolution)
end while
return bestSolution
end function
```

During the construction phase, the solution is built step by step, one item at a time. For each problem to be solved, it should be clearly defined that it is an element of the solution. In each iteration, the next element to be added to the solution is determined by ordering the candidate elements according to a greedy criterion. This criterion measures the incremental cost that each candidate element contributes to the solution. The heuristic is adaptive because in each iteration this incremental cost is updated to reflect the contribution of the last added element. The probabilistic component of GRASP is characterized by randomly choosing one of the best elements, but not necessarily the best of all. For this, a list with the best elements is created in each iteration, called the Restricted Candidates List (RCL). The length of the RCL list is a parameter to determine in each particular problem. The general outline of the construction phase is presented with Algorithm 4. There may be an intermediate repair phase in case the construction stage returns non-feasible solutions. In this phase the necessary elements of the solution are modified to make it feasible. The solutions obtained in the construction phase may not be locally optimal with respect to some neighborhood criterion [49].

Algorithm 4. Construction Phase

```
function CONSTRUCTGREEDYRANDOMIZEDSOLUTION
    initialize solution s
    initialize the candidate element set
    evaluate the incremental cost of each element
    while solution s incomplete do
        construct RCL list
        e ← random element of RCL
        s ← s ∪ {e}
        update the candidate element set
        recalculate the incremental cost of each element
    end while
    return s
end function
```

The neighborhood $N(s)$ is a set of solutions neighboring s, which means that the solutions of $N(s)$ can be obtained by changing some element of s or performing some operation on s. In the same way, the solution s must be able to be obtained from the solutions of $N(s)$ by carrying out the inverse process. All this will depend on each particular problem. After the construction phase of GRASP, a local search algorithm is applied, in which the solution obtained is iteratively replaced by another neighboring solution with a better objective function value, until it is locally optimal. The general scheme of local search is presented in Algorithm 5. The local search implemented will depend on each particular problem. For example, the replacement solution may be the one with the best objective function value in all $N(s)$ or the first to be found that exceeds the one to be replaced. Or, it could happen that if the defined neighborhood is too big, not everything is explored due to the computational cost that it would imply and only a subset of it is revised.

Algorithm 5. Local Search Phase

```
procedure LocalSearch(s)
    while exists t ∈ N(s) which f(s) < f(s) do
            choose a solution t ∈ N(s) which f(s) < f(s)
    s ← t
    end while
end procedure
```

If the solution obtained at the end of the local search process exceeds the best one obtained so far, it replaces it and the entire procedure is repeated, provided the GRASP stop criterion is not known. The stop criterion can be, for example, exceeding a maximum number of total iterations, a maximum number of iterations without improving the best solution found, reaching a known height, or any other criterion depending on the problem [50].

7.11.1 GRASP for Spare Capacity Allocation Problem (SCA)

We assume that in addition to the input data we have a set P of candidate cycles from which a solution must be constructed.

Construction Phase.

The SCA solution is a set (with repeats) of cycles, or a set of $\langle cycle, number of copies \rangle$ pairs. Therefore, the elements to construct the solution iteratively will be copies of the candidate cycles of the set P. The criterion used to arm the RCL list with the best candidates will be the real efficiency. In each iteration, the K cycles with the highest real efficiency are saved in the RCL list (The K parameter will be determined based on experimentation.) Then one of these cycles is chosen at random and added a copy of it to the solution. The $BestCycles$ function returns the K best cycles of P according to the past efficiency as a parameter.

There is a first drawback with the SCA problem. The set of feasible solutions is not finite. We can make it finite if we restrict feasible solutions to be minimal. This means that if a copy of any cycle is removed from the solution, it will no longer be feasible. This is accomplished with the $RemoveRedundancies$ procedure which iteratively reviews the cycles of a solution and evaluates whether they can be removed, and in that case removes a copy. Repeat the process until the solution is minimal. Doing so also returns a higher quality solution.

Local Search Phase

Another drawback that arises is the difficulty of defining an appropriate neighborhood for this problem considering that the elements of the solution are copies of candidate cycles.

The options considered were many, but none was promising. Then we replace the local search with a simple improvement algorithm whose objective is to replace cycles of the solution with others that are in the set of candidates P, again following the criterion of real efficiency. For each cycle c of the solution, a copy is removed and the solution is rebuilt, prohibiting the use of cycle c. The original solution is replaced by the least expensive solution found in this way. The process is repeated until the cost is not reduced.

Final Scheme

Its complexity depends on the $ListCycles$ procedure, which is external to the proposed algorithm. As a stopping criterion, we established a maximum number of total iterations and iterations without improving the solution (the algorithm ends when one of these limits is exceeded). These parameters are adjusted during experimentation.

7.11.2 GRASP Optimization for the Multi-Level Capacitated Minimum Spanning Tree Problem

The GRASP employs heuristic rules used both to do a partition of $V - \{r\}$ and to decide whether to call an exact method within the construction and the local search phases. A GRASP iteration consists of two subsequent phases: construction phase and local search phase. The construction phase builds a feasible solution. The local search starts with the solution built in the former phase and tries, by investigating neighborhoods, to achieve improvements until local minima. The procedure returns the best solution found after Max_It iterations.

Construction Phase

The construction phase builds in two steps a feasible solution to the MLCMST (Multi-Level Capacitated Minimum Spanning Tree) problem. It is used a greedy randomized heuristic to do a partition of $V - \{r\}$ in R_k, where $k = 1, ..., K$ are subsets. This step has an input parameter $w \geq z^L$ that limits the cardinality of each subset in the partition.

$$K = \lceil \frac{V - \{r\}}{w} \rceil \tag{7.58}$$

where K, represented in Eq. 7.58, is in the unitary node weights case. We build one subset of the partition at a time. Let S denote the candidate nodes to be inserted in subset R_k being built, initially we set $S = V - \{r\}$ and $k = 1$. The procedure starts an iteration by removing a node i at random from S and inserting it in R_k. While $|R_k| < w$ and $S = \emptyset$, the procedure moves one node after the other from S to R_k. Therefore, all nodes remaining in S are candidates to be inserted in R_k. Then, we create a RCL formed by the best elements given by a greedy evaluation function. The RCL is formed by those nodes whose incorporation to R_k results in the smallest incremental cost according to Prim's algorithm to compute a MST. Let d_j, for a node $j \in S$, be a label defined as $d_j = min\left\{c^1_{ij} : i \in R_k\right\}$. Moreover, we set d_{min} and d_{max} respectively to the minimum and maximum values of d_j, j ∈ S. Given a parameter $\alpha \in [0, 1]$, RCL is defined in Eq. 7.59.

$$RCL = \{j \in S : d_j \leq d_{min} + \alpha(d_{max} - d_{min})\} \tag{7.59}$$

The element to be moved from S to R_k is randomly selected from those in the RCL, and labels d_j are updated for the nodes remaining in S following Prim's algorithmic. When w nodes are inserted in R_k, we increment k and proceed until a partition of $V - \{r\}$ is made.

There are K subproblems consisting each of an independent MLCMST instance on the subgraph induced in G by $R_k \cup \{r\}$, $k = 1, ..., K$. Then, we solve independently each of the K subproblems to optimality, and we have thus a feasible solution to the original MLCMST instance. We solve the sub-problems, nodes grouped in one subset R_k of the partition which forms two or more components connected to r.

Local Search Phase

Here, we improve a feasible solution by re-arranging nodes of different components connected to r. Let us consider a spanning tree $T = \left(V, \hat{E}\right)$ feasible to the MLCMST problem. We designate a connected component of the forest obtained by the elimination of r and its incident edges from T as a component of T. The value of applying a move to T in order to generate a neighbor $T = \left(V, \hat{E}'\right)$ is described in Eq. 7.60.

$$\Delta = \sum_{(i,j)\in\hat{E}'} c^{\hat{l}'}_{i,j} - \sum_{(i,j)\in\hat{E}'} c^{\hat{l}}_{i,j} \tag{7.60}$$

where $c^{\hat{l}}_{i,j}$ is the cost of capacity $z^{\hat{l}}_{i,j}$ installed on edge $(i, j) \in \hat{E}$. The value of Δ is computed by obtaining the best way to connect to r the nodes of the components involved. As mentioned before, evaluate this kind of move means solving a smaller sized MLCMST instance in the worst-case. Thus, we limit the size of the sub-problem induced in G by the components defining a move. We make within the local search calls to an optimization package to solve exactly each MLCMST instance modeled with the capacity-indexed formulation. We propose thus heuristic rules to form the sub-problems.

The rationale of the heuristic is to identify a local gain in connecting two nodes that are in different components of T. Given a node i, let us define by V_i the set of nodes in the component containing i, and we also denote by $c(T_i)$ the cost of the subgraph induced in T by $V_i \cup \{r\}$ with the respectively installed capacities. We consider the set S of non-leaf nodes in T as candidates to be reference nodes to form a sub-problem. At each iteration of the local search, a node i is chosen at random from S. The procedure starts to build a subset of nodes P which may induce a sub-problem in G. Now, P is set to $V_i \cup \{r\}$, and the gain γ is set to $c(T_i)$. The set $\overline{S}$ contains the nodes that connect the other components to r. The parameter h, which limits the cardinality of P, is chosen at random from the interval $[\underline{h}, \overline{h}]$, where $\underline{h}$ and $\overline{h}$ are positive integers. We add the nodes of one or more components to P to build a sub-problem. A node j is chosen at random from $\overline{S}$, and the procedure looks for a node u belonging to V_j that could be connected to i with the capacity $\hat{l}$ installed on edge $(p(u), u)$ at a smaller cost. The component V_j is included in the sub-problem being built if both there exists such a node and the cardinality of $V_j \cup P$ does not exceed h. The gain is increased by adding $c(T_i)$ to γ. Node j is removed from $\overline{S}$ and the procedure continues trying to enlarge the sub-problem until either the cardinality of P is h or $\overline{S}$ is empty.

A sub-problem has been built if P contains also nodes other than the ones in $V_i \cup \{r\}$. We then solve to optimality the subproblem induced in G by P. We try to re-arrange in an optimal manner the components of T selected to be included in P. Let us suppose that the optimal solution of the sub-problem has a value of φ, note that φ cannot be greater than γ. If φ is smaller than γ, an improving move with value $\Delta = \varphi - \gamma$ has been found, and the current solution T is updated. The set S of candidates to be reference nodes is also reconfigured. We include in S the non-leaf nodes of the optimal solution of the sub-problem that are not already in it. The choice of a node i from S at the beginning of an iteration may not lead to an improving move for two reasons: the move value Δ is zero or none of the components has been added to P. In the first case, the optimal solution of the sub-problem has the same configuration the components involved already had in T, and we remove from S all the nodes of component V_i. On the other hand, the procedure iterates while S in not empty [51].

7.12 Applications

Within metaheuristics, we can distinguish two types of search strategies. On the one hand, we have the "smart" extensions of local search methods (path-based metaheuristics). The goal of these strategies is to avoid local minima somehow and move to other promising regions of the search space. This type of strategy is followed by Tabu Search, iterated local search, variable neighborhood search, and simulated cooling. These metaheuristics work on one or more neighborhood structures imposed by the search space. Another type of strategy is that followed by ant colonies or evolutionary algorithms. These incorporate a learning component in the sense that, implicitly or explicitly, they try to learn the correlation between the variables of the problem to identify the regions of the search space with high-quality solutions (population-based metaheuristics). These methods perform, in this sense, a biased sampling of the search space.

The planning of cellular wireless networks involves making decisions based on different characteristics and parameters. One of the most critical objectives of planning for mobile telecommunications systems is to design the configuration necessary to provide a service optimally concerning some performance criterion while satisfying a set of restrictions. The performance criteria can be, for example, the cost associated with the topology or the quality of the service offered. One of the problems associated with the planning stage of wireless systems is coverage of a given area using the minimum number of radio base stations [52].

Tasks inherent in the design of wireless networks such as cell phone networks give rise to a variety of problems that must be solved in order to provide quality service to users. Many of these problems have the characteristic of being highly complex and combinatorial problems.

Evolutionary Algorithms (AEs) encompass a set of very popular metaheuristics and spread over the last thirty years, due to their versatility in solving a large number of problems. The ae base their operation on an emulation of the process of the natural evolution of living beings, applying the neo-Darwinian concepts of natural selection, the survival of the fittest individuals, and genetic diversity to solve search, optimization, and learning problems [53].

Evolutionary algorithms have proven to be a valid alternative in solving combinatorial problems. Mainly, within the set of evolutionary algorithms, the Distribution Estimation Algorithms (DEA) constitute a set of methods that replace the classic crossover and mutation operators, by an estimate of the underlying probability distribution of the population of potential solutions and subsequent sampling of it. After the generation of an initial population and subsequent evaluation of all its members, the best individuals are selected to build a probabilistic model of the population. From it, a sampling is performed to generate a new set of individuals. This process is repeated until some stopping criteria is reached [54].

Algorithms that come from the field of evolutionary computing have proven to be useful in solving combinatorial problems of complexity [55]. An evolutionary algorithm simulates the adaptability of species to the environment in which they live. Given a population of individuals, those who are fittest have a better chance of survival, while those who are less fit tend to perish. Beyond terminology and analogies with natural evolution, AEs are stochastic procedures that maintain a population of individuals $P(t)$ for each iteration t. Each individual constitutes a potential solution to the problem to be treated, which is represented by a particular structure. The solution is evaluated using a measure of the fitness function. Subsequently, a new population is formed (iteration $t + 1$) selecting the best individuals of $P(t)$. Some members of the new population undergo unary transformations m_i (mutation), which create a new individual from itself, and higher-order transformations, such as crossing, which create new individuals by combining parts of two or more individuals. The algorithm runs until a certain number of generations have evolved or until some stopping criteria are satisfied. The best individual is considered a close to the optimal solution to the problem. Distribution estimation algorithms are heuristics that share characteristics of evolutionary algorithms but where the potential solutions to the problem that make up the population are considered as realizations of a multidimensional random variable, whose joint probability distribution can be estimated and updated to through different mechanisms.

Among the most relevant application areas in science, industry and commerce, the following can be highlighted [56]:

- Automatic and control systems: metaheuristics have been applied as efficient learning methods to reduce the need for human presence in automated tasks.
- Bioinformatics: Metaheuristics have been positioned as effective methods to solve complex problems that handle large volumes of data for the study of protein structure, drug design, sequence alignment, and other relevant problems.
- Engineering: Metaheuristics are useful methods for the design of systems with many components and complex functions such as the design and optimization of aerodynamic profiles, signal and image processing, design of machinery, electronics, hydraulic networks and other problems [57].
- Information processing: metaheuristics are useful methods to solve problems of characteristic selection, classification and grouping of data, natural language processing and others.
- Manufacturing and industry: metaheuristics improve competitiveness in relevant problems of the globalized economy, including optimization of assembly and packaging lines, optimization of department stores, etc.
- Planning and scheduling: metaheuristics positively influence productivity and service quality, when used as an aid to decision-making on the allocation of resources to important tasks, in reduced execution times.
- Routing and Logistics: Metaheuristics are useful methods to manage and optimize the flow of resources in large complex scenarios.
- Telecommunications: metaheuristics solve problems related to the design and optimization of wired, wireless and mobile networks.

7.13 Conclusions

The objective of this chapter has been to carry out a compilation and bibliographic review of the main heuristic methods of solving optimization problems, which we find in real life in areas such as telecommunications. Due to the high number of variables and restrictions in these problems, it is sometimes challenging to solve them accurately, so these techniques are of vital importance to obtain right solutions in acceptable times.

The telecommunications industry has provided and continues to provide a host of optimization problems that arise from the design of the communication system itself to some aspects of its operation. The resolution of these problems has undoubtedly played a vital role in the development and use of this type of system. However, as they have become more popular and their market penetration is greater, the size of telecommunications systems has grown and, therefore, the problems they pose are so high that they are unapproachable with exact techniques. Metaheuristic algorithms are one of the best options in this context since they are capable of finding quality solutions in acceptable times.

The evolution of metaheuristics during the last 25 years has had an almost exponential behavior. In the time that elapses from the first reluctance (due to its supposed lack of scientific rigor) until today, very high-quality solutions have been found to problems that long ago seemed unapproachable.

A particular type, known as optimization inspired by nature or "metaheuristic" is gaining substantial popularity in the research community due to its advantages, which are applicable in computational intelligence, data mining, and its applications. Taken from the wonders of nature, such algorithms computationally optimize complex search problems, with superior search performance and efficiency compared to previously used optimization techniques. It should be added that, based on many literary studies, it is clear that some are more efficient and popular than others. Therefore, it would be useful to carry out more studies for each type. Metaheuristic algorithms manage to explore a wider range of possible solutions and return higher-quality solutions to given problems, especially in the instances with the largest number of nodes in a different type of networks.

References

1. Bernard MS, Pei T, Nasser K (2019) QoS strategies for wireless multimedia sensor networks in the context of IoT at the MAC Layer, Application Layer, and Cross-Layer Algorithms. J Comput Netw Commun
2. Aswale P, Shukla A, Bharati P, Bharambe S, Palve S (2019) An overview of internet of things: architecture protocols and challenges. In: Information and communication technology for intelligent systems. Springer, Singapore, pp 299–308
3. Guleria K, Verma AK (2019) Comprehensive review for energy efficient hierarchical routing protocols on wireless sensor networks. Wireless Netw 25(3):1159–1183
4. Nayak AK, Mishra BSP, Das H (2019) In: Mishra BB, Dehuri S, Panigrahi BK (eds) Computational intelligence in sensor networks. Springer, Berlin
5. Kaveh A (2014) Advances in metaheuristic algorithms for optimal design of structures. Springer International Publishing, Switzerland, pp 9–40
6. Xing H, Zhou X, Wang X, Luo S, Dai P, Li K, Yang H (2019) An integer encoding grey wolf optimizer for virtual network function placement. Appl Soft Comput 76:575–594
7. Clausen T, Jacquet P (eds) (2003) RFC3626: optimized link state routing protocol (OLSR)
8. Boushaba A, Benabbou A, Benabbou R, Zahi A, Oumsis M (2015) Multi-point relay selection strategies to reduce topology control traffic for OLSR protocol in MANETs. J Netw Comput Appl 53:91–102

9. García-Nieto JTJM, Alba E. (2010) Configuración Óptima del Protocolo de Encaminamiento OLSR para VANETs Mediante Evolución Diferencial, Conference: Congreso Español de Metaheurísticos, Algoritmos Evolutivos y Bioinspirados, (MAEB'10), Valencia
10. Price KV, Storn RM, Lampinen JA (2005) The differential evolution algorithm. Differential evolution: a practical approach to global optimization, pp 37–134
11. Lobato FS, Steffen V Jr, Neto AS (2012) Estimation of space-dependent single scattering albedo in a radiative transfer problem using differential evolution. Inverse Probl Sci Eng 20(7):1043–1055
12. Dorigo M (2007) Ant Colony Optimization. Scholarpedia 2(3):1461
13. Okdem S, Karaboga D (2009) Routing in wireless sensor networks using an ant colony optimization (ACO) router chip. Sensors 9(2):909–921
14. Rodríguez AI (2013) Algoritmos Inspirados En Swarm Intelligence Para El Enrutamiento En Redes De Telecomunicaciones
15. Kaveh A, Ghobadi M (2020) Optimization of egress in fire using hybrid graph theory and metaheuristic algorithms. Iranian J Sci Technol Trans Civil Eng 1–8
16. Li M, Hao JK, Wu Q (2020) General swap-based multiple neighborhood adaptive search for the maximum balanced biclique problem. Comput Oper Res 104922
17. Resende MG, Ribeiro CC (2003) Greedy randomized adaptive search procedures. In: Handbook of metaheuristics. Springer, Boston, pp 219–249
18. Feo TA, Resende MG, Smith SH (1994) A greedy randomized adaptive search procedure for maximum independent set. Oper Res 42(5):860–878
19. Mirjalili S, Mirjalili SM, Lewis A (2014) Grey wolf optimizer. Adv Eng Softw 69:46–61
20. Faris H, Aljarah I, Al-Betar MA, Mirjalili S (2018) Grey wolf optimizer: a review of recent variants and applications. Neural Comput Appl 30(2):413–435
21. Zhao X, Zhu H, Aleksic S, Gao Q (2018) Energy-efficient routing protocol for wireless sensor networks based on improved grey wolf optimizer. KSII Trans Internet Inf Syst 12(6)
22. Shah-Hosseini H (2009) The intelligent water drops algorithm: a nature-inspired swarm-based optimization algorithm. Int J Bio-Insp Comput 1(1–2):71–79
23. Clerc M (2010) Particle swarm optimization, vol 93. Wiley, New York
24. Kennedy J, Eberhart R (1995) Particle swarm optimization. In: Proceedings of ICNN'95-international conference on neural networks, vol 4. IEEE, pp 1942–1948, Nov 1995.
25. Lee A (2013) Particle swarm optimization (PSO) with constraint support. Python Software Foundation, Accessed 18 Apr 2018
26. Glover F, Laguna M (1998) Tabu search. In: Handbook of combinatorial optimization. Springer, Boston, pp 2093–2229
27. Laguna M, Kelly JP, González-Velarde J, Glover F (1995) Tabu search for the multilevel generalized assignment problem. Eur J Oper Res 82(1):176–189
28. Glover F (1986) Future paths for integer programming and links to artificial intelligence. Comput Oper Res 13(5):533–549
29. Gopakumar A, Jacob L (2009) Performance of some metaheuristic algorithms for localization in wireless sensor networks. Int J Netw Manage 19(5):355–373
30. Batista BM, Glover F (2006) Introducción a la búsqueda Tabu, vol 3, pp 1–36
31. Yang XS, He X (2013) Firefly algorithm: recent advances and applications. arXiv preprint arXiv:1308.3898
32. Nayak J, Naik B, Pelusi D, Krishna AV (2020) A comprehensive review and performance analysis of firefly algorithm for artificial neural networks. In: Nature-inspired computation in data mining and machine learning. Springer, Cham, pp 137–159
33. Bui DK, Nguyen TN, Ngo TD, Nguyen-Xuan H (2020) An artificial neural network (ANN) expert system enhanced with the electromagnetism-based firefly algorithm (EFA) for predicting the energy consumption in buildings. Energy 190:116370
34. Mcclelland JL, Rumelhart DE, PDP Research Group et al (1987) Parallel distributed processing, vol 2. MIT press, Cambridge
35. Rojas Delgado J, Trujillo Rasúa R (2018) Algoritmo meta-heurístico Firefly aplicado al pre-entrenamiento de redes neuronales artificiales. Revista Cubana De Ciencias Informáticas 12(1):14–27

36. Glover F, Laguna M, Martí R (2000) Fundamentals of scatter search and path relinking. Control Cybern 29(3):653–684
37. Glover F (1998) A template for scatter search and path relinking. Lect Notes Comput Sci 1363:13–54
38. Glover F, Laguna M, Martí R (2003) Scatter search. In: Advances in evolutionary computing. Springer, Berlin, pp 519–537
39. Nebro AJ, Luna F, Alba E, Dorronsoro B, Durillo JJ, Beham A (2008) AbYSS: adapting scatter search to multiobjective optimization. IEEE Trans Evol Comput 12(4):439–457
40. Herrera F, Lozano M, Molina D (2006) Continuous scatter search: an analysis of the integration of some combination methods and improvement strategies. Eur J Oper Res 169(2):450–476
41. Luna Valero F (2008) Metaheurísticas avanzadas para problemas reales en redes de telecomunicaciones
42. Melián Batista MB (2003) Optimización metaheurística para la planificación de redes WDM
43. Deb K (2001) Multi-objective optimization using evolutionary algorithms, vol 16. Wiley, New York
44. Van Veldhuizen DA, Lamont GB (1998) Multiobjective evolutionary algorithm research: a history and analysis, pp. 1–88. Technical Report TR-98-03, Department of Electrical and Computer Engineering, Graduate School of Engineering, Air Force Institute of Technology, Wright-Patterson AFB, Ohio
45. Zitzler E, Deb K, Thiele L (2000) Comparison of multiobjective evolutionary algorithms: empirical results. Evolut Comput 8(2):173–195
46. Deb K, Pratap A, Agarwal S, Meyarivan T (2002) A fast and elitist multiobjective genetic algorithm: NSGAII. IEEE Trans Evol Comput 6(2):182–197
47. Zitzler E, Laumanns M, Thiele L (2001) SPEA2: Improving the strength pareto evolutionary algorithm. Technical Report 103, Swiss Federal Institute of Technology (ETH), Zurich, Switzerland
48. Delgadillo E (2013) Modelos y algoritmos para diseno de redes de comunicaciones con requisitos de supervivencia (Doctoral dissertation, Tesis de Licenciatura, Departamento de Computación, Facultad de Ciencias Exactas y Naturales, Universidad de Buenos Aires)
49. Festa P, Resende MG (2002) GRASP: an annotated bibliography. In: Essays and surveys in metaheuristics. Springer, Boston, pp 325–367
50. Barros B, Pinheiro R, Ochi L, Ramos G (2020) A GRASP approach for the minimum spanning tree under conflict constraints. In: Anais do XVI Encontro Nacional de Inteligência Artificial e Computacional. SBC, Jan 2020, pp 166–177
51. Gamvros I, Raghavan S, Golden B (2003) An evolutionary approach to the multi-level capacitated minimum spanning tree problem. In: Telecommunications network design and management. Springer, Boston, pp 99–124
52. Martins AX, de Souza MC, Souza MJ, Toffolo TA (2009) GRASP with hybrid heuristic-subproblem optimization for the multi-level capacitated minimum spanning tree problem. J Heurist 15(2):133–151
53. Glover F (1977) Heuristics for integer programming using surrogate constraints. Decis Sci 8(1):156–166
54. Baluja S (1994) Population-based incremental learning. A method for integrating genetic search based function optimization and competitive learning (No. CMU-CS-94–163). Carnegie-Mellon Univ Pittsburgh Pa Dept Of Computer Science
55. Goldberg DE (1989) Genetic algorithms in search, optimization, and machine learning. Addison-Wesley, Reading. NN Schraudolph and J, 3(1)
56. para el Diseño CDE, de Redes Inalámbricas OM. Mecánica Computacional, vol XXVIII, Number 31. Optimization and Control (A)
57. Odili JB, Noraziah A, Ambar R, Wahab MHA (2018) A critical review of major nature-inspired optimization algorithms. In: The Eurasia proceedings of science technology engineering and mathematics, vol 2, pp 376–394

Chapter 8
Metaheuristic Algorithms Applied to the Inventory Problem

8.1 Introduction

Almost all industries and businesses make purchases; they usually have more than one possible supplier; in most cases, they have many possible suppliers, since suppliers have different prices, policies, different levels of quality. Making wise decisions may save money, which has an impact on the final price of products, and therefore, on the company's competitiveness. The problem can be solved using mathematical models to represent the cost of the inventory management, and finally, applying techniques to minimize the cost. Mathematical models of the inventory management problem may be complex and NP-hard, and as a result, evaluating all possible solutions to find the cheapest one may be unfeasible, even with a computer. When that happens, metaheuristic algorithms may be used to find a reasonable solution in a reasonable amount of time. This chapter deals with this topic. This chapter presents an example problem, in order to illustrate the complexity of the inventory management and how a large number of possible solutions may arise, then it solves the problem with traditional and metaheuristic algorithms to demonstrate the advantages of metaheuristic algorithms. Finally, other problems are discussed, and further information about algorithms is provided.

8.1.1 *The Inventory Example Problem*

In this section, an example problem is presented, along with an explanation designed to answer the following questions: what kind of mathematical models represent the cost or the objective optimization function in the inventory management? Why do they lead to non-linear equations? Why is the number of possible solutions often so large, even infinite? Why do local optimal points appear?

© The Author(s), under exclusive license to Springer Nature Switzerland AG 2021
E. Cuevas et al., *Recent Metaheuristic Computation Schemes in Engineering*,
Studies in Computational Intelligence 948,
https://doi.org/10.1007/978-3-030-66007-9_8

The example problem presented here has also been studied in several scientific publications [1–3], where metaheuristic algorithms and other types of algorithms were used to reach a solution.

A manufacturing company that will be called "*the customer*", needs to purchase certain items in order to produce the final product. For example, it could be a company that makes doors. This company may purchase the knobs from a wide variety of other companies, which will be called "*the suppliers*". One or more suppliers must be chosen to purchase from, as well as how many items (knobs) to order each time, and how often to make orders from the (or each) supplier.

Manufacturers usually need to make these decisions for many different items. Final retailers also decide how many products to purchase in order to have them available for the final user.

In our example, the *customer* calculated that 1000 units a month were needed, setting the demand at $d = 1000$. The parameters of this problem will be introduced throughout the example in order to save space and make it easier for the reader to get involved with the problem and its characteristics. A summary of the parameters is presented in Table 8.1, and they will be explained in detail.

Having a constant demand applies to manufacturers that have production lines, meaning that the number of final products manufactured in a certain amount of time can be determined. The items can be purchased from three different suppliers. This parameter will be referred to as $r = 3$. Since the cost of interrupting the production is very high, shortages are not allowed, meaning that purchasing fewer items than required is not an option.

Each supplier sells the items at a different price. The ***price of the items*** is designated as p_i, where i indicates which of the three suppliers we are dealing with; for example, the price given by the first supplier is $p_1 = 20$ dollars per item, while the second supplier sells the item at $p_2 = 24$ dollars each, and the third one offers a price of $p_3 = 30$ dollars per item.

Each time an order is placed, suppliers charge a setup or ***ordering cost***, which may also be different from one supplier to another one; in our example, this is designated as k_i. The first supplier charges 160 dollars per order ($k_1 = 160$), the second 140 ($k_2 = 140$), and the third 130 dollars ($k_3 = 130$). This setup cost is independent of the number of items purchased, and therefore, it is not convenient to order a small number of items.

The lead time l_i, indicates how long we must wait from the allocation of the order until the items are received. This must be considered during the planning of the purchasing, but it does not have an associated cost. In the example problem, the respective lead times for each supplier are $l_1 = 1$ day, $l_2 = 3$ days, and $l_3 = 2$ days.

Each supplier has a different monthly capacity, c_i. In this specific example, none of the suppliers can produce the demand for $d = 1000$ units a month. Their respective capacities are $c_1 = 700$, $c_2 = 800$, and $c_3 = 750$, units per month. This forces us to choose at least two suppliers and makes the problem more interesting.

Another parameter, which is not a cost but it is important for the problem, is the perfect rate of suppliers, q_i, which indicates the percentage of non-defective parts over the total purchased units guaranteed by the supplier. $q_1 = 0.93$ means that

Table 8.1 Parameters of the example problem [2]

Data	Definition
$r = 3$	Number of available suppliers
$d = 1000$	Demand in units a month
$w = 16$	Weight of an item shipped (lbs)
$h = 10$	Inventory holding cost in dollars per unit a month
$k_1 = 160$	Setup cost of the first supplier in dollars per order
$k_2 = 140$	Setup cost of the second supplier in dollars per order
$k_3 = 130$	Setup cost of the third supplier in dollars per order
$p_1 = 20$	Price of items of the first supplier in dollars per unit
$p_2 = 24$	Price of items of the second supplier in dollars per unit
$p_3 = 30$	Price of items of the third supplier in dollars per unit
$l_1 = 1$	The lead time of the first supplier in days
$l_2 = 3$	The lead time of the second supplier in days
$l_3 = 2$	The lead time of the third supplier in days
$q_1 = 0.93$	The perfect rate of the first supplier
$q_2 = 0.95$	The perfect rate of the second supplier
$q_3 = 0.98$	The perfect rate of the third supplier
$q_a = 0.95$	The minimum required perfect rate
$Y = 1$	Time length of the planning scenario in months
$c_1 = 700$	Production capacity of the first supplier (units per month)
$c_2 = 800$	Production capacity of the second supplier (units per month)
$c_3 = 750$	Production capacity of the third supplier (units per month)

supplier 1 guarantees that 93 items (from every 100 items purchased) are flawless and can be used in production. The two other suppliers offer a perfect rate of $q_2 = 0.95$, $q_3 = 0.98$, respectively. The customer also has a minimum required perfect rate of $q_a = 0.95$ for the total items purchased, which means that they have specified that they can accept up to 5% of defectives parts. Observe that this does not mean the customer cannot purchase items from supplier 1, whose perfect rate is 0.93, but if it does it, it would have to also purchase items from supplier 3, in order to have an average perfect rate (from all items purchased), of 0.95 (at least).

Back to the topic of the cost, there is a cost related to storing the items, the ***inventory holding cost***, referred to using the letter h, given in dollars per unit per month. In the example case, $h = 10$ dollars per unit per month, which means that storing one item costs 10 dollars. This holding cost can be extended to the on transit inventory in the ***transit inventory cost***. The cost function equation will be reviewed later in this section.

Finally, the last cost is ***transportation cost***. The example problem considers that the manufacturer pays for the transportation costs and owns the inventory in transit. This is known as a FOB (free-on-board) policy [4]. In the example problem, the transportation cost contains quantity discounts, which introduces a non-linear behavior of the problem and leads to a locally optimal solution. Quantity discounts may be present not only in the transportation cost but also in the price of items. In all cases, the cost behavior is similar, and this deserves special attention. The following sub-sections will discuss this topic.

8.1.2 Behavior of a Cost Function Under Quantity Discounts

The transportation discounts contained in this example are one example of cost functions under quantity discounts. The cost is provided by a table, and it is a non-linear function of the shipment weight. It may also be different from one supplier to another one. The weight of each item is represented by $w = 16$ lb in the example case. Table 8.2 shows the shipping cost of supplier 3 in the illustrative example. The cost is given in dollars per hundredweight (cent weight, CWT), which is in this case dollars per hundreds of pounds, and it contains what we call *breakpoints*, in which, when the weight of shipments surpass the *breakpoint*, a new *discounted* or *lower* cost applies.

Figure 8.1 shows the transportation cost of a certain shipment versus the weight shipped using the information in Table 8.2, which is also called *"nominal"* freight rates. Notice that transportation companies offer discounts to motivate larger shipments.

It may be observed that due to the volume discounts, an interesting behavior occurs; the cost of shipping 1000 lbs is \$611.4, the same cost as shipping 815.85 lbs. The 815.85 lbs weight is called *indifference-point*. Moreover, the total transportation

Table 8.2 Nominal freight rates of supplier [2]

Shipped weight range (lbs)	Supplier 3
1–499	\$81.96/CWT
500–999	\$74.94/CWT
1000–1999	\$61.14/CWT
2000–4999	\$49.65/CWT
5000–9999	\$39.73/CWT
10,000–19,999	\$33.44/CWT
20,000–29,999	\$18.36/CWT
30,000–40,000	\$5030

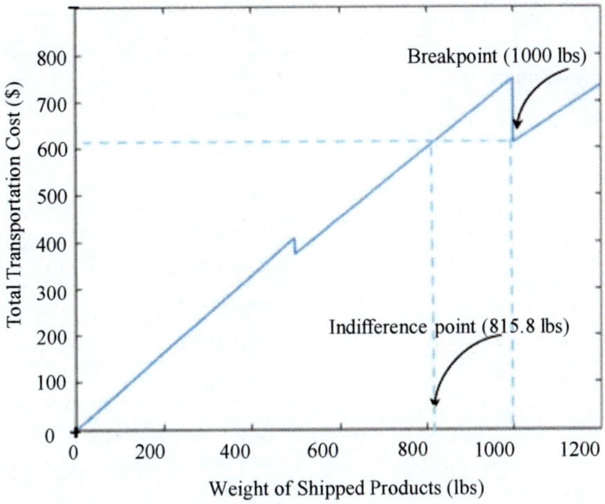

Fig. 8.1 Total transportation cost function

cost corresponding to any weight in the range of 815.85–999 lbs would be larger than that number.

A practice called over-declaring is motivated by this situation. In this case, any order in this range of weights is declared as a 1000 lbs shipping. Another option is to *jump* to a 1000 lbs shipping whenever the shipment is in the range of 815.85–999 lbs. The real or actual cost between the *indifference-point* and the *breakpoint* can be calculated by dividing the cost of a 1000 lbs shipment over the weight. When over-declaring is considered, the shipping cost function takes the stepped form shown in Fig. 8.2.

Shipping costs in Table 8.2 can be modified if the actual freight rate is calculated considering the over-declaring; the result is shown in Table 8.3.

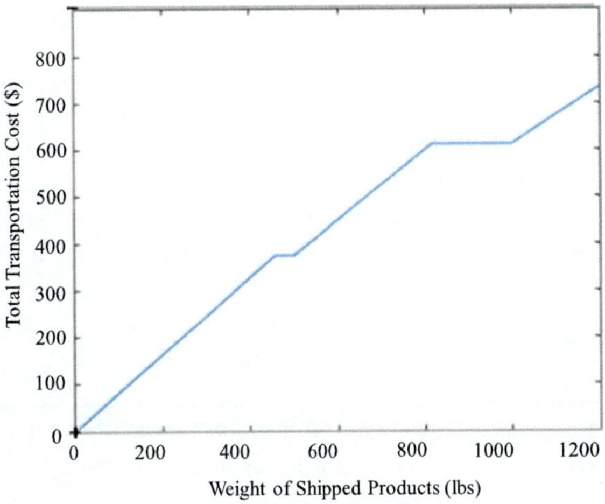

Fig. 8.2 Transportation cost with over-declaring [2]

Table 8.3 Actual freight rates considering over-declaring

Weight range (lbs)	Freight rate
1–428	$81.96/CWT
429–499	$374.7
500–771	$74.94/CWT
772–999	$611.4
1000–1803	$61.14/CWT
1804–1999	$993
2000–4070	$49.65/CWT
4071–4999	$1,986.5
5000–7682	$39.73/CWT
7683–9999	$3,344
10,000–13,702	$33.44/CWT
13,703–19,999	$3,672
20,000–27,383	$18.36/CWT
27,384–40,000	$5,030

A table similar to Table 8.3 is used for each supplier. The fact that a cost is given by a table based on different linear rates and breakpoints creates a cost that includes quantity discounts, non-linear, and introduced local minimum points. In terms of cost, remember that the objective will be minimizing the cost of a function composed of the different discussed costs.

8.1.3 The Solution, Format, and Parameters

A solution to the problem contains the information given in the set of variables indicated in Table 8.4.

The solution indicates how many items are purchased from each supplier during the full order cycle. For example, a solution may be $j_1 = 2$, $j_2 = 3$, and $j_3 = 4$, which means that during the full order cycle, we will make 2 orders to supplier 1, 3 orders to supplier 2, and 4 orders to supplier 3. This can also be expressed in vector form $j = [2, 3, 4]$. If the vector contains zeros, that means those suppliers are not selected. We must also indicate the order quantity, Qi, which indicates how many

Table 8.4 Solution variables

Variable	Description
j_i	Number of orders placed to supplier i per order cycle period, ∀ i = 1,…,r.
Q_i	Ordered quantity for orders placed to supplier i (in units), ∀ i = 1, …, r
T_C	Order cycle period (in months)

items are contained in each order. It is expected that all orders to the same supplier are the same size, but the order size may differ from one supplier to another one. An example of the solution may be $Q = [500, 500, 100]$. In the order cycle period T_C, in months, we can place the orders distributed in this period. An example could be $T_C = 3.5$ months.

There are two variables that are part of the solution but are auxiliary variables that will help us to express some characteristics of the solutions. One of them is the total number of orders per cycle (sum of elements in vector j).

$$M = \sum j_i, \quad \forall i = 1, \ldots r, \tag{8.1}$$

In the example previously discussed of solution $M = 9$. The other auxiliary variable is R_i, which represents how many items we purchased from each supplier during the full order cycle:

$$R_i = j_i Q_i, \quad \forall i = 1, \ldots r, \tag{8.2}$$

This can help to determine which supplier is reaching its maximum production capacity. In the example solution $R = [1000, 1500, 400]$. The sum of R_i represents the number of total units purchased from all suppliers during the order cycle. This, along with the order cycle period T_C, can be used to determine whether demand d at the manufacturer is being fulfilled.

8.1.4 Testing if a Possible Solution Is Feasible

Before talking about the correct solution, we will discuss the feasibility of a solution. For example, not purchasing anything ($j = [0, 0, 0]$, $Q = [0, 0, 0]$, and $T = 0$) is one solution, but it is not feasible because we cannot maintain the manufacturer production using this solution. Therefore, we can define a solution as feasible, if it makes it possible to maintain the production of the customer.

Among all the feasible solutions, some of them will lead to a smaller average total cost than others. If the objective is to minimize the cost, the cheapest cost may be considered the right solution. However, as readers may know, in this type of problem, in which the number of solutions is infinite (the solution is composed of positive numbers unbounded in size), since it is impossible to evaluate an infinite number of solutions, determining the cheapest solution (globally) is complicated. The complexity of this problem is outside the scope of this book, so we will go back to focusing on the feasibility of one solution to the discussed problem.

Once a solution is found or proposed, the total average cost is important, but there are constraints that must be fulfilled. First of all, shortages are not allowed. The number of purchased units must be equal or larger than the demand during the full order cycle. This can be ensured if Eq. (8.3) holds true:

$$\sum_{i=1}^{r} R_i \geq dT_C \tag{8.3}$$

The right side of (8.3) indicates the number of units required to satisfy the demand during the order cycle period; this is stated as the monthly demand d multiplied by the order cycle period T_C in months.

Equation (8.3) guarantees there is no shortage, but another equation is required to ensure the perfect rate of purchased units do not lead to a shortage of non-defective parts. That is to say, a large number of defective parts may lead to a shortage even when the demand d is covered. This can be ensured with Eq. (8.4):

$$\sum_{i=1}^{r} R_i q_i \geq dq_a T_C \tag{8.4}$$

Where the left side of (8.4) calculates the total number of non-defective parts purchased during the full order cycle period, and the right side of (8.4) shows the number of non-defective parts that the customer needs to produce. T_C (Eq. 8.4) will be used in the objective function.

And finally, suppliers cannot offer more monthly item units than what they can provide. This can be ensured using the next equation:

$$\frac{R_i}{T_C} \leq c_i \tag{8.5}$$

Equation (8.5) shows that the units purchased from supplier i throughout the full order cycle period, divided over the order cycle period in months, must be equal or smaller than the maximum capacity of supplier i. (8.5) must be evaluated for each of the r suppliers. Any proposed solution must comply with the set of Eqs. (8.3)–(8.5) to be considered a feasible solution.

8.1.5 The Total Average Cost Function

The total average cost function, the function we are trying to minimize, is:

(C) Min

$$Z_T = \frac{1}{T_C}\left[\sum_{i=1}^{r} j_i k_i + \sum_{i=1}^{r} R_i p_i + \frac{h}{2d}\sum_{i=1}^{r}\frac{R_i^2}{j_i} + \frac{h}{Y}\sum_{i=1}^{r} R_i l_i + \sum_{i=1}^{r} j_i TC_i\right] \tag{8.6}$$

The model is a summation of five terms divided over the order cycle period. The first term is the ordering cost of all orders made during an order cycle, the second term is the cost of all purchased items during the order period, the third term represents the cost of inventory on hand, and the fourth term represents the cost of inventory in transit, where (l_i/Y) is the proportion of time that a shipment spends in transit. Finally, the fifth term represents the transportation cost; it must be obtained from tables like the one shown in Table 8.3.

To have an idea of how this function looks, we can use a simplified example. Let's imagine there is only one supplier (with enough capacity to provide the demand). The question is now the number of items in an order. The function looks like Fig. 8.3, in which only the ordering cost and the holding (or inventory storage) cost are considered. There are no local optimal points, and the optimization of the cost has relatively low complexity. Figure 8.4 shows a similar example, but in this case, the transportation cost is considered with quantity discounts, like in the example problem. We can

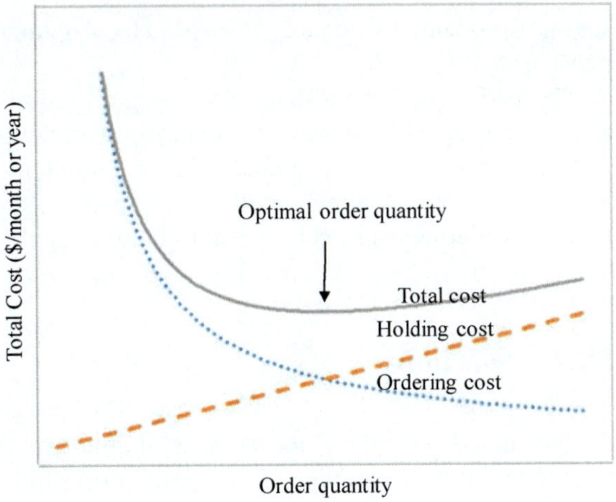

Fig. 8.3 Total cost considering ordering and holding cost

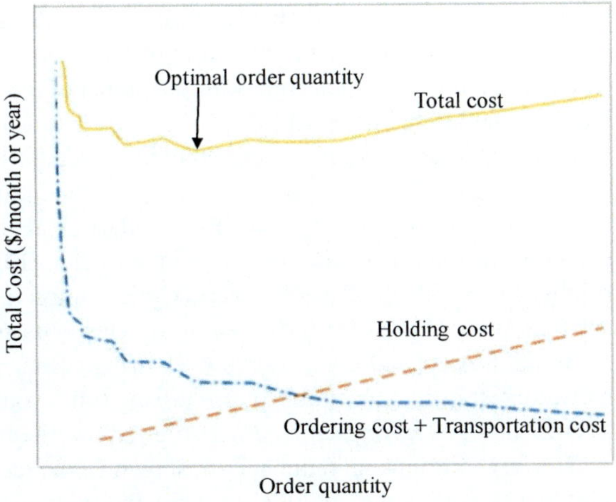

Fig. 8.4 Total cost considering quantity discounts

Table 8.5 Nominal Freight rates for suppliers [2]

Shipped weight range (lbs)	Supplier 1	Supplier 2	Supplier 3
1–499	$107.75/CWT	$136.26/CWT	$81.96/CWT
500–999	$92.26/CWT	$109.87/CWT	$74.94/CWT
1000–1999	$71.14/CWT	$91.61/CWT	$61.14/CWT
2000–4999	$64.14/CWT	$79.45/CWT	$49.65/CWT
5000–9999	$52.21/CWT	$69.91/CWT	$39.73/CWT
10,000–19,999	$40.11/CWT	$54.61/CWT	$33.44/CWT
20,000–29,999	$27.48/CWT	$48.12/CWT	$18.36/CWT
30,000–40,000	$7525	$13,200	$5030

see that the quantity-discounts introduced local optimal points and make the problem non-linear.

The reader can imagine than considering several suppliers with quantity discounts and limited capacity increases the complexity of the problem.

Continuing with the example problem, in addition to Table 8.1, the nominal freight rate from all three suppliers is shown in Table 8.5, and the actual freight rates (considering over-declaring) for all three suppliers are shown in Table 8.6.

8.1.6 Solutions

As mentioned previously, the number of solutions for the illustrative example is infinite because the solution is composed of positive numbers unbounded in size. This makes the problem interesting. Several scientific articles have dealt with it using different methods (analytics and metaheuristics). Table 8.7 presents a list of several solutions, some of them published in scientific journals, and Table 8.8 presents the cost of each of them. The optimization of this kind of problem is a challenging task since evaluating many possible solutions, most of the time, results in an excessively large computational time period.

Different alternatives have been studied in order to solve the supplier selection and order quantity allocation problem, aiming to obtain the optimal solution. The illustrative problem has been studied and solved using analytic methods [1, 2], where the solution was obtained using commercial software (LINGO) and evolutionary algorithms [3], specifically Particle Swarm Optimization (PSO), Genetic Algorithm (GA) and Differential Evolution (DE), providing a lower cost in shorter computational time.

Solution A [1] and B [2] were solved using commercial software. The use of this software results in many hours of processing time. Solution B was obtained by Lingo after 20 h of processing time. Other solutions were obtained by different methods.

The last solutions in Table 8.7 are shown for a reason. There may be an intuitive idea that extending the possible solutions to larger numbers does not lead to cheaper

Table 8.6 Actual freight rates considering over-declaring [2]

Weight range (lbs)	Freight rate	Weight range (lbs)	Freight rate
Supplier 1			
1–428	$107.75/CWT	4071–4999	$2610.5
429–499	$461.3	5000–7682	$52.21/CWT
500–771	$92.26/CWT	7683–9999	$4011
772–999	$711.4	10,000–13,702	$40.11/CWT
1000–1803	$71.14/CWT	13,703–19,999	$5496
1804–1999	$1282.8	20,000–27,383	$27.48/CWT
2000–4070	$64.14/CWT	27,384–40,000	$7525
Supplier 2			
1–403	$136.26/CWT	4400–4999	$3495.5
404–499	$549.35	5000–7811	$69.91/CWT
500–833	$109.87/CWT	7812–9999	$5461
834–999	$916.1	10,000–17,623	$54.61/CWT
1000–1734	$91.61/CWT	17,624–19,999	$9624
1735–1999	$1589	20,000–27,431	$48.12/CWT
2000–4399	$79.45/CWT	27,432–40,000	$13,200
Supplier 3			
1–428	$81.96/CWT	4071–4999	$1986.5
429–499	$374.7	5000–7682	$39.73/CWT
500–771	$74.94/CWT	7683–9999	$3344
772–999	$611.4	10,000–13,702	$33.44/CWT
1000–1803	$61.14/CWT	13,703–19,999	$3672
1804–1999	$993	20,000–27,383	$18.36/CWT
2000–4070	$49.65/CWT	27,384–40,000	$5030

Table 8.7 Solutions

Sol	J_1	J_2	J_3	Q_1	Q_2	Q_3	Order cycle period
A	3	0	4	625	0	313	3.13 months
B	2	1	0	625	625	0	1.85 months
C	9	4	0	626	635	0	8.05 months
D	9	4	0	625	633	0	8.03 months
E	9	4	0	625	633	0	8.03 months
F	16	7	0	625	643	313	14.29 months
G	29	13	0	625	627	313	25.89 months
H	776	349	0	625	625	313	692.91 months
I	73739	33155	0	625	625	0	65838.50 months
J	8170568	3673684	0	625	625	0	7295150.03 months

Table 8.8 Solutions

Sol	Total Cost ($/month)
A	$33,680.18
B	$32,912.08
C	$32,786.39
D	$32,778.12
E	$32,778.12
F	$32,792.68
G	$32,768.06
H	$32,765.23
I	$32,764.88
J	$32,764.87

solutions, and therefore, the number of solutions that need to be evaluated is limited. However, this kind of problem may show behavior in which extending the number of evaluated solutions leads to cheaper and cheaper solutions. The next section explains in more detail the use of metaheuristics algorithms to solve the supplier selection and inventory problems, and specifically discusses the example problem.

8.2 Solving Using Metaheuristics Algorithms

This section describes some of the most representative metaheuristic algorithms used to solve the lot sizing and inventory problems. The optimization of these problems represents the process of finding the "best solution" to the problem among a large set of possible solutions.

Let us define metaheuristic as an iterative process, with the potential to avoid a local solution (which may be found using a heuristic method), using a neighborhood search space. This search strategy allows us to find a possible solution, and an iterative generation process conducts a subordinate heuristic to explore the search space and to find a near-optimal solution (if it exists), within a reasonable period of time.

The optimization methods apply an iterative process to explore the search space. There are two kinds of algorithms [5]: classical methods and evolutionary methods. The classical methods use the function gradient to generate new possible solutions. The most important aspect of the classical methods is that the function must be both differentiable and unimodal.

Most real lot-sizing problems cannot be solved with classical methods, especially when non-linear quantity discounts are considered. Unlike the classical methods, evolutionary methods do not use the function gradient to find a solution. These methods apply heuristic methods to search. Sometimes these methods are based on nature and social processes. These methods usually are stochastic, which means they use stochastic processes to determine the search direction.

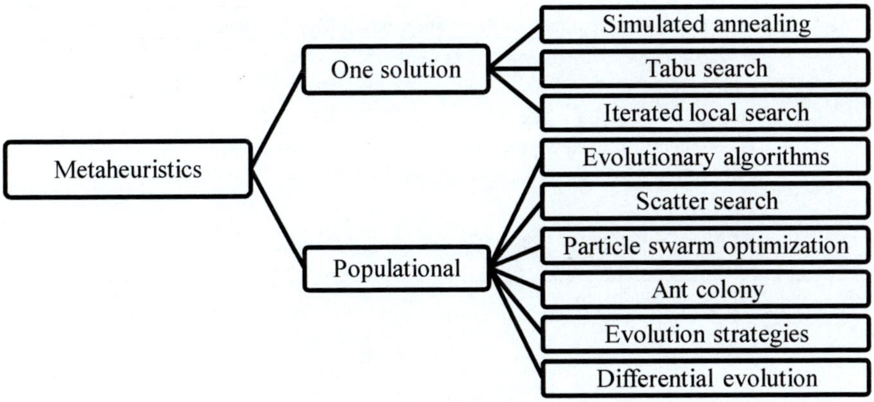

Fig. 8.5 Metaheuristic algorithms

Sometimes the evolutionary methods are separated from the swarm algorithms. Swarm algorithms are based on the collective behavior of groups of animals or insects. Two types of swarm algorithms are Ant Colony Optimization and Particle Swarm Optimization. The mechanism of these swarm algorithms is similar to evolutionary methods, and they are populational. This chapter considers them within the evolutionary algorithms.

Metaheuristic algorithms can be classified into two main classes, as shown in Fig. 8.5: (i) Single solution based metaheuristics, sometimes physics-based techniques, and (ii) Population-based metaheuristics.

The algorithm efficiency and effectiveness rely on the correct selection of certain parameters, which are tuned for allowing more flexibility and robustness.

Performance parameters may be classified into three groups [6]:

- Solution quality. Defined in terms of precision, and usually based on measuring the distance or the deviation percentage of the candidate solution.
- Computational effort. CPU time with or without input/output and preprocessing/postprocessing time.
- Robustness. Defined insensitivity against small deviations in the input instances or the parameters of the metaheuristic.

This chapter explores mainly population-based metaheuristic algorithms, such as Ant Colony Optimization, Evolutionary Algorithms, Particle Swarm Optimization. Algorithm 8.1 shows the high-level template of population-based metaheuristics [7].

Algorithm 8.1. High-level template of Population-based Metaheuristics

```
P=Po; /*Generating the initial population*/
t=0;
Repeat
Generate (Pt'); /*New population*/
Pt+1= Pt υ Pt'; /*selecting new population*/
t=t+1
Until    Stopping criteria
Output: Best solution so far
```

8.2.1 Particle Swarm Optimization

The purpose of this section is to show the main principles of the Particle Swarm Optimization algorithm. Also, an illustrative example presented in Sect. 8.1.1 is solved using the metaheuristic algorithm, and a solution is presented for the problem.

The illustrative example has an infinite number of solutions. The number of possible solutions can be restricted by controlling the number of orders per supplier (M) in Eq. (8.1) and using the constraints presented in Eqs. (8.3)–(8.5).

The search space cannot be infinite, but it may seem contradictory to restrict the number of possible solutions in order to restrict the searching space, especially after explaining that this was a disadvantage of traditional or commercial computer programs. However, after working on a specific problem, it is found that the searching space can be constrained to a much larger size compared to commercial computer programs, and better solutions are found in a very short time, in a matter of seconds rather than the hours needed for traditional algorithms.

The objective is to obtain a good solution in a reasonable processing time period. PSO is easy to implement, and its operators are very intuitive. PSO makes it possible to overcome local solutions and obtain better solutions.

8.2.1.1 Procedure for PSO

Particle Swarm Optimization, proposed in [8], is inspired by the natural behavior of flocks of birds. The algorithm has a particle population, which works along with a defined search space. An iterative process is developed where a new position for each particle is calculated. The new position is determined using a calculated velocity, the global solution, and a better local solution (if it exists) of each particle. Each particle must be evaluated in the objective function with the goal of obtaining the function value (the total cost in the discussed example). The Particle Swarm Optimization process is described as follows:

Initialization

As with all metaheuristic algorithms, the search process of PSO begins by initializing the population and its velocities. Particles (candidate solutions) are randomly generated and dispersed over the search space, which is limited by the lower and upper bounds (parameters). After the initialization, the population is evaluated in the objective function, and the best local and global particles are calculated. Then, an iterative process begins, where the velocity and position updating operators are applied to the population.

It is necessary to define the parameters of the problem, such as the dimensions (which depend on the number of decision variables), the bounds for each dimension, and the constraints. Particles are randomly generated for the first time. This may be done by using Eq. (8.7).

$$x_i^k = l_i + rand(u_i - l_i), x_i^k \in \mathrm{x}^\mathrm{k} \tag{8.7}$$

Where x_i^k represents the particle i of the iteration k. l_i, and u_i are the lower and upper bounds for each dimension (decision variable) of the search space; *rand* is a random value in the interval of [0,1].

Update Velocity

The velocity is updated based on the global solution (so far) and the best local solution (so far). These solutions have an effect on the next particles. These effects can be controlled with the cognitive and social factors, respectively. The velocity v_i^k of each particle x_i^k is updated by calculating Eq. (8.8).

$$v_i^{k+1} = v_i^k + c_1\left(r_1^k\left(p_i^k - x_i^k\right)\right) + c_2\left(r_2^k\left(g^k - x_i^k\right)\right) \tag{8.8}$$

Where k represents the current iteration, v_i^{k+1} is the updated velocity for the next generation (k + 1), of the particle x_i^k, v_i^k is the current velocity, p_i^k is the local best so far of x_i^k, g^k is the global best so far. Furthermore, r_1^k and r_2^k are random values in the interval of [0,1], while the parameters c_1, and c_2 are the cognitive and social factors, respectively.

Update Particles

Particles are dispersed throughout the search space to find new candidate solutions. The movement of each particle is calculated according to its updated velocity. The position of each particle is also updated, as shown in Eq. (8.9):

$$x_i^{k+1} = x_i^k + v_i^{k+1} \tag{8.9}$$

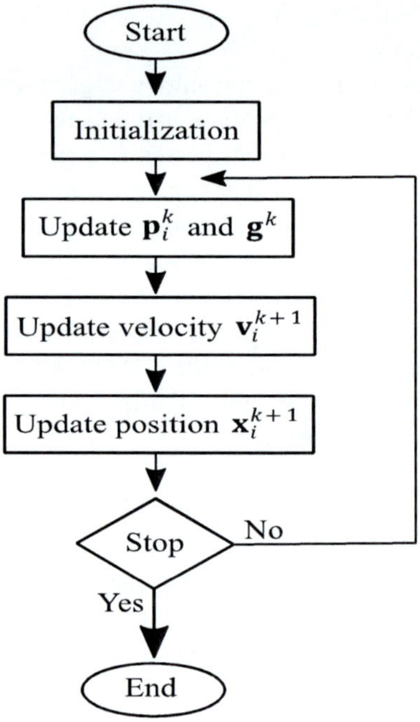

Fig. 8.6 Flowchart of particle swarm optimization [3]

Where x_i^{k+1} is the updated position of the particle x_i^k for the new iteration, where its updated velocity is v_i^{k+1}.

Then, the updated particles (candidate solutions) are evaluated in the objective function, and the best local and global particles so far are calculated and updated. The process is repeated until the fixed maximum number of iterations is reached. The flowchart for the PSO algorithm is shown in Fig. 8.6.

8.2.1.2 PSO Application

As mentioned before, the illustrative example presented in Sect. 8.1.1 was solved using the PSO algorithm, following the procedure explained in Sect. 8.2.1.1.

Parameters and Results

The parameters of the Particle Swarm Optimization must be set for the optimization of the problem. Initial parameters of PSO are configured with the following values: the population size m has been set to 200 particles, and the maximum number of iterations k_{max} has been set to 300. The number of dimensions, n, is 6, corresponding to the decision variables of the problem: j_1, j_2, j_3, Q_1, Q_2, and Q_3.

Table 8.9 Some solutions from the 30 independent executions

Sol	J_1	J_2	J_3	Q_1	Q_2	Q_3	Total cost ($/month)	Order cycle period
A	9	4	0	626	635	0	$32,786.39	8.05 months
B	10	4	2	638	631	170	$32,890.66	9.12 months
C	10	4	1	629	630	345	$32,840.12	9.03 months
D	10	4	0	625	706	0	$32,891.34	8.94 months
E	9	0	8	637	0	316	$33,052.03	8.22 months

Table 8.10 Results of PSO algorithm, contemplating 30 independent executions

Lower cost (per month)	Z_l	$32,786.39
Higher Cost (per month)	Z_h	$33,049.33
Average Cost (per month)	Z_a	$32,897.22
Computational time (seconds)	ct	3.11
Average computational time (seconds)	ct_a	3.77

The cognitive and social factors used have been fixed, according to the best-reported values in the literature, at $c_1 = 2$ and $c_2 = 2$.

After the parameters are established, the PSO algorithm is executed to solve the numerical example presented in Sect. 8.1.1. Since the PSO algorithm is a stochastic strategy, the optimization process is repeated in 30 independent executions to verify the consistency of the results. 30 solutions and the objective function values are obtained (evaluated in Eq. 8.6), which represent the best-found solutions. Some indicators were calculated, such as: the lowest cost Z_l, highest cost Z_h, average cost Z_a, the computational time for the best solution *ct*, and average computational time ct_a. Indicators Z_l, Z_h, and Z_a evaluate the accuracy of the solution. Finally, *ct*, and ct_a evaluate the speed of the algorithm. Some experimental results are presented in Table 8.9. Table 8.10 summarizes the indicators for the 30 executions.

The convergence of the best solution presents a total cost of $32,786.39, and the algorithm reaches the best result in iteration 255, the computational time was 3.11 s.

8.2.2 *Genetic Algorithm (GA)*

The purpose of this section is to present the main characteristics of the Genetic Algorithm. Also, the illustrative example presented in Sect. 8.1.1 is solved by using the metaheuristic algorithm, and a solution is presented for the example problem.

The search space was limited using lower and upper bounds for the decision variables. The number of orders per supplier was controlled (Eq. 8.1), and all constraints of the problem were considered (Eqs. 8.3–8.5). The objective is to obtain a good solution in a reasonable processing time period. GA allows local solutions to be overcome in order to obtain better solutions.

8.2.2.1 GA Procedure

The Genetic Algorithm was developed by J. Holland in 1970. This algorithm is based on natural selection in the biological evolution process [9]. Among the evolutionary algorithms, this method is very popular. In this chapter, a discrete and non-binary version of GA will be executed. A description of the algorithm is shown in the following sub-section.

Initialization

The first step in the search process is to randomly initialize the population (solutions) within the lower and upper bounds and to distribute over the search space. The population is of size m. m candidate solutions (individuals). After initialization, the population must be evaluated in the objective function. Then, an iterative process begins, and crossover, mutation, and selection operators are applied.

Crossover

The purpose of the crossover is combining genotype information. This exchange is simulated using information from the solutions vectors. Sets of two individuals are randomly chosen from the population, and an offspring of individuals are produced by mixing the individuals from the population.

Genetic algorithms use several methods for selecting the parents for the crossover, such as the Proportional Selection Method, Tournament Selection, and Rank-Based Selection.

There are several kinds of crossover, such as n-point or uniform. The example developed in this section is solved using n-point due to its simplicity.

The number of sets is determined according to a crossover probability CR. These parents (two individuals) are chosen to be the creators of new individuals by applying the crossover operation. In this process, the genes of each pair of parents are combined, as shown in Eqs. (8.10) and (8.11).

$$y_{c,j}^{k} = \begin{cases} x_{a,j}^{k}, & j < CP, \forall j = 1, \ldots, n \\ x_{b,j}^{k}, & otherwise \end{cases} \tag{8.10}$$

$$y_{d,j}^{k} = \begin{cases} x_{b,j}^{k}, & j < CP, \forall j = 1, \ldots, n \\ x_{a,j}^{k}, & otherwise \end{cases} \tag{8.11}$$

where an individual (genotype of the possible solution) x_a^k is combined with an individual x_b^k to produce the new individuals y_c^k, and y_d^k the constant CP defines how many genes from x_a^k and x_b^k are being taken to create the offspring. The total number

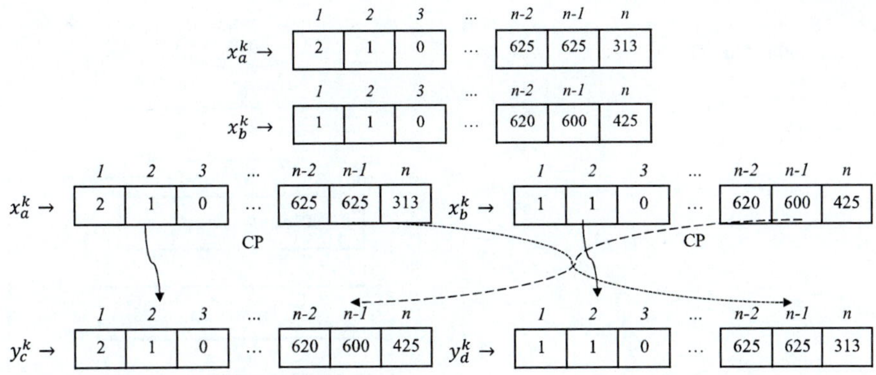

Fig. 8.7 Crossover process of GA

of genes n corresponds to the dimensions of the problem. The process is graphically explained in Fig. 8.7.

Mutation

When the offspring is generated, the mutation operator is applied to the individuals in order to generate the diversity of the population. In the process, there is a mutation probability of MP to control the genes of the offspring to be mutated. This procedure is shown in Eq. (8.12):

$$y_{i,j}^{k} = \begin{cases} rand(lb, ub), & rand(0, 1) < MP \\ y_{i,j}^{k}, & otherwise \end{cases} \tag{8.12}$$

where lb and ub are the lower and upper bounds of the search space.

Re-selection of Population

Following the crossover and mutation operators, the offspring must be evaluated in the objective function. Then, the best individuals from the offspring and the population are chosen to be the new generation.

The GA algorithm can be reviewed in the flowchart shown in Fig. 8.8.

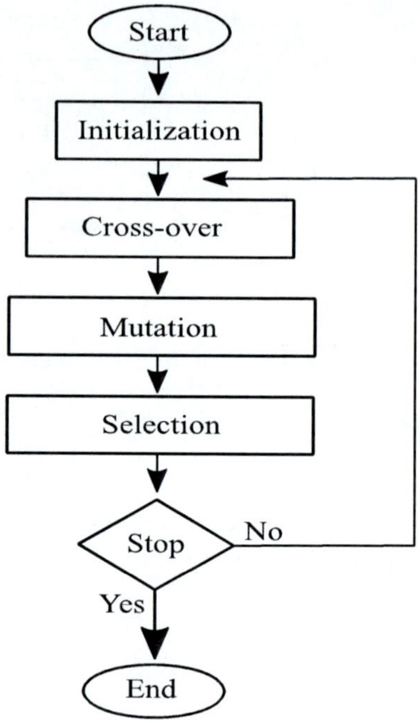

Fig. 8.8 Flowchart of the genetic algorithm [3]

8.2.2.2 GA Application

As mentioned before, the purpose of solving the illustrative example presented in Sect. 8.1.1 was to explain the application of Genetic Algorithm in lot-sizing and inventory problems, following the procedure presented in Sect. 8.2.2.1.

Parameters and Results

Parameters of the Genetic Algorithm must be set for the optimization of the problem. The initial parameters of GA are configured with the following values: the population size m has been set to 200 individuals, and the maximum number of iterations k_{max} has been set to 300. The number of dimensions, n, is 6, corresponding to the decision variables of the problem: j_1, j_2, j_3, Q_1, Q_2, and Q_3.

The selecting of parents is established using the Proportional Selection Method, and the cross over was made by combining n-genes.

The constant CP (how many genes from x_a^k and x_b^k are being taken to create the offspring) is fixed at 3. The mutation probability is established (MP) at 0.2 and CR at 0.9.

After the parameters are established, the GA algorithm is executed to solve the numerical example. The process is repeated in 30 independent executions to verify the consistency of the results. 30 solutions and the objective function values (fitness) were

Table 8.11 Some solutions of GA from the 30 independent executions

Sol	J_1	J_2	J_3	Q_1	Q_2	Q_3	Total cost ($/month)	Order cycle period
A	9	4	0	625	633	481	$32,778.12	8.03 months
B	10	4	1	625	625	313	$32,797.14	8.94 months
C	8	3	1	625	625	367	$32,817.20	7.14 months
D	8	3	1	625	633	339	$32,815.16	7.14 months
E	9	4	0	635	643	85	$32,824.68	8.16 months

Table 8.12 Results of GA, contemplating 30 independent executions

Lower cost (per month)	Z_l	$32,778.12
Higher cost (per month)	Z_h	$32,817.21
Average cost (per month)	Z_a	$32,796.74
Computational time (seconds)	ct	2.30
Average computational time (seconds)	ct_a	2.27

obtained, which represent the best-found solutions. Some indicators were calculated, such as the lowest cost Z_l, highest cost Z_h, average cost Z_a, computational time ct, and average computational time ct_a. Indicators Z_l, Z_h, and Z_a evaluate the accuracy of the solution. Finally, ct, and ct_a evaluate the speed of the algorithm. Some experimental results are presented in Table 8.11. Table 8.12 summarizes the indicators for the 30 executions.

The convergence of the best solution has a total cost of $32,778.12, and the algorithm reaches the best result in iteration 145, the computational time was 2.30 s.

8.2.3 *Differential Evolution (dE)*

DE is a population-based stochastic algorithm that is able to find a global optimum solution in multimodal, non-differentiable, and non-linear functions. The algorithm is described as being easy, efficient, and fast. The algorithm was proposed by Kenneth Price and Rainer Storn [10].

This algorithm employs mutation, crossover, and selection operations. In the population, each solution is considered a candidate solution vector that evolves along with the iterations. The mutation is the process where a new vector is generated for the sum of the weighted difference between two random solutions.

8.2.3.1 DE Procedure

After initializing the population, the search process of classical DE uses operators such as mutation, crossover, and selection.

Initialization

Fixing the lower and upper bounds of the land space is an important part of initializing the population. An easy way to initialize the population is to consider two vectors (d-dimensional) called b_{low} and $b_{up,}$ where the subscripts *low* and *up* represent the lower and upper bounds for each decision variable (dimension). The population can be randomly generated over the search space by using Eq. (8.13).

$$x_{i,j}^{k} = b_{j,low} + rand(0,1)(b_{j,up} - b_{j,low}) \tag{8.13}$$

Where k represents the current iteration, j, and i are the number of parameters and the number of individuals, respectively.

The individuals are created and distributed following uniform probability distribution. After initialization, the individuals are evaluated in the objective function. Then, an iterative process begins by applying mutation, crossover, and selection operations to the population.

Mutation

This operator provides exchange information between different candidate solutions to combine the population in order to generate a new population.

This operator generates mutant vectors by calculating the sum of a random candidate solution and the weighted difference between two random candidate solutions. The mutation operator is generated by the following equation:

$$v^{k} = x_{r3}^{k} + F(x_{r1}^{k} - x_{r2}^{k}) \tag{8.14}$$

Where, v^k is the mutant vector, $r1, r2, r3 \in \{1, 2, 3, \ldots, i-1, i+1, \ldots, N\}$ without repetition, then x_{r1}^{k},x_{r2}^{k} and x_{r3}^{k} are three different random solutions selected from the population, and F is a number in the interval of [0,2] for scaling differential vectors.

Crossover

DE implements the crossover operation to generate new solutions and increase the diversity of the population. This operator combines the mutant vector v^k with a solution x_i^k to create a trial vector $u^{k,}$ which is generated by the following Eq. (8.15):

$$u_{i,j}^{k} = \begin{cases} v_{i,j}^{k}, & rand(0,1) < CO \\ x_{i,j}^{k}, & otherwise \end{cases} \tag{8.15}$$

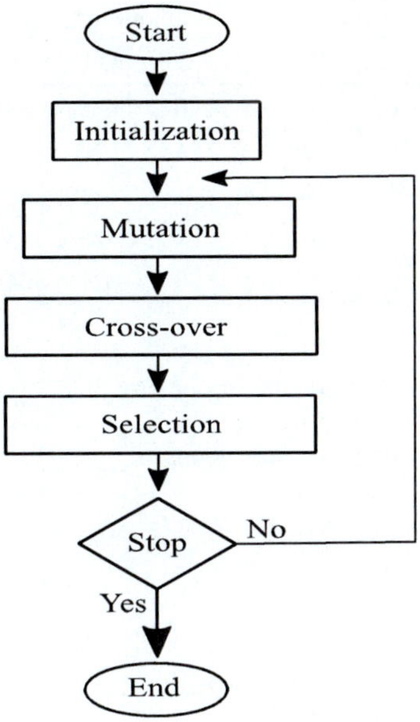

Fig. 8.9 Flowchart of the differential evolution algorithm [3]

Where *rand*(0,1) is a random number between 0 and 1, *CO* represents a constant parameter defined by the user ($0.0 \leq CO \leq 1.0$), which specifies the probability of crossover.

Selection

A Greedy method is used to find the best solutions. In the selection process, the purpose is to decide if the trial vector u^k generated in the crossover operation is better than the current solution x_i^k. The solution will be replaced if the objective value is better than the current solution. In a minimization problem, the selection process is defined as shown in Eq. (8.16):

$$x_i^{k+1} = \begin{cases} u^k, & f(u^k) < f(x_i^k) \\ x_i^k, & otherwise \end{cases} \tag{8.16}$$

Where x_i^{k+1} is the new solution for the next generation $k + 1$, $f(u^k)$, and $f(x_i^k)$ are the fitness values of u^k and x_i^k respectively. The process is repeated until the maximum number of iterations is reached. The DE algorithm can be illustrated in the flowchart of Fig. 8.9.

Table 8.13 Some solutions of DE, contemplating 30 independent executions

Sol	J_1	J_2	J_3	Q_1	Q_2	Q_3	Total cost ($/month)	Order cycle period
A	9	4	0	625	633	0	$32,778.12	8.03 months
B	7	3	1	630	643	80	$32,879.28	6.32 months
C	10	4	1	631	666	200	$32,900.52	9.04 months
D	8	3	1	625	635	348	$32,829.56	7.15 months
E	9	4	0	627	644	292	$32,820.06	8.1 months

Table 8.14 Results of DE, contemplating 30 independent executions

Lower cost (per month)	Z_l	$32,778.12
Higher Cost (per month)	Z_h	$32,800.36
Average Cost (per month)	Z_a	$32,788.58
Computational time (seconds)	ct	3.01
Average computational time (seconds)	ct_a	3.06

8.2.3.2 DE Application

As mentioned before, the purpose of solving the illustrative example presented in Sect. 8.1.1 was to explain the application of Differential Evolution in lot-sizing and inventory problems, following the procedure presented in Sect. 8.2.3.1

Parameters and Results

The parameters of the Differential Evolution must be set for the optimization of the problem. The initial parameters of the DE are configured with the following values: the population size m has been set to 200 individuals, and the maximum number of iterations k_{max} has been set to 300. The number of dimensions, n, is 6, corresponding to the decision variables of the problem: $j_1, j_2, j_3, Q_1, Q_2,$ and Q_3.

The constant $\boldsymbol{F}$ for scaling differential vectors was fixed at 0.2, and the constant CO for the crossover probability is fixed at 0.5.

After the parameters are established, the DE algorithm is executed to solve the numerical example presented in Sect. 8.1.1. The process is repeated in 30 independent executions to verify the consistency of the results. 30 solutions and the objective function values (fitness) are obtained, which represent the best-found solutions. Some indicators were calculated, such as: the lowest cost Z_l, highest cost Z_h, average cost Z_a, computational time ct, and average computational time ct_a. Indicators Z_l, Z_h, and Z_a evaluate the accuracy of the solution. Finally, ct, and ct_a evaluate the speed of the algorithm. Some experimental results are presented in Table 8.13. Table 8.14 summarizes the indicator for the 30 executions.

The best solution has a total cost of $32,778.12; when evaluating it in the objective function, and the algorithm reaches the best result in iteration 234, the computational time is 3.01 s.

8.2.4 Tabu Search

8.2.4.1 Tabu Search Procedure

Tabu Search is the metaheuristic algorithm proposed in [11]. The algorithm is frequently used for solving combinatorial problems, and it uses a heuristic local search procedure to explore the solution space.

There are several techniques that use a local search for optimizing functions. The basic procedure of the local search is as follows:

1. To generate an initial solution X.
2. To select a neighboring solution that corresponds to the neighborhood $N(x)$.
3. Evaluating the solution X and a solution of $N(x)$, if: value for $N(x) <$ value for X; $X = N(x)$.
4. Repeat steps 2 and 3.

Tabu Search uses some key elements:

- Tabu constraints: limit the search by classifying certain movements as prohibited to avoid obtaining recently generated solutions.
- Aspiration criteria: allows storing a solution if it is better than the current solution even if it is a movement tabu, using the short-term memory function.

Let us explain a little more about the Tabu Search procedure and how it overcomes local solutions. In order to avoid going back to past solutions and looping, Tabu Search uses a temporary memory, which is called *tabu list*. It stores the recent history of the search, and n iterations are used. Therefore, Tabu Search is a local search which uses a short-term memory. Tabu Search usually finds good solutions and overcomes the existence of locally optimal solutions.

The search region analyzed by Tabu Search is the space of all candidate solutions than can be explored during the search. The Tabu Search explores a neighborhood search using an iterative procedure to move from a candidate solution (x) to an improved solution until the stopping-criteria has been reached. The iterative procedure allows the application of the local transformation of the candidate solution (x). These solutions define a new neighborhood in the search space, $N(x)$. Therefore, $N(x)$ is a subset of the search space, the neighborhood of x.

As mentioned before, the tabu list is the main characteristic of Tabu Search. The list is used to prevent cycling when overcoming a local solution using non-improving movements. For instance, if an airliner pilot is moved from $Line_1$ to $Line_2$, it can be defined as tabu moving from $Line_2$ to $Line_1$. Therefore, the Tabu list involves several prohibited movements that help to filter which solutions will be admitted in $N(x)$.

Short-term memory makes it possible to store the tabus, and the information (limited) is recorded. Usually, the user decides to apply the tabus function of recording the last transformation performed on the candidate solution and prohibiting reverse transformations. Multiple tabu lists can be employed. This action is sometimes advisable when different kinds of movements are utilized to generate the neighborhood.

Tabus are very useful, but one disadvantage is that they may prohibit useful movements. For this reason, it is advisable to use aspiration criteria, which allow us to cancel tabus, and consist of permitting to a movement (even if this movement is contained in tabu) when this solution results in a better objective function value than the current solution.

A good search intensification must be made using the solutions and exploring more thoroughly the regions to make sure the better solutions. Intensification uses intermediate-term memory and considers the number of successive iterations that several elements of the solution have been explored without interruption in the candidate solution.

Diversification attempts to obtain solutions not previously found. It is made storing the total number of times that several solution elements have been present in the current solution.

8.2.4.2 Tabu Search Application

The Tabu Search allows the local search to overcome the optimal local solution [12]. The basic characteristic is to pursue a local search to find a local optimum and overcome it in order to improve the solution and obtain a better result.

As was mentioned, the purpose of solving the illustrative example presented in Sect. 8.1.1 was to explain the application of simple Tabu Search in lot-sizing and inventory problems, following the procedure presented in Sect. 8.2.4.1.

Table 8.15 presents the variables used to apply the TS algorithm presented in Sect. 8.2.4.1.

Design of Solution

Solution X is coded as a vector of size m that will contain, in each position, the value of each decision variable (Fig. 8.10). For instance: if the problem is to be solved considering 3 suppliers, then the first positions represent the number of orders assigned to each supplier, and the last positions represent the order quantity for each supplier.

Table 8.15 Variables used in the template for a basic Tabu Search

Variable	Description
X	Current solution vector
X_o	Initial solution vector
f	Total Cost of the evaluated neighboring solution
F	Best total cost of the neighborhood
f*	Best global total cost
N(x)	Solution vector of the neighborhood
X*	Best-know solution vector
T	Vector for controlling the values of prohibit movements for the element i
T_l	Number of prohibit movements
Y	New random value for the element i_{th} of the vector X
I	Element i_{th} of the vector X that generated the best neighboring

Fig. 8.10 Solution *X*

J_1	J_2	J_3	Q_1	Q_2	Q_3

Search Space

Since the initial solution X_0 is randomly generated, the search space will consider visiting feasible solutions and allowing the search for movements of non-feasible solutions (diversification).

Figure 8.11 shows the search space region of the function (which will be minimized) for a single supplier (for illustrative purpose) and a randomly generated initial solution. So, the generated solution can explore any area of the function without restricting the search to only feasible solutions. The limits of the variables were fixed at: $L_b = [0, 0, 0, 0, 0, 0]$, $U_b = [10, 10, 10, 800, 800, 800]$, where L_b represents the lower bound and U_b represents the upper bound for each variable.

Tabu List and Diversification

In the Tabu Search, diversification seeks to lead the search to regions of the search space not previously visited and to generate new solutions that differ significantly from solutions already evaluated. This is achieved through the tabu list.

A vector called tabu list (*T*) will control the movements to the neighboring solutions. This vector has the same dimension as solution *X*, and it records how many movements back that element of the solution was changed. For this example, the tabu list is initialized at zero, and it is established as follows:

$$T(i) = k, \quad \forall i \tag{8.17}$$

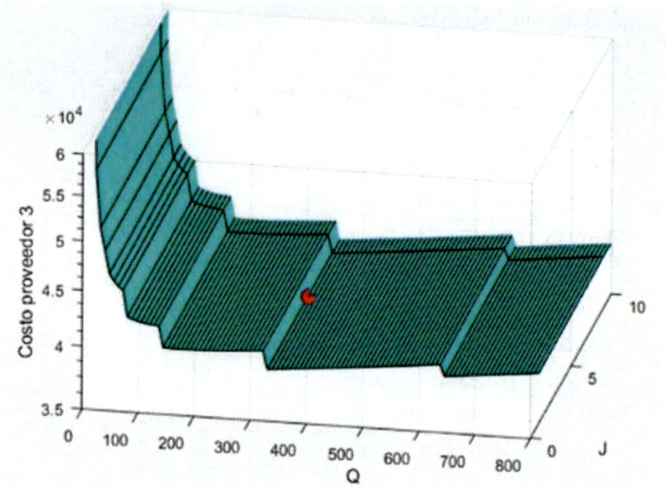

Fig. 8.11 Search space

Where k represents the number of movements that element i (of the solution vector X) will not be moved. If $k = 0$, it means that element i can be changed to generate a neighboring.

Strategy for Intensification and Finding the Neighborhood

In order to fulfill the purpose of choosing good solutions, a local transformation is applied to the current solution X; a set of neighboring solutions is defined in the search space. For the illustrative problem, the set of neighboring solutions will be defined as follows: if $k = 0$ (for each element in the tabu list), the neighboring will be of length X. The algorithm is presented in Algorithm 8.2.

Each neighboring solution is randomly generated with a new value for position i of the solution vector X. It must be ensured that the new value for position i is different from the current value. The neighboring solutions are evaluated in the objective function (Eq. 8.6), and the best one is chosen (the value of the objective function and the vector solution are stored in F and in $N(x)$, respectively).

Algorithm 8.2: Process for generating the neighborhood

for *X //∀ i//*
  if *(T(i) = 0)*
  Repeat
  Y = random value( ) //considering L_b, U_b//
if *(X(i) ≠ Y)*
  break;
end
end
X(i) = Y
  f = Objective_function(X)
if *(f < F)*
F = f
N(x) = X //∀ i//
I = i
  end if
end for

Once the best neighboring solution is chosen, the position that was changed (I) is stored and registered in the *tabu_list*, and it will not be moved in n iterations, (T_I). If some element i of vector X has been recently moved but not in this iteration, ($T(i) \neq 0$), then *tabu_list* is decreased by one unit.

The best solution and its total cost generated from the neighborhood process ($N(x)$, F) must be compared in each iteration to allow us to record the best result of all the iterations so far (X^*, f^*).

After completing the current iteration, the element i that was moved to generate the best neighboring must be updated in the *tabu_list* (T) with the maximum value (T_I); for this illustrative example, it was considered equal to four.

The procedure to generate the global solution, global cost, and controlling the tabu list are presented in Algorithm 8.3.

Table 8.16 Solutions using TS

Sol	J_1	J_2	J_3	Q_1	Q_2	Q_3	Total cost ($/month)
A	6	2	1	625	660	336	$32,879.0
B	6	2	1	639	642	447	$32,949.9
C	9	3	1	627	640	631	$32,893.1
D	10	4	2	631	630	182	$32,897.6
E	8	2	3	628	655	322	$32,932.4

Table 8.17 Results of TS, contemplating 30 independent executions

Lower cost (per month)	Z_l	$32,879.0
Higher Cost (per month)	Z_h	$32,949.9
Average Cost (per month)	Z_a	$32,946.2
Computational time (seconds)	ct	3.0
Average computational time (seconds)	ct_a	3.14

Algorithm 8.3:

Process for generating the global solution, total global cost, and tabu list

```
if (F < f*)
f* = F
   X* = N(x),  //∀i//
   I = i
end
for X //∀i//
if (T(i) ≠ 0)
T(i) = T(i) – 1
end
T(I) = Tl
```

Parameters and Results

The purpose of this section is to apply the TS algorithm to the illustrative problem presented in Sect. 8.1.1. In this algorithm, a different way to execute the process was explored. In this case, a number of iterations were not considered as stopping criteria. The stopping-criteria is to overcome the total cost of $33,000 since this cost was a bound in the previous results.

Some experimental results are presented in Table 8.16.

Table 8.17 summarizes some indicators of 30 executions. The algorithm achieves a good total cost for the function in a reasonable elapsed time. The computational time for the best solution was 3 s.

8.2.5 *Simulated Annealing*

8.2.5.1 Simulated Annealing Procedure

The simulated annealing algorithm [13] is an optimization technique that simulates the cooling process in metallic materials. The main idea is to cool the material in a controlled way, with the purpose of avoiding defects in the metallic structures. The algorithm was proposed by Kirkpatrick, Gelatt, and Vecchi in 1983. This algorithm is able to avoid local solutions. The characteristic of the algorithm is a gradual reduction in the atomic changes that decrease the density of lattice defects until the lowest-energy state is achieved.

The objective function (to optimize) is similar to the energy in a thermodynamic system. In high temperatures, the algorithm can explore far-away points in the search space; also, the acceptance probability is greater to non-improving solutions. By using lower temperatures, it is possible to generate nearby points, and the acceptance probability is lower to non-improving solutions. The algorithm uses only one candidate solution along the process (x^k), which is changed in each iteration using a similar procedure of random search, where each point is modified using a random vector Δx. An advantage of the algorithm is to allow not only changes that improving the objective function but also applying a stochastic procedure for considering the worst solutions. The algorithm is able to escape from local solutions.

If the optimization function is a minimization problem, a new solution x^{k+1} will be accepted considering two options:

1. If the objective function value is better, lower than the last solution x^k. $f(x^{k+1}) < f(x^k)$.
2. Under an acceptance probability p_a.

The second alternative determines that if the objective function value for x^{k+1} is not better than the objective value for x^k, ($f(x^{k+1}) > f(x^k)$), the solution x^{k+1} can be accepted using an acceptance probability $p_{a,}$ which is calculated as:

$$p_a = e^{-\Delta f/T} \tag{8.18}$$

Where T represents the temperature for controlling the cooling process, Δf defines the difference of energy between x^{k+1} and x^k, which is calculated as:

$$\Delta f = f(x^{k+1}) - f(x^k) \tag{8.19}$$

The acceptance of the solution x^{k+1} is made following the next procedure. First, it is obtained a random value r_1 in the interval of [0,1]; if $r_1 < p_a$ then, the solution x^{k+1} is stored as the new solution.

If T is a large value, then $p_a \to 1$, which indicates that almost all x^{k+1} will be accepted. If T is a small value, then $p_a \to 0$, which means that when x^{k+1} improves the solution, then it will be accepted. For this reason, the slow cooling of metals decreasing temperature is very important to obtain good solutions. There are several ways to control the cooling process from an initial temperature T_{ini} to final temperature T_{end}, linear and geometric. In the cooling process following a geometric temperature reduction scheme, the temperature decreases by applying a cooling factor β in the interval [0,1].

$$T(k) = T\beta \tag{8.20}$$

The simulated annealing method, shown in algorithm 8.4, begins by setting up the parameters T_{ini}, T_{end}, β, and the number of iterations *Niter.* After that, a random solution is generated between the lower bound and upper bound into the search space, x. Then, the evolutive process is started and finished until that last iteration is made or until the last temperature T_{end} is reached.

Algorithm 8.4: Process for simulated annealing

Setting T_{ini}, T_{end}, β, σ, *Niter*, L_b, U_b

$k \leftarrow 0$, $T = T_{ini}$

$x^k \leftarrow random[x]$ //considering L_b, U_b//

While $((T > T_{end})$*and* $(k < Niter))\{$

While $(x^{k+1} \notin x)\{$

$\Delta x \leftarrow N(0, \sigma T)$

$x^{k+1} \leftarrow x^k + \Delta x\}$

If $(f(x^{k+1}) < f(x^k))\{$

Update x with the value of x^{k+1} and f with the objective function value $f(x^{k+1})$

Else

$\Delta f \leftarrow f(x^{k+1}) - f(x^k)$

$p_a \leftarrow e(-\Delta f/T)$

If $(rand() < p_a)$

Update x with the value of x^{k+1} and f with the objective function value $f(x^{k+1}))\}$

$k = k+1$

$T = \beta T\}$

During the process, a candidate solution x^{k+1} is generated using a random value, Δx. The objective function value for x^{k+1} is evaluated, and if this value is better than the value for x^k then, x^{k+1} is accepted as a new solution. If solution x^{k+1} is not better than x^k, then the acceptance probability is considered, $p_{a.}$ If the value for p_a is larger than a random value r_1, solution x^{k+1} is accepted as a new solution. Along with the iterations, the temperature is decreased by a factor β with the purpose of reducing the acceptance percentage of new solutions x^{k+1} that do not represent a better solution than x^k.

Table 8.18 Solutions using Simulated Annealing

Sol	J_1	J_2	J_3	Q_1	Q_2	Q_3	Total cost ($/month)
A	10	5	0	625	629	0	$32,927.39
B	10	3	1	634	800	441	$33,076.23
C	7	3	1	682	637	541	$33,280.76
D	11	1	4	520	798	458	$34,209.65
E	17	2	3	160	424	127	$34,895.03

Table 8.19 Results of SA, contemplating 30 independent executions

Lower cost (per month)	Z_l	$32,927.39
Higher Cost (per month)	Z_h	$34,895.03
Average Cost (per month)	Z_a	$33,793.59
Computational time (seconds)	ct	3.4
Average computational time (seconds)	ct_a	3.27

8.2.5.2 Simulated Annealing Application

Parameters and Results

The purpose of this section is to apply the Simulated Annealing algorithm for the illustrative problem presented in Sect. 8.1.1. Setting $T_{ini} = 10$, $T_{end} = 0$, $\beta = 0.95$, $\sigma = 2.5$, $Niter = 300$. The number of dimensions is 6, corresponding to the decision variables of the problem: j_1, j_2, j_3, Q_1, Q_2, and Q_3. The limits of the variables were fixed to: $L_b = [0, 0, 0, 0, 0, 0]$, $U_b = [10, 10, 10, 800, 800, 800]$, where L_b represents the lower bound and U_b represents the upper bound for each variable.

Some experimental results are presented in Table 8.18. Table 8.19 summarizes some indicators of 30 executions.

Figure 8.12 shows how the temperature begins in T_{ini} value and decreases along with the iterations, similar to the slow cooling process of the metals, which is made by a gradual reduction in the atomic movements that decrease the density of defects until a lowest-energy state.

8.2.6 Grey Wolf Optimizer

8.2.6.1 Grey Wolf Procedure

The Grey Wolf Optimizer (GWO) algorithm is a new metaheuristic proposed by Mirjalili et al. [14]. This metaheuristic is inspired by the behavior of the grey wolf in nature. Generally, they live in groups of 5–12 grey wolves and form a pack. The algorithm is based on the social hierarchy behavior of the wolves and their mechanism

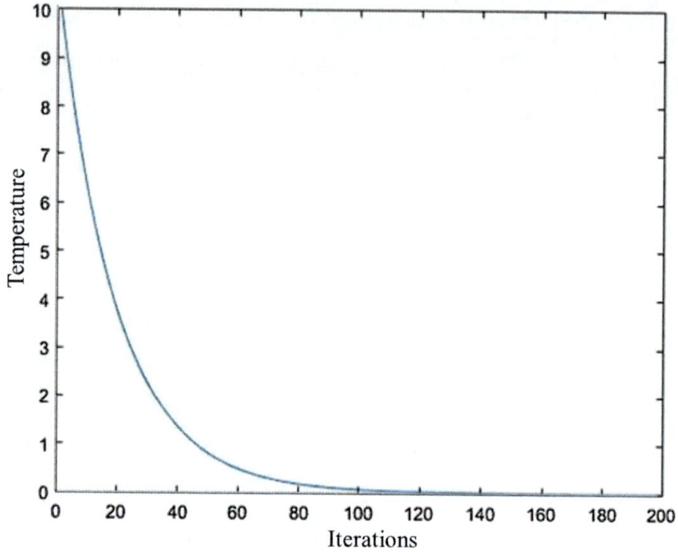

Fig. 8.12 Temperature along iterations

of obtaining prey (hunting). The wolves pack has several hierarchical levels: the alpha wolf (α) who is responsible for making decisions about when sleeping or how hunting, they lead the herd, and the members follow the decisions of alpha wolves. The Beta wolf (β) is who helps the alpha wolf, coordinating and collaborating with the management of the herd. They are subordinate to alpha wolves. They are the second hierarchy level of the structure. The other level is called Delta wolves (δ) who helps to alpha, beta wolves managing the herd. The Omega wolves (Ω) are the lowest level of the hierarchy. They must obey to alpha, beta, and delta wolves.

The GWO algorithm considers the position of the prey as the optimal solution of optimization. Then, using the natural behavior of the grey wolves, the algorithm tries to obtain the position of the prey. There are four stages in the hunting process of the grey wolves:

- Encircling prey
- Hunting
- Attacking prey
- Search for prey

Encircling Prey

The grey wolves begin the hunting process by surrounding the prey. They encircle their prey. This action is determined using the following formulations (8.21), (8.22) to update the position of the wolves in the encircling action:

$$\vec{D} = \vec{C}\vec{X}_p(t) - \vec{X}(t) \tag{8.21}$$

$$\vec{X}(t+1) = \vec{X}_p(t) - \vec{A}\vec{D} \tag{8.22}$$

Where $\overrightarrow{X}_p$ is the position of the prey, $\overrightarrow{X}$ indicates the position of the wolves, t represents the current iteration, $\overrightarrow{C}$ and $\overrightarrow{A}$ are the coefficients, $\overrightarrow{A}$ determines the search radius of the hunting. $\overrightarrow{C}$ and $\overrightarrow{A}$ coefficients are calculated as follows:

$$\vec{A} = 2\vec{a}\vec{r}_1 - \vec{a} \tag{8.23}$$

$$\vec{C} = 2\vec{r}_2 \tag{8.24}$$

Where $\overrightarrow{a}$ is linearly decreased from 2 to 0 along the course of iterations, $\overrightarrow{r}_1$ and $\overrightarrow{r}_2$ are random values in the range [0,1].

Hunting

In the real process of hunting, the Alpha wolf determines the position of the prey, and the Beta and Delta wolves follow to the alpha wolf and participate in the hunting. The positions of Alpha (best candidate solution), Beta, and Delta have a better understanding of the potential location of prey; they save the first three best solutions obtained so far and forcing the other search agents (including omegas) to update their positions according to the position of the best search agents.

$$\vec{D}_\alpha = \left|\vec{C}_1\vec{X}_\alpha - \vec{X}\right|, \ \vec{D}_\beta\left|\vec{C}_2\vec{X}_\beta - \vec{X}\right|, \ \vec{D}_\delta = \left|\vec{C}_3\vec{X}_\delta - \vec{X}\right| \tag{8.25}$$

$$\vec{X}_1 = \vec{X}_a - \vec{A}_1\vec{D}_\alpha, \vec{X}_2 = \vec{X}_\beta - \vec{A}_2\vec{D}_\beta, \vec{X}_3 = \vec{X}_\delta - \vec{A}_3\vec{D}_\delta \tag{8.26}$$

$$\vec{X}(t+1) = \frac{\vec{X}_1 + \vec{X}_2 + \vec{X}_3}{3} \tag{8.27}$$

Attacking Prey

Wolves capture prey when it stops moving. This action is modeled decreasing the value of the $\overrightarrow{a}$ along the course of iterations from 2 to 0, then $\overrightarrow{A}$ is also decreased. $\overrightarrow{A}$ is a random value in $[-2a, 2a]$. If random values $\overrightarrow{A}$ are in $[-1, 1]$, the next position of a search agent may be in any position between the position of the prey and its position. When $|A| < 1$, the grey wolves are forced to attack the prey. With the use of these operators, the algorithm allows the search agents to update their position based on the position of the alpha, beta, and delta. Only using these operators, the algorithm is susceptible to stay in local solutions; for this reason, more operators are needed.

Search for Prey

The search is done according to the position of the wolves (alpha, beta, delta). The wolves diverge from each other with the purpose of searching for prey and converging to attack it. The divergence is reached using random values, $\vec{A} > 1$ or $\vec{A} < 1$ to force the search agent to diverge from the prey. This process helps at exploration and allows finding a global solution.

8.3 More Information About the Inventory Problem in the State of Art and in History

Solving the inventory problem can be a complex task, and almost all manufacturer companies and retailers (practically the whole industry) are involved in the inventory problem. Activities that seem remarkably simple conceal a process of inventory management. A real example is a supermarket where products are presented to the customer. These products were selected from a wide variety of potential products. The company chose the best products with the goal of satisfying the customer's requirements and increasing profit. Competitive markets pose important challenges to supply chain management. One of the most critical challenges is the minimization of production costs [15] and meeting customer requirements. Companies need to decide how many items to purchase, how often to place an order, how to select the correct supplier in order to minimize costs while also meeting demand.

Supplier selection is a complicated process because several criteria must be considered, such as prices, volume discounts, reliability, and quality [16]. Companies explore and apply different methods or decision models to select final suppliers [17].

This section presents a literature review of how different techniques have been proposed to solve supplier selection and order quantity allocation problems. More information about inventory behavior and the evolution in the state of the art of this kind of problem will be explained in order to better guide the reader.

8.3.1 How Has Lot-Sizing, Supplier Selection, and Inventory Problems Been Solved Over the Years?

Supplier selection has a big impact on the purchasing process. An appropriate choice of suppliers is crucial since it improves the competitive advantages of industrial companies [18]. Deciding how many items must be ordered from the right suppliers and how frequently an order must be placed are important tasks.

The lot-sizing problem and supplier selection are some of the most important activities for the company. The lot-sizing problem requires several activities to determine the suppliers, average inventory, and cycle order period, all focused on meeting demand. Several costs are involved in these activities. The setup cost, holding cost, purchasing costs, and in some models, the transportation cost is considered.

The lot-sizing problem arises from the Economic Order Quantity (EOQ) model, which is one of the most important theories in production. The EOQ model was developed by Harris [19]. The main objective is to minimize the total inventory costs, where the mathematical model determines the optimal order quantity of an item [20].

EOQ has received widespread attention over the years, and it is an interesting research field. Reference [21] presented a survey and described the results of a study of literature in the lot-sizing problem.

Reference [22] studied supplier selection and examined how purchasing strategies influence supply management activities. They developed a supplier evaluation based on operational and strategic criteria, with the goal of ensuring better purchasing, quality, delivery, flexibility, and innovation. Other authors have examined different applications for supplier selection, such as [23–25].

The simplest way of solving the lot-sizing problem is to only consider a single item, single supplier, constant demand, no shortages, a single time period, and not consider any discount. But this situation may be unrealistic. More realistic problems consider other aspects, such as multi-period problems [26, 27] considered different types of discount (all-unit cost, incremental discount, and total business volume discount) through multi-objective formulation for the single item purchasing problem. Single-item complexity can increase depending on the criteria considered. [28] presented four different mathematical programming formulations of the lot-sizing classical problem. They discussed different extensions for real-world applications of this problem.

Other research that studied the lot-sizing problem and inventory costs for supplier selection, and explored the complexity presented by single product structure and larger size problems were presented by [29–31].

Over the years, the complexity of the lot-sizing problem has evolved, and numerous models and solution strategies have emerged with the goal of making this problem more realistic. The lot-sizing problem for multiple items increases the complexity of the model considerably. For example, [32] presented a mixed-integer programming model based on a piecewise linear approximation of the number of orders. They considered multi-product, multi-constraint inventory systems from suppliers, as well as incremental quantity discounts. Bohner and Minner [33] considered supplier selection and the order allocation problem for multiple products. They presented a mixed-integer linear programming model to minimize the total cost. The suppliers offer quantity discounts (all-units and incremental quantity discounts).

As a result of the lot-sizing complexity for single-item, and more frequently for multiple-item, when several criteria are considered, heuristics and metaheuristics have been proposed in the literature, as a result of reducing the elapsed processing time and finding better solutions. Research studies have applied Artificial Intelligence methods such as ant-colony optimization, simulated annealing, particle swarm optimization, and differential evolution, among others.

The following are some examples of heuristics and metaheuristics for a single item: Mahdavi Mazdeh et al. [34] analyzed a lot-sizing problem with supplier selection and proposed a new heuristic, which is based on the Fordyce–Webster Algorithm [35]. In the first case, quantity discounts are not taken into account; in the second case, incremental and all-unit quantity discounts are implemented. Lee et al. [36] presented a mixed integer programming (MIP) model to solve the lot-sizing problem with multiple suppliers, multiple periods, and quantity discounts. They proposed a Genetic Algorithm (GA) to solve the problem when it becomes too complicated. Their model minimizes total costs, where the costs include the ordering cost, holding cost, purchasing cost, and transportation cost; shortage is not allowed.

For the case of lot-sizing considering multiple-item, several heuristics and metaheuristics have been applied due to the great complexity of the models. Alfares and Turnadi [37] presented a mixed-integer programming model for a lot-sizing problem with multiple suppliers, multiple products, multiple periods, quantity discounts, and back-ordering of shortages. Due to a large number of variables, they proposed two heuristics, the first method is based on the Silver-Meal heuristic [38], and the second is solved by applying a GA.

Metaheuristic algorithms have been successfully applied to solve the supplier selection problem [39]. Mousavi et al. [40] presented the optimization of a two-echelon distribution supply chain network with multiple buyers and vendors. They considering ordering, holding, and purchasing costs. Some other applications of metaheuristic algorithms in supply chain models can be found in [41–45].

All authors mentioned here have shown the lot-sizing problem and its complexity. Strategies to solve the problem must be proposed, such as mathematical models, heuristics, and metaheuristics. In some cases, these models are solved using commercial software, but, when the lot-sizing and inventory problem becomes too complex (too many decision variables and several constraints), analytics methods and commercial software are no longer enough to solve the problem since this can consume a lot of time. Also, in some cases, a feasible solution cannot be found. For this reason, some metaheuristic algorithms must be explored in order to find a good solution in a reasonable amount of time.

In earlier sections, this chapter presented the use of some metaheuristic algorithms to solve this kind of problem, and an illustrative example was explored with the purpose of explaining how some metaheuristics are being applied.

8.4 Conclusions

Purchasing is a key activity in almost all industries. Making wise decisions may save money, which has an impact on the final price of products, and therefore, on the company competitivity. This chapter has covered some metaheuristic algorithms which are used in industrial engineering, and their use on the inventory management problem.

The current chapter first introduced an example problem, which is a state-of-the-art problem, studied in several references, from several points of view. The example problem provides us with a perspective of the complexity and, in some cases, the impossibility of finding the optimal global solutions, for example, when the number of possible solutions is infinite.

Some metaheuristic algorithms are then briefly explained, such as Particle Swarm Optimization PSO, Genetic Algorithm GA, Differential Evolution DE, Tabu Search Algorithm TS, and Simulated Annealing SA. After a brief explanation of these algorithms, this chapter introduced one way the example problem can be solved and provided several solutions. The main objective of this chapter is that readers can introduce themselves to other problems of the industrial engineering and to apply the algorithms discussed herein in order to solve them, especially in the field of supply chain management.

References

1. Mendoza A, Ventura JA (2013) Modeling actual transportation costs in supplier selection and order quantity allocation decisions. Oper Res Int J 13(1):5–25
2. Alejo-Reyes A, Mendoza A, Olivares-Benitez E (2019) Inventory replenishment decisions model for the supplier selection problem facing low perfect rate situations. Optim Lett. https://doi.org/10.1007/s11590-019-01510-0
3. Alejo-Reyes A, Olivares-Benitez E, Mendoza A, odriguez A (2020) Inventory replenishment decision model for the supplier selection problem using metaheuristic algorithms. Math Biosci Eng 17:2016–2036. https://doi.org/10.3934/mbe.2020107
4. Bowersox DJ, Closs DJ (1996) Logistical management: the integrated supply chain process. McGraw-Hill, New York, NY
5. Yang X-S (2010) Engineering optimization. Wiley, Hoboken
6. Barr BS, Golden BL, Kelly JP, Resende MGC, Stewart WR (1995) Designing and reporting on computational experiments with heuristic methods. J Heuristics 46:9–32
7. Talbi E-G (2009) Metaheuristics: from design to implementation. Wiley, Hoboken, pp 54–67
8. Eberhart R, Kennedy J (1995) A new optimizer using particle swarm theory. In: Proceedings of the sixth international symposium on micro machine and human science, pp 39–43
9. Sampson JR (1975) Adaptation in natural and artificial systems (John H. Holland). The University of Michigan Press, Ann Arbor
10. Price KV, Storn RM, Lampinen JA (2005) The differential evolution algorithm. Differential evolution: a practical approach to global optimization, pp 37–134
11. Glover F, Laguna M (1998) Tabu search. In: Handbook of combinatorial optimization. Springer, Boston, pp 2093–2229
12. Gendreau M, Potvin JY (2019) Handbook of metaheuristics. Operations research and management science. Springer, Berlin
13. Kirkpatrick S, Gellat C, Vecchi P (1983) Optimization by simulated annealing. Science 220(4598):671–680
14. Mirjalili S, Mirjalili SM, Lewis A (2014) Grey wolf optimizer. Adv Eng Softw 69:46–61. ISSN 0965-9978. https://doi.org/10.1016/j.advengsoft.2013.12.007

15. Meindl SCP (2016) Supply chain management: Strategy, planning, and operations. Tsinghua University Press, Beijing
16. Verma R, Pullman ME (1998) An analysis of the supplier selection process. Omega 26:739–50
17. Boer LD, Labro E, Morlacchi P (2001) A review of methods supporting supplier selection. Eur J Purch Supply Manag 7:75–89
18. Zhang D, Zhang J, Lai K, Lu Y (2009) An novel approach to supplier selection based on vague sets group decision. Expert Syst Appl 36:9557–9563
19. Harris FW (1913) How many parts to make at once. Mag Manag 10:135–152
20. Kundu A, Guchhait P, Pramanik P, Maiti MK, Maiti M (2016) A production inventory model with price discounted fuzzy demand using an interval compared hybrid algorithm. Swarm Evolut Comput. https://doi.org/10.1016/j.swevo.2016.11.004
21. Glock CH, Grosse EH, Ries JM (2014) The lot sizing problem: a tertiary study. Int J Prod Econ 155:39–51. ISSN 0925-5273. https://doi.org/10.1016/j.ijpe.2013.12.009
22. Nair A, Jayaram J, Das A (2015) Strategic purchasing participation, supplier selection, supplier evaluation and purchasing performance. Int J Prod Res 53(20):6263–6278. https://doi.org/10.1080/00207543.2015.1047983
23. Mafakheri F, Breton M, Ghoniem A (2011) Supplier selection-order allocation: a two-stage multiple criteria dynamic programming approach. Int J Prod Econ 132(1):52–57. ISSN 0925-5273. https://doi.org/10.1016/j.ijpe.2011.03.005
24. Chang C-T, Chen H-M, Zhuang Z-Y (2014) Integrated multi-choice goal programming and multi-segment goal programming for supplier selection considering imperfect-quality and price-quantity discounts in a multiple sourcing environment. Int J Syst Sci 45(5):1101–1111. https://doi.org/10.1080/00207721.2012.745024
25. Batuhan M, Huseyin A, Kilic S (2015) A two stage approach for supplier selection problem in multi-item/multi-supplier environment with quantity discounts. Comput Indus Eng 85:1–12. ISSN 0360-8352. https://doi.org/10.1016/j.cie.2015.02.026
26. Choudhary D, Shankar R (2011) Modeling and analysis of single item multi-period procurement lot-sizing problem considering rejections and late deliveries. Comput Indus Eng 61(4):1318–1323. ISSN 0360-8352. https://doi.org/10.1016/j.cie.2011.08.005
27. Mohammad Ebrahim R, Razmi J, Haleh H (2009) Scatter search algorithm for supplier selection and order lot sizing under multiple price discount environment. Adv Eng Soft 40(9):766–776. ISSN 0965-9978. https://doi.org/10.1016/j.advengsoft.2009.02.003
28. Brahimi N, Dauzere-Peres S, Najid NM, Nordli A (2006) Single item lot sizing problems. Eur J Oper Res 168(1):1–16. ISSN 0377-2217. https://doi.org/10.1016/j.ejor.2004.01.054
29. Chen S, Feng Y, Kumar A, Lin B (2008) An algorithm for single-item economic lot-sizing problem with general inventory cost, non-decreasing capacity, and non-increasing setup and production cost. Oper Res Lett 36(3):300–302. https://doi.org/10.1016/j.orl.2007.09.005
30. Massahian Tafti MP, Godichaud M, Amodeo L (2019) Models for the single product disassembly lot sizing problem with disposal 52(13):547–552. https://doi.org/10.1016/j.ifacol.2019.11.215
31. Ghaniabadi M, Mazinani A (2017) Dynamic lot sizing with multiple suppliers, backlogging and quantity discounts. Comput Ind Eng 110:67–74. https://doi.org/10.1016/j.cie.2017.05.031
32. Haksever C, Moussourakis J (2008) Determining order quantities in multi-product inventory systems subject to multiple constraints and incremental discounts. Eur J Oper Res 184(3):930–945. https://doi.org/10.1016/j.ejor.2006.12.019
33. Bohner C, Minner S (2017) Supplier selection under failure risk, quantity and business volume discounts. Comput Ind Eng 104:145–155. https://doi.org/10.1016/j.cie.2016.11.028
34. Mahdavi Mazdeh M, Emadikhiav M, Parsa I (2015) A heuristic to solve the dynamic lot sizing problem with supplier selection and quantity discounts. Comput Indus Eng 85:33–43. https://doi.org/10.1016/j.cie.2015.02.027
35. Fordyce JM, Webster FM (1984) The Wagner, Whitin algorithm made simple. Product Invent Manage 25(2):21–30
36. Lee AH, Kang H-Y, Lai C-M, Hong W-Y (2013) An integrated model for lot sizing with supplier selection and quantity discounts. Appl Math Model 37(7):4733–4746. https://doi.org/10.1016/j.apm.2012.09.056

37. Alfares H, Turnadi R (2018) Lot sizing and supplier selection with multiple items, multiple periods, quantity discounts, and backordering. Comput Ind Eng 116:59–71
38. Silver EA, Meal HC (1973) A heuristic for selecting lot size quantities for the case of a deterministic time-varying demand rate and discrete opportunities for replenishment. Product Invent Manage 14(2):64–74
39. Eydi A, Fazli L (2016) Asia-Pac J Oper Res 33(06)
40. Mousavi SM, Bahreininejad A, Musa SN, Yusof F (2017) A modified particle swarm optimization for solving the integrated location and inventory control problems in a two-echelon supply chain network. J Intell Manuf 28(1):191–206
41. Li Y, Ding K, Wang L, Zheng W, Peng Z, Guo S (2018) An optimizing model for solving outsourcing supplier selecting problem based on particle swarm algorithm. J Ind Prod Eng 35(8):526–534
42. Kang H, Lee AHI, Wu C, Lee C (2017) An efficient method for dynamic-demand joint replenishment problem with multiple suppliers and multiple vehicles. Int J Prod Res 55(4):1065–1084
43. Wang Y, Geng X, Zhang F, Ruan J (2018) An immune genetic algorithm for multi-Echelon inventory cost control of IOT based supply chains. IEEE Access 6:8547–8555
44. Xiong F, Gong P, Jin P, Fan JF (2018) Supply chain scheduling optimization based on genetic particle swarm optimization algorithm. Cluster Comput 1–9
45. Fallahpour A, Olugu EU, Musa SN, Khezrimotlagh D, Wong KY (2016) An integrated model for green supplier selection under fuzzy environment: application of data envelopment analysis and genetic programming approach. Neural Comput Appl 27:707–725